New Techniques
in Astronomy

New Techniques
in Astronomy

English Version Edited by

Hector C. Ingrao

Optics and Microwave Research Laboratory
NASA Electronics Research Center
Cambridge, U.S.A.

and

Harvard College Observatory
Cambridge, U.S.A.

GORDON AND BREACH SCIENCE PUBLISHERS
New York   London   Paris

Copyright © 1971 by

    Gordon and Breach, Science Publishers, Inc.
    440 Park Avenue South
    New York, N. Y. 10016

Editorial office for the United Kingdom

    Gordon and Breach, Science Publishers Ltd.
    12 Bloomsbury Way
    London W. C. 1

Editorial office for France

    Gordon & Breach
    7—9 rue Emile Dubois
    Paris 14^e

This is an English version of the original Russian *Novaya Tekhnika v Astronomii* (Volumes I & II) published by the USSR Academy of Sciences in February 1963 and June 1965, respectively.

Library of Congress catalog card number 74–162553. ISBN 0 677 13540 8. All rights reserved. No part of this book may be reproduced or utilized in any form or by any means, electronic or mechanical, including photocopying, recording, or by any information storage and retrieval system, without permission in writing from the publishers. Printed in east Germany

Preface to the English Version

Three conferences on astronomical techniques were organized by the Commission on Instrumentation, Astronomical Council, USSR Academy of Sciences, and held in Moscow April 18–20, 1961; Kazan May 12–14, 1964; and Pulkovo, November 27–30, 1967. Proceedings of the Moscow conference were published in 1963 (Vol. I), Kazan in 1965 (Vol. II), and Pulkovo in 1970 (Vol. III), under the title *New Techniques in Astronomy*. This English Version of *New Techniques in Astronomy* spans the Moscow (Section I) and Kazan (Section II) conferences. We anticipate that these conferences will continue to be held in the future, at approximately the same time intervals, and trust that the proceedings will follow. The final decision to have *New Techniques in Astronomy* translated into English was made because our colleagues in the USSR had exhibited new approaches to the development of astronomical instrumentation (especially in the use of automatic control) although sometimes using only very basic technology. This decision was also prompted by a series of events occurring between 1964 and 1967. In November 1964, at the Harvard College Observatory I was the host for a group of astronomers and engineers visiting the United States under the National Academy of Sciences. The group was integrated by Dr. V. B. Nikonov, Head of Stellar Physics at the Crimean Astrophysical Observatory; Dr. B. K. Ioannisiani (Lenin Laureate), Section Head of the State Optical Instrument Plant (GOMZ) and designer of the 6-M altazimuth optical telescope; Dr. N. N. Mikhelson, Senior Scientific Associate of the Main Astronomical Observatory; and Dr. A. R. Gorshkov, consultant to GOMZ.

During their stay, we had the opportunity to discuss, at length, two (at the time) controversial subjects concerning large astronomical telescope design: Altazimuthal Mounts vs. Equatorial Mounts, and Telescope Drives using Digital Control. These discussions were continued during my visit and lecture at the Main Astronomical Observatory in Pulkovo (August, 1967) where I had the pleasure of being the guest of Dr. Mikhelson and had the opportunity of seeing the working model (1/10 scale) of the controversial 6-M altazimuth telescope.

I met a skillful Russian translator, Mrs. Ruth Feinstein, who was a visitor in Cambridge, upon my return. We discussed the possibility of translating the two volumes of *New Techniques in Astronomy* and decided to plunge in. Unfortunately, she had to return to London before the translation was complete; thus, I had to complete the first draft.

To make the English version of *New Techniques in Astronomy* useful, several editorial changes were incorporated. For example: Captions have been added to almost all figures and photographs. Original drawings were checked and corrections introduced as necessary. Photographs and sketches have been added to better illustrate some contributions. Footnotes have been introduced in places where the editor thought it could assist the reader.

The authors affiliation was introduced; in some instances this required quite a bit of library search. A subject and name index has been added. A complete review of the references was carried out, and when possible, author's name, complete title and the name of the publication were incorporated.

Contribution I-5: "The Control System of the PM-700", by Yu. A. Belyaev, *et al.* was not very clear and some drawings were not fully explained. For this reason, this was replaced by a more comprehensive one, I-5a, "The Electronic Digital Control System of the PM-700 Telescope" by Yu. A. Belyaev. Again, for the same reason, Contribution II-6 "Experimental Methods for the Study of Telescope Components and Assemblies" by Ye. G. Grossvald was replaced by II-6a, "Investigation of Models of the Large Telescope Primary Mirror by Photoelectric Methods" and II-6b, "Experimental Model Investigation of the Large Hyperbolic Mirror", by Ye. G. Grossvald and K. S. Tavastsherna.

These changes and additions suggested to me that it was better to identify this book as an English *version* rather than as just an English translation.

In the past, transliterations of Russian names from the Cyrillic to the Latin alphabet were done using different criteria and to date they still lack consistency. For this book, I followed the recommendations of the U.S. Government Printing Office Style Manual, January 1967, with two exceptions: The character ь which is transliterated as an apostrophe (') and the character ъ which is given as a double apostrophe ("). In both cases, the apostrophes were eliminated for the sake of easy reading in English at the expense of distorting the original Russian sound. (I apologize to the linguisticians for this choice.)

A more serious problem is the transliteration into the Latin alphabet of the non-Russian names that have been previously transliterated from the

Latin into the Cyrillic alphabet in a rather free fashion. The results can be rather catastrophic in some cases. I recall reading the book *Technologie der Astronomischen Optik*, the German translation of the Russian book by D. D. Maksutov. In Chapters I and II there is a reference to the work done by an optician named Ricci. Since I never heard of Ricci or read of him, I decided to go to the library to search for Ricci's work. After an unsuccessful search, especially in the Italian literature, I reached the conclusion, with the help of some clues, that Maksutov referred to the famous American optician Ritchey. To avoid a similar occurrence in this book, all the non-Russian names have been checked for proper spelling in the original papers or reliable catalogues; I hope that I have succeeded.

I would like to thank several people who encouraged me to undertake this English version and those who helped me during the various phases: Professor Donald H. Menzel, Paine Professor Practical Astronomy and Professor of Astrophysics at Harvard University, who gave me the encouragement and moral support required to undertake this task (and with whom I was professionally associated from 1957 to 1968); Professor A. Edward Lilley, Professor of Astronomy at Harvard University for his interest and frequent discussions on some of the papers; Dr. Jay Pasachoff, my tutee as an undergraduate at Harvard College and at present Research Fellow in Astrophysics at the California Institute of Technology, for reading contributions I-26 and I-28; Mr. Dino Argentini of the Itek Corporation for assisting me on some aspects of translation of contribution I-4; and to Mrs. Thomasine C. Brooks, Librarian of the Phillips Library of the Harvard College Observatory, for her enthusiastic cooperation in the literature search. Finally, I am particularly grateful to Miss Elaine M. Ciccarelli for her invaluable assistance throughout the preparation of this book.

Wellesley Hills, Massachusetts HECTOR C. INGRAO
USA

Contents

Preface to the English version v

Section I

Preface 3

I-1 The new soviet telescope, the largest in Europe (*B. K. Ioannisiani, A. B. Severny* and *P. I. Shelovitelev*) 6

I-2 The Shain 2.6-m reflector (*B. K. Ioannisiani*) . . . 11

I-3 The control system of the 2.6-m telescope (*N. S. Zhurkin, V. M. Konshin* and *G. L. Bruk*) 20

I-4 The manufacture and preliminary tests results of the primary mirror (2.6-m) for the Crimean Astrophysical Observatory telescope (*V. V. Oshurko*) 24

I-5 The electronic digital control system of the PM-700 telescope (*YU. A. Belyaev*) 32

I-6 An automatic guiding and setting system for altazimuth telescopes (*YU. A. Sabinin* and *P. V. Nikolayev*) . . . 73

I-7 Automatic control systems for astronomical instruments using a spherical coordinates converter (*Y. P. Yegorov* and *YU. A. Sabinin*) 85

I-8 Altazimuthal and positional drive mechanisms in a combined control system for altazimuth instruments (*V. P. Yegorov* and *YU. A. Sabinin*) 107

I-9 Some principles for the design of digital control systems for altazimuth instruments (*S. V. Korotkov, V. A. Myasnikov* and *YU. A. Sabinin*) 119

I-10 Automatic tube flexure compensation in stellar telescopes (*G. L. Goreva, YU. A. Sabinin, P. V. Nikolayev* and *A. V. Shumakher*) 143

I-11 A photoelectric method to measure the "seeing" of stellar images (*I. P. Rozhnova* and *YU. A. Sabinin*) . . . 161

Contents

I-12 Programmed control of a telescope for limited observing programs (*Z. H. Kubeva* and *YU. A. Sabinin*) . . . 169

I-13 The electronic drive of the 48-cm reflector of the Vilnius Astronomical Observatory (*V. A. Yasevichus*) . . . 179

I-14 Performance of an iris photometer (*V. S. Avedisova*) . . 181

I-15 A camera with an optical compensator for observations of artificial earth satellites (*KH. I. Potter* and *YU. S. Streletskii*) . 185

I-16 Observational results from the new transit instrument of the Pulkovo Observatory (*N. N. Pavlov*) 195

I-17 A standard wedge level-tester (*L. A. Sukharev*) . . . 199

I-18 A reflecting astrolabe for research in fundamental astrometry (*D. D. Polozhentsev, KH. I. Potter* and *YU. S. Streletskii*) . 207

I-19 Automation of the operation of a photographic zenith tube (*D. N. Ponomarev* and *YE. M. Lapkin*) 213

I-20 The control system of the Pulkovo photographic zenith tube (*V. A. Naumov*) 217

I-21 The automatic stellar photoelectric photometer of the Latvian Astrophysical Laboratory (*A. P. Kundzin*) 223

I-22 An automatic photoelectric polarimater and other instruments (*L. V. Ksanfomaliti*) 231

I-23 A stellar photometer and a spectrometer for the 1–2.5 micron region (*V. I. Moroz*) 243

I-24 A Fabry-Pérot etalon used with an image intensifier for the observation of faint emission objects (*P. V. Shcheglov*) . . 255

I-25 A recording intensity microphotometer based in the MF-4 type (*E. V. Kononovich*) 259

I-26 A spectrograph with a photocontact image converter for the observation of nebulae (*V. F. Yesipov*) 269

I-27 A vacuum chamber for testing precision mirrors for astronomical instruments (*V. A. Savin*) 279

I-28 An interference-polarization filter for astrophysical studies of the sun in the K-line of ionized calcium (*S. B. Ioffe* and *N. M. Drichko*) 287

I-29 A slitless stellar spectrograph with guiding and spectral reference lines (*V. P. Linnik*) 289

Section II

Preface 309

II-1 Fast optical systems developed at the Crimean Astrophysical Observatory and their application to astrophysics (*G. M. Popov*) 311

II-2 A catadioptric telescope (*P. P. Argunov*) 317

II-3 An astrometric study of the 70-cm meniscus telescope at the Abastumani Observatory (*A. SH. Khatisov*) 331

II-4 A study of the optical systems of the PM-700 telescope (*G. I. Bolshakova* and *A. V. Shumakher*) 345

II-5 A study of the deformation of the PM-700 telescope mount (*G. I. Bolshakova* and *N. N. Mikhelson*) 353

II-6a Investigation of models of the large telescope primary mirror by photoelastic methods (*YE. G. Grossvald* and *K. S. Tavastsherna*) 359

II-6b Experimental model investigations of the large telescope hyperbolic mirror (*YE. G. Grossvald* and *K. S. Tavastsherna*). 367

II-7 On the design selection for the support system of an astronomical mirror (*N. N. Mikhelson*) 377

II-8 Electrical control for the telescope mount APSH-6 (*U. K. Veisman* and *T. E. Kyubar*) 385

II-9 A photoelectric photometer with digital printout (*U. K. Veisman*) 389

II-10 Image converters in astronomical research (*P. V. Shcheglov*) . 393

II-11 A low-order Fabry-Pérot etalon used in astronomical research (*P. V. Shcheglov*) 395

II-12 Rowland ghosts in double pass diffraction monochromators and their compensation in photoelectric spectrophotometry (*P. P. Kozak*) 399

II-13 An iris microphotometer based on the MF-2 microphotometer (*L. A. Urasin*) 405

II-14 Time marking in high-speed solar cinematography (*U. I. Ilyasov*) 411

II-15 The use of a horizontal long focal length telescope with coelostat for positional observations of the Moon using photography (N. G. *Rizvanov*). 421

II-16 The use of short exposure photography for meteors (*YE. N. Kramer*) 425

II-17 An astronomical dome made of plastic (*T. E. Kyubar*) . . 431

Subject Index 435

Name Index 443

SECTION I

Preface to Section I

The modern equipment required by astronomical observatories for different observations can now be provided by the Soviet optical and mechanical industry.

A 500 mm aperture telescope has been constructed for a university, and a 450 mm telescope is in the process of completion. Several 700 mm reflectors and one 125 cm reflector have been installed and are in operation, including the original meniscus telescope at the Abastumani Astrophysical Observatory and the telescope at the Main Astronomical Observatory (Pulkovo) which has a metallic mirror.* The 1 m Schmidt telescope at the Byurakan Observatory has been completed, installed and is operating successfully. Most recently, a 2.6 m aperture reflector has been designed, built and installed at the Crimean Astrophysical Observatory.

In the last century telescopes used gravity drives for guiding on stars; in the first quarter of this century they used gravity drives with electrical winding and in the second quarter, electric drives using synchronous motors. Today telescopes make extensive use of automatic electric drives (in combination with computers) including compensation for mean refraction and in some cases simultaneous synchronization of the dome rotation and slit opening.

In addition to the above mentioned stellar and other telescopes, the industry has produced relatively small-aperture special-purpose cameras observing artifical earth satellites, various telescopes (and accessories) for solar observations and in particular the large Crimean solar telescope, a stellar interferometer, and several astrometric telescopes with unusual drive mechanisms.

Laboratory instruments in Soviet observatories include special-purpose microphotometers and spectrophotometers, photocomparators and spectrocomparators, magnetographs, spectrographs, monochromators, etc., but considerable improvements need to be made to these instruments by introducing automatic control techniques.

* Recently the metallic mirror has been replaced by a glass mirror (Ed. English version).

A great achievement has been the development and manufacture of diffraction gratings and in particular replica, echelets, and echelles gratings, wide-band interference and narrow-band interference-polarizing filters, photocells and photomultiplier tubes, image converters and special-purpose television image pick-up tubes, opticalacoustic and other thermal radiation detectors, etc. Observational methods have been greatly improved by the gradual transition from photographic to television techniques and electronic methods in general. The portion of the electromagnetic spectrum over which observations can be made has increased considerably. Observations of soft and hard X-rays have become possible as a result of the development of high-altitude observations (high-altitude balloons, rockets, artificial earth satellites), while observations of extremly long waves are possible due to advances in radioastronomy.

At present observations are possible in a spectral region which covers about forty octaves; however, the receivers are not sufficiently sensitive particularly in the wavelength region of a fraction of a millimeter (up to 1 mm).

The prospect for future studies of outer space is very promising. Observational astronomers, following the successful development of powerful rocket technology since 1957, are gradually becoming experimental astronomers.

The articles in this book (Section I) are part of the Proceedings of the First Conference on New Techniques in Astronomy which took place in Moscow April 18 and 20 in 1961. The conference was organized by the Astrosoviet and the Commission on Instrumentation of the Astronomical Council of the Academy of Sciences, with N. N. Mikhelson as Chairman. Unfortunately, lack of time has prevented the inclusion of the following contributions:

B. K. Ioannisiani *The Design of the 2.6-M Shain Reflecting Telescope (ZTSh)*. (In view of the importance of the subject we are replacing this report, with the author's permission, by two articles, the first from *Pravda*, February 20, 1961 and the second from *Optiko-mekhanicheskaya promyshlennost*, 1958 (No. 4, p. 25).

V. B. Nikonov, I. M. Kopylov, K. K. Chuvayev *Performance of the ZTSh Telescope.*

P. V. Dobychin *Astronomical Instrumentation Work at GOMZ.*

V. V. Aleksandrov, P. S. Konev *A Brief Note on the Present Status of the AST-452 (0.35 m aperture) and the AZT-3.*

D. D. Maksutov *A New* 700 mm *Meniscus Telescope for Astrometry, and a Telescope for Amateurs.*

Yu. S. Streletskii, V. G. Ilin, T. S. Gerasimova, I. F. Saksina, A. V. Shumakher *The Design of the PM-700 Telescope.*

M. K. Abele *An Automatic Three-Axis Camera for Observations of Artifical Earth Satellites.*

N. D. Kalinenkov *The Lunar Camera at the Engelgardt Astronomical Observatory.*

P. M. Afanaseva, V. B. Sukhov, Yu. P. Platonov *A Device for Calculating the Mean Transit Times of Stars Using Magnetic Tape and Punched-Card Output.*

N. D. Kalinenkov *An Oscillographic Microphotometer and Its Performance.*

N. A. Dimov *Integration Methods in Astrophysical Instruments.*

N. M. Shakhovskii *The Stellar Polarimeter at the Crimean Astrophysical Observatory.*

F. M. Gerasimov *New Development in Diffraction Gratings.*

V. N. Karpinskii *A Photoelectric Monochromator with Double Diffraction.*

A. V. Bruns *A Spectrocomparator.*

Main Astronomical Observatory   O. A. MELNIKOV
Pulkovo, USSR

CHAPTER **I-1**

The new soviet telescope, the largest in Europe*

B. K. IOANNISIANI†, A. B. SEVERNY‡ and P. I. SHELOVITELEV

† *State Optical and Mechanical Plant*
Leningrad

‡ *Crimean Astrophysical Observatory*
Nauchny

From the earliest times, man has sought to understand the secrets of the Universe. The stars were studied by the Ancient Chinese, Hindus, Egyptians and Greeks. But, lacking optical instruments, the human eye could not penetrate the depths of boundless outer space. Rudimentary techniques and the consequent lack of accurate observations were insurmountable obstacles to the scientists striving for knowledge.

An important stage in the development of astronomy was the discovery of the telescope at the beginning of the 17th century. At the moment the great Italian scientist Galileo Galilei first pointed his telescope at the sky, the Universe, so to speak, came nearer to man and the curtain hiding the secrets of life in distant worlds was partly opened. From that time, there has been a constant improvement in telescopes, they have become the eyes of the astronomer and the basic instrument of astronomical research with which many remarkable discoveries have been made.

The "sharp eye" of the modern telescope can penetrate space to distances calculated in millions of light years. Astronomers are studying the furthermost corners of the Universe, discovering its secrets and increasing the knowledge of outer space.

To study the farthest parts of the Universe by optical methods, the astronomer needs telescopes with a large light-gathering power and equipped with the proper accessories.

* This article was written for *Pravda* (*20 February, 1961*). In the English version we tried to keep only the relevant technical aspects. (Ed. English version.)

The telescope must gather as much light as possible from celestial bodies that are at great distances from us. This capability depends on the diameter of the mirror objective which collects the light. The greater the mirror diameter is, the brighter the image will be, the greater the detail that can be observed on a given object and the larger the number of stars that can be seen. The construction of telescopes with mirrors of large diameters is a complex task.

The development of the Soviet optical and mechanical industry which can now produce large glass blanks has made possible the construction of large astronomical instruments.

Several such instruments including reflectors with mirror diameters of up to 1.2 m have been built since World War II but, until now, astronomers have not had telescopes large enough for certain research programs. The government therefore requested that a 2.6-m telescope be constructed in the shortest possible time.

This telescope, named after G. A. Shain, has now been installed at the Crimean Astrophysical Observatory and its use has been approved by the State Committee. The construction of this highly accurate optical and mechanical instrument is an achivement of Soviet science, technology and industry.

Apart from such general considerations as convenience in use and ease of control, the design of the telescope had to provide exceptional accuracy of operation and to allow for the large changes of load in the telescope structure during operation. Also, it was necessary to combine mechanical rigidity and temperature stability, both of the individual components and of the system as a whole.

At the State Optical and Mechanical plant (GOMZ) in Leningrad, the leading astronomical instrument building factory in the Soviet Union, together with several other organizations succeeded in solving the complex technical problems in a simple and original way. About forty organizations participated in the construction of the telescope itself, the building and the dome, the different mechanisms and the required special purpose auxiliary equipment. The coordinated operation produced and completed this unique installation in four years.

The specifications for the reflector determined on the basis of the scientific requirements were laid down by a committee of the Academy of Sciences, led by Prof. V. B. Nikonov.

The basic component of the optical system of the telescope is a mirror of 2.6 m diameter and 10 m focal length. Casting a blank for a mirror

1*

weighing more than five tons of special glass with a small coefficient of expansion is not an easy task. A special furnace with programmed temperature control had to be built to achieve the fine annealing of the blank in order to minimize internal strain. The annealing operation lasted five months.

The reflecting surface of the mirror is a very exact paraboloid, with deviations from a strictly mathematical surface not permitted to exceed hundredths of a micron. To produce a surface of such precision requires great technical skill, but it was done by GOMZ in record time, not much more than six months*, using a special grinding and polishing machine, devised and manufactured by the Gorky Milling Machine Factory.

The aluminium film was deposited over the glass surface by means of a vacuum evaporator.

The completed mirror weighs 4 tons, and so a special support system is required. Without such a system, the mirror would become deformed under its own weight and the shape of the reflecting surface would change whenever the telescope changed attitude. Such a change, however small it might appear at first glance, would distort the telescope images.

Besides the basic optical system, the telescope has a number of auxiliary systems which makes changes in the focal length possible, from 10 to 100 m, and allows use of the telescope for photography, photometry and spectral analysis.

The fact that the rotating parts of the telescope weighs 62 tons and that this large mass must turn around the Earth's axis to track stars with the accuracy of a first-class drive mechanism, indicates the difficulties which had to be overcome in the construction.

The front north pivot of the polar axis, which is the base of the rotating part of the whole instrument, is an accurately-ground massive cylinder 5.5 m in diameter. The south support of the axis, a hemispherical pivot of 1.4 m diameter, is finished with near-optical precision. The polar axis turns on oil pads; the oil film is kept under a pressure of about 40 atmospheres. This bearing surface ensures exceptional smoothness and ease of rotation. It is hard to imagine a mass of 62 tons being moved by hand, but that is what happens.

The telescope is as automatic as it could be made; more than 160 electric motors are used. Many of the operations previously performed by the

---

* For assessment of the technique used, it would be helpful to know the actual number of hours spent in the manufacturing. (Ed. English version.)

astronomer are done by computers, coordinate converters and servo-mechanisms.

Tracking of a star, control of the photographic shutters, focusing and change of the plateholders are among the operations done from the central and auxiliary consoles. The telescope can track the observed objects by means of an electric motor driven by a crystal frequency generator. Excellence in optical quality requires that, during observation, the detrimental effects of the Earth's atmosphere and the turbulence of the air in the telescope building be reduced as much as possible. These factors cause random motion of the stellar images. The design of the telescope tube, the dome and the building is such as to minimize the effect of these interferences.

The area under the dome is equipped with a special air-conditioning system, the telescope tube is isolated from the effects of the air turbulence under the dome. There is an insulated closed space around the mirror in which a selected temperature, corresponding to that expected during the night, is maintained before observation. The dome has two walls with insulation between them. The walls of the telescope building are covered by thin metallic slats to reduce the heating by the sun; leafy trees are planted around the building to improve the local "seeing". The aperture of the Soviet telescope is the biggest in Europe. A better idea of its light-gathering power is given by the following example. If we light a candle at Moscow the telescope could detect the light generated by it (in the absence of any atmosphere) at Vladivostok.

The actual power, which is determined not only by the amount of light which the mirror gathers, but also by the way in which this light is concentrated and used, depends on the light-gathering efficiency of the mirror. A detailed photograph of the night sky made with the new teescope and its accessory equipment has demonstrated the high quality of the optics.

The quality of the stellar images produced by the telescope (diameter of star 2″) are as good as the ones obtained by the largest telescope in the world, the 200-inch Hale telescope on Mount Palomar (U.S.A.). We have already been able to photograph very faint stars, practically the faintest given in the Palomar photographic star atlas.*

For the first time in the Soviet Union we are able to study processes which are occurring in the furthest parts of the Universe and at distances

---

* We should indicate that the National Geographic Society–Palomar Observatory Sky Survey was obtained with the Schmidt 48-72-inch telescope at Mount Palomar. (Ed. English version.)

several times exceeding those previously accessible. The volume of outer space that can be studied has been greatly increased. The new telescope also greatly assists the study of celestial bodies near the Earth, such as the Moon, the planets and interplanetary matter. Its "gaze", penetrating to the depths of the Universe, will be able to follow the flight of a space ship on course to a planet to a distance of several millions of miles.

All the potentialities created for science by the new telescope will be realized in full measure as its experimental program is accomplished, but it has already become clear that its construction will permit Soviet astronomers to penetrate deeper into the secrets of the Universe.

CHAPTER **I-2**

The Shain 2.6-m reflector*

B. K. IOANNISIANI

State Optical and Mechanical Plant
Leningrad

1 INTRODUCTION

The need to expand the knowledge of the Universe requires the construction of powerful telescopes with varied accessory equipment. The use of a telescope of large light-gathering power is a prerequisite for all astrophysical research, although differences exist among telescopes due to methods and types of observations.

In 1954 GOMZ began the manufacture of the mirrors for the 2.6-m Crimean Astrophysical Observatory telescope. The complexity of the task is obvious: we might just briefly discuss some of the basic questions of design and the considerations which determined the choice of technical solutions.

While it is desirable to have a telescope which is suited to various modes of operation, it is clear that special-purpose telescopes, which are designed for specific tasks, are the most efficient. The deciding factor at any point was the request of the astronomers, who laid down several specific requirements.

2 THE OPTICAL SYSTEMS

The main mirror of the reflector is a paraboloid, with a clear aperture of 2.6 m, $f/4$. It can be used at the prime focus and for Cassegrain, Nasmyth and Coudé optical systems.

* This article appeared in *Optiko-mekhanicheskaya Promyshlennost* No. 4 (1958).

In addition to the usual mirror systems to increase the usable field and to have different aperture ratios, three correcting lens systems for the prime focus are planned. These are:

(1) A system with optical components made of clear fused silica and lithium fluoride, giving an aperture ratio of 3.9 and a field of view of $2\omega = 12'$,

(2) A meniscus system giving an aperture ratio of 4 and a field of view of $2\omega = 28'$,

(3) A wide-angle system with an aperture ratio of 2.6 and a field of view $2\omega = 45'$.

The Cassegrain and Nasmyth systems have an aperture ratio of approximately 16. The Coudé system, to observe the sky between declinations of $+45°$ and $-30°$, has a single flat mirror operating at half of the declination angle. For observations in the near polar region, a system with four flat mirrors will be used in the plane containing the declination and the polar axis. The aperture ratio of this Coudé system is 40.

Calculations of aberrations for an on-axis point source had to satisfy the Rayleigh condition ($\lambda/4$). For off-axis points this condition was relaxed to give a confusion disk of $1''$.

3 MOUNTING

The types of mounts considered were the fork mount and the English off-axis mount (as used for the 84-inch telescope at the McDonald Observatory, USA), with the most suitable outlet for the Coudé focus.

The final design is a somewhat modified fork mount. The base is a weldment with three-point support resting on a concrete foundation. Two of the supports are articulated stands which can depart from the vertical position to allow for azimuthal adjustments. The third support is a large hydraulic screw jack. The polar axis consists of the following main components: a platform, a cone, a worm wheel and a hemispherical pivot. The platform has a diameter of 5.5 m and rotates on two self-adjusting hydrostatic pads. The pivot of the polar axis is a spherical surface of 700 mm radius and its position is defined by three hydrostatic pads. All five pads are fixed to the telescope base (see Figure 2-1).

The arms of the fork are bolted to the platform. They carry the trunnions of the declination axis which are mounted in ball bearings. To secure uniform distribution of the load on both arms of the fork, the bearings are preloaded to roughly 15,000 kg.

To minimize deformations of the central part of the telescope tube, the declinations trunnions are linked to the tube by means of Cardan (gimbal) couplings. Rotation of the polar axis is accomplished by two worm-gears, coarse and fine, with 720 teeth and a 12 mm pitch. The gears are rigidly connected to one another. The worms are driven at the same rate, but due to the use of a special device either worm is operated in the power mode. The worms are floating to allow full engagement with the gears. Backlash in the polar axis drive is removed by a counterweight which maintains constant contact between the worm and the gears of the drive.

Rotation around the declination axis is accomplished by a worm-gear pair similar to the one in the polar axis. The gear of the declination is rigidly connected to the central part of the telescope tube and the worm

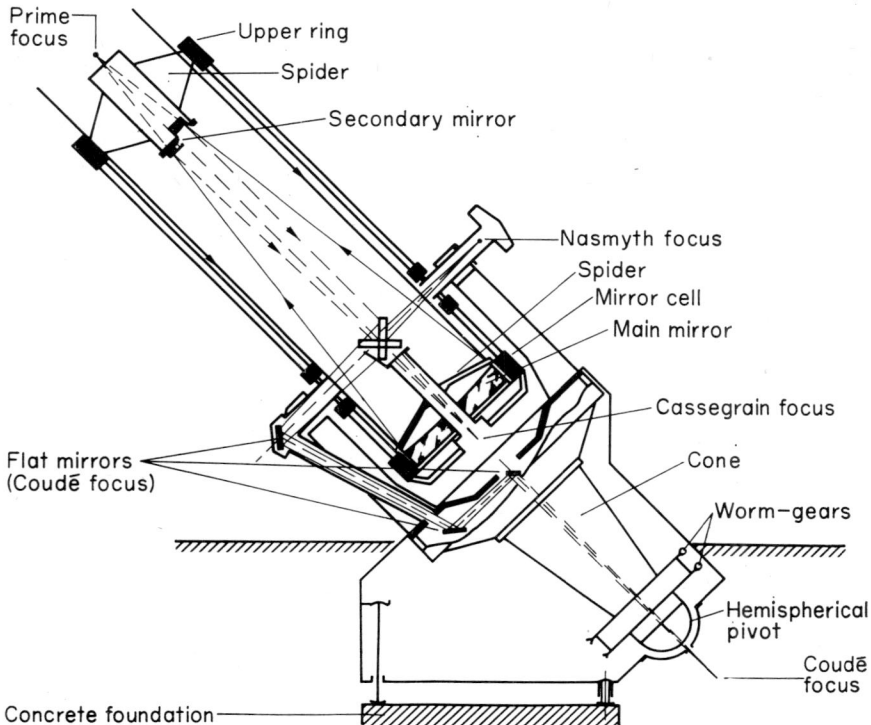

Figure 2-1 Cross section of the Shain 2.6-m reflector at the Crimean Astrophysical Observatory (courtesy of Prof. B. K. Ioannisiani).

to an arm of the fork. Backlash in the declination drive is removed by a cable which is secure from one end to the hub of the worm and from the other to a drum rotating with the same angular velocity as the gear. To ensure full engagement between the worm and the gear, an elastic element (a torsion shaft) is incorporated in the kinematic circuit.

4 THE TUBE

The design principle of the telescope tube is to minimize transversal and angular relative displacements of the tube ends for any attitude of the telescope tube (see Figure 2-2).

This principle* was first used in the 200–inch Hale telescope (Palomar Observatory, USA) and then in the 120-inch telescope (Lick Observatory, USA). To satisfy the flexure equalization principle, the cross sections of the tubular trusses of the telescope are chosen on the basis of the weight of the upper and lower ends of the telescope tube.

Complete flexure equalization of the ends of the tube has not been accomplished. The magnitude of the total residual uncompensated flexure is 0.4 mm and the combined maximum relative angle of displacement of the upper and lower ends of the tube is 2″. All parts of the tube are made of the same material. The upper and lower rings, the mirror cell and the central part of the telescope tube are welded structures.

The next important problem was how to support the main mirror. The support has to be made so that the deformations resulting from the weight of the mirrors, whatever its attitude, do not cause distortions of the reflected wave front by more than $\lambda/8$. If we accept deformation of the mirror surface by half of this amount, then the deviation of the surface from the mean shape shoud not exceed $0.035\,\mu$. It therefore seemed advisable to locate the support units in sockets in the back of the mirror. These sockets are bored along two concentric circles, eight in the first and sixteen in the second. The units are designed to provide radial and axial reactions. There are three fixed defining pads in the front part of the mirror which determines the plane perpendicular to the optical axis. The main mirror is centered by a sleeve inserted in the central hole.

According to our calculations, with this support system the maximum flexure of the mirror due to its own weight cannot exceed $0.01\,\mu$, allowing for the effects of friction in the support units themselves. A device is

---

* This design is due to Mark Serrurier of the California Institute of Technology. It was conceived for the 200-inch Hale telescope. (Ed. English version.)

Figure 2-2 Photograph of the Shain 2.6-m reflector at the Crimean Astrophysical Observatory (courtesy of the USSR Academy of Sciences).

being manufactured at present to determine the actual magnitude of the friction at different angles of inclination. The cell for the main mirror is a weldment, at the center of which is attached a tube carrying the diagonal mirror of the Nasmyth–Coudé system. The tube is kept in position by a four-legged spider. This type of mount keeps the position of the diagonal mirror unchanged regardless of the attitude of the telescope.

The cell at the prime focus is connected to the upper ring of the telescope tube by means of a four-legged spider.

The focusing unit is the same for the prime focus and the other systems using the secondary mirror. With an instrument of this size it is difficult to operate at the prime focus owing to inaccessibility. The use of a Ritchey-type plateholder, remotely controlled using the guiding telescopes mounted on the main telescope tube is exceptionally complicated and inconvenient to use. In large telescopes, due to changes in the position of the focal plane, a Foucault test is necessary before an exposure starts. As far as we know, the plateholder in the 200-inch Hale telescope has to be removed and a Foucault tester introduced every ten to fifteen minutes.

We intend to use the classical Ritchey plateholder, and in addition, at the prime focus the guiding will be done with a guide telescope which has an aperture of 320 mm and a focal length of 9300 mm. The position of the image on the photographic plate during exposure will be monitored by a special optical system, the eyepiece of which is placed near the eyepiece of the guider. We discuss below the scheme of this system. When direct photography is carried out at the Cassegrain focus, guiding will be done with the classical Ritchey plateholder. In other cases, we propose to use photoelectric guiding. The experimental mounts at the Crimean Astrophysical Observatory are already operating successfully. It should be noted that three blanks made of Pyrex-type glass with an internal strain not exceeding $5 \, m\mu \, cm^{-1}$ have been cast at one of our glass works. The sockets in the back of the mirror were bored basically using hard-alloy cutting tools followed by polishing, a method which reduces production time. Thus it took from five to six hours to bore one socket 180 mm in diameter and 235 mm in depth.

5   THE CONTROL SYSTEM

The control system provides for:
   (a) automatic and semi-automatic setting for given coordinates; (b) sidereal motion; (c) introduction of the necessary corrections; (d) obser-

vation of the moon and planets; (e) guide motions; (f) automatic cutoff of the drives at the final positions of the telescope tube; (g) control of the diagonal mirror in the Coudé system; (h) synchronization between, the telescope tube and the dome windscreen.

Manual setting of the telescope is done from a central control console and from a console in the observing area. Automatic setting is done from the central control console, and semi-automatic from both consoles. The slew rate for both coordinates is $1°$ sec^{-1}. The accuracy of the setting is $\pm 10''$ (static error). The transients in both modes are critically damped. Two identical guiding mechanisms are provided for the two axes. The control elements are sine-cosine rotary transformers which drives the amplidynes controlling the servomotors of the telescope.

Semi-automatic setting is done by the same servomechanisms at the coarse and fine setting speeds. The slew speed is $1°$ sec^{-1} and that of fine setting varies from $12''$ to $600''$ sec^{-1} The attitude of the telescope tube (α, δ) is monitored on dials.

Visual guiding is accomplished by using a pushbutton control at setting and guide speeds. The guide speeds in declination varies from $0''.25$ to $2''$ sec^{-1}. For the hour angle the guide speed is a function of declination, varying from $0''.25$ to $12''$ sec^{-1} according to the secant law; the second value corresponds to $80°$ declination. The drive in hour angle at guiding speeds is done by a synchronous motor fed from a crystal frequency generator.

6 FOCUSING MECHANISM

The purpose of the focusing mechanism is to ensure that the photographic plate at the prime focus remains in the best focus plane during exposure.

The plateholder portion facing the primary mirror is marked with a line. If the line is on the focal plane, reflection on the primary mirror will give a parallel beam. This beam is intercepted by two mirrors; from the right-hand mirror the light beam goes through the partially reflecting left-hand mirror. The latter sends both beams in the same direction. A plane mirror reflects these coincident beams to an achromatic objective and the images formed on its focal plane are seen through the eyepiece.

Poor focusing causes the left and right beams to be nonparallel and can be recognized at the eyepiece as a relative displacement of the line images on the left and right side of the plateholder. By periodic illumination of the line, the coincidence of the respective images can be observed through the eyepiece.

When relative displacements are detected, coincidence is regained by resetting the focusing mechanism. For a normal eye with an acuity of 6 (resolve 10″) a displacement of 0.015 mm in the position of the focal plane can be detected.

7 TELESCOPE AND BUILDING TEMPERATURE CONSIDERATIONS

The maintenance of the required temperature conditions for the optical components and the elimination of the detrimental thermal effects in the telescope tube and in the dome are important problems which have to be carefully studied, and have already received much consideration. The causes which affect the image quality to varying degrees, can be separated into climatic and geographic conditions (about which little can be done) and those inherent to the instrument (see Figure 2-3).

Serious attempts have been made recently to reduce the causes of the latter kind. The work of Couder, to elimate slow turbulent currents in the telescope tube by generating an artificial current in the tube, has aroused our interest. We propose to use this principle; at present we are designing a detachable double-walled casing for the telescope tube.

To reduce the effect of the difference in temperature between the inside and the outside of the dome, the end of the tube will be connected to a ring in the dome windscreen by means of a bellow made of fabric. The motions of the dome and the telescope tube will be synchronized both in direction and in speed by a special-purpose computer which will control the dome azimuth, the zenith angle of the ring on the windscreen. It is difficult to estimate at this stage of the project how effective these measures will be, and so we have allowed for the possibility of using the telescope either with an open tube or with a closed tube connected to the dome. In time, we can then obtain a comparative quantitative estimate of the image quality. The temperature of the main mirror will be controlled by insulating the mirror cell. The sectors of the mirror cover are also insulated. A closed space can then be formed around the mirror and before observations begin, the temperature of the mirror can be set to the one expected during the night of observation. Twelve thermostats mounted inside the mirror will be used to control its temperature.

To conclude, it should be said that fairly lengthy and thorough experiments will probably have to be carried out on the complete instrument, to finalize the operation of the different units in the system.

Figure 2-3 Shain 2.6-m telescope building at the Crimean Astrophysical Observatory (photograph courtesy of the USSR Academy of Sciences).

CHAPTER **I-3**

The control system of the 2.6-m telescope

N. S. ZHURKIN, V. M. KONSHIN and G. L. BRUK

State Optical and Mechanical Plant
Leningrad

The main emphasis in the design of the control system for the 2.6-m telescope was to achieve the maximum possible automation. The basic requirements were worked out in 1959 under the guidance of the Crimean Astrophysical Observatory. This was the first time that a control system has been developed for such a large telescope; several organizations participated in the solution of the complex problems involved.

The control system is integrated by the following basic units:

(1) A central control console (TsPU) which provides automatic and semi-automatic setting of the telescope in right ascension α and declination δ, control of the operating mode (i.e. operation from the TsPU or from other consoles), control of the flat mirror mechanisms for the Coudé (direct and elbowed) and Nasmyth systems, control of a stable frequency generator, of the exposure time, of focusing and other operations. The TsPU is also used to control the operation of the photoguider, the mechanisms of the covers for the main mirror, the guider mechanisms of the plateholder at the main focus and the cover of the correcting lens at the prime focus.

(2) A computer which calculates the components for mean refraction correction, tube and mount flexure with respect to hour angle and declination, transforms equatorial coordinates (hour angle t and declination δ) into horizontal coordinates (azimuth A and zenith distance z) to synchronize the motion of the telescope and the dome; computes the positional angle, which must be known to rotate the ring on the dome windscreen.

(3) A stable frequency generator for sidereal motion of the telescope, as well as guidance of the telescope using a lunar-planetary drive.

(4) An auxiliary console with all the semi-automatic controls for the dome mechanisms.

(5) Auxiliary controls in the observation area, at the Coudé, Cassegrain foci, etc.

Automatic setting of the telescope with respect to the given coordinates α and δ is done from the TsPU by means of a two-channel system of sine–cosine rotary transformers (SKVT) in a differential circuit and with a ratio of 132 between the coarse and fine readings. The transmitting SKVT are installed in the TsPU and the receiving SKVT for hour angle and declination in the drives of the telescope. The guidance servomechanisms of the telescope consists of systems with amplidyne and dc servomotors (EMU-12 A and MI-32 T).

The telescope is set on α and δ by slew motion at a rate of $1°$ $\sec^{-1}$. The accuracy of the setting in both coordinates is $\pm 15''$ and the motion is aperiodic.

Semi-automatic setting of the telescope for coordinates α and δ is made using a three-reading Selsyn indicator transmission from the TsPU, and also from the consoles in the observing area, at the Coudé focus and the hand switch box at the Cassegrain focus. Semi-automatic setting is made by a tachometric drive, in which the positional sensor is a potentiometer, and the speed sensor is a tachometer generator.

Semi-automatic setting operates at the rate of $3600''$ $\sec^{-1}$ for coarse setting and from 600 to $12''$ $\sec^{-1}$ for fine setting; fine correction can also be adjusted continuously from $12''$ to $0.5''$ $\sec^{-1}$.

Coarse and fine setting of the telescope is done by the same drives as in the automatic mode.

Similar drives, but of smaller power (an amplidyne EMU-5 A and motor SL-569), are used for the guide motions.

The accelerations at high speeds and during braking of the telescope drives are limited by flywheels on the shafts of the servomotors MI-32 T. The mechanisms to drive the diagonal mirrors are controlled by ac servomechanisms according to the data from transmitting and receiving SKVTs.

The components of the mean atmospheric refraction and mount flexure corrections respect to hour angle and declination are calculated using two conoids (three-dimensional cams). The input data to the conoids mechanism are the hour angle, transmitted as rotation to the conoids, and the declination, proportional to the conoid probes displacements. The corrections Δt and $\Delta \delta$ are proportional to the angle of rotation of the respective probes of the conoid mechanism. The rotation proportional to Δt and $\Delta \delta$ is transmitted to the shaft of the transmitting SKVT and the differential Selsyns. The scales of the conoid mechanism are chosen so that in the future it will be possible to use the conoids to compute the correction component of tube and mount flexure as well as refraction. After the actual values of the flexure errors have been measured on the telescope, we shall make new conoids to introduce the combined corrections. To design the conoid we assumed that the correction for flexure with respect to the zenith distance follows the $\sin z$ law.

A computer converts the equatorial coordinates t and δ, used for setting the telescope, into the horizontal coordinates A and z, used for the positioning of the dome slit and the windscreen which are synchronized with the movements of the telescope. DC functional potentiometers and magnetic amplifiers are used in the coordinate converter. The coordinates required (A and z) are obtained by using a contact-relay type servomechanism. Magnetic amplifiers are also used in the servomechanisms. In both servomechanisms amplification is regulated automatically. For the zenith distance, this is done by multiplying the output signal (the error signal) by the quantity $1/\sin z$. The automatic amplification regulation for the azimuth signal is obtained by a servomechnism.

The input data to the computer, as we have said, are the hour angle t and the declination δ. The hour angle is transmitted to the computer by a shaft. The declination angle is fed to the computer by a servomechanism using a two-channel Selsyn arrangement.

The computed azimuth and zenith distance are transmitted to the servomotor by a two-reading system in the SKVT; the servomotor drives the dome in azimuth and the windscreen in zenith distance. The slew motion of the telescope when it passes through a near-zenith zone, where the rate of change in azimuth exceeds the limiting value of $2°$ $\sec^{-1}$, is limited by a relay acting at the inputs of the amplifiers of the telescope servomechanisms. The data unit for the slew motion of the dome is a tachometer generator set up in the kinematic circuit of the computer which calculates the azimuth.

Thermal air currents inside the telescope tube and originated inside the dome will be minimized. This will be accomplished by connecting the upper end of the telescope tube and the window in the windscreen of the dome by means of a bellow. A cylindrical ring connects the bellow to the window. To prevent twisting of the bellow, the ring rotates proportionally to the positional angle thus following the rotation of the telescope tube relative to the window. The positional angle is obtained from a phantom telescope in the computer to which the hour angle and the declination are fed from the main telescope. Lack of synchronization between the phantom dome and the dome, when the telescope passes the zenith, is prevented by feeding the azimuth data from the computer to the shaft of the phantom dome.

The positional angle output from the phantom telescope is transmitted to a small Selsyn in the form of angular rotation. The positional data transmitted by the Selsyn is received and processed by the servomechanism to drive the ring.

A 50-V two-phase synchronous motor driven by a power amplifier and fed by a crystal frequency generator in the control system drives the telescope in right ascension. The crystal oscillator is of the type Sch-VII; its frequency stability is between 2 to 3 parts into 10^7.

The variable rate of the telescope motion around the t and δ axes required for the observation of the planets and the moon is achieved by adjusting the frequency of the RC-generators (with smooth variation in frequency). An auxiliary control console next to the central console contains the units for controlling the dome, the windscreen, mirror cover and the ring (semi-automatic control).

The observing platform allows the astronomer easy access to any part of the telescope. It moves in azimuth and height, and can rotate relative to the vertical support and also relative to the arm. An operator at the observing platform can control its movement from the console using semi-automatic drives.

The central and the auxiliary control consoles are attended by a single operator. Both of the consoles and the observing platform are interconnected by telephone.

CHAPTER **I-4**

The manufacture and preliminary tests results of the primary mirror (2.6-m) for the Crimean Astrophysical Observatory telescope

V. V. OSHURKO

State Optical and Mechanical Plant

Leningrad

The mirror blank was casted of LK-5 (Pyrex type) glass with a coefficient of linear expansion of 32.7×10^{-7} °C^{-1}. The fine annealing took about seven months and was done under special conditions. The blank which was casted in the form of a solid disk and weighing 5700 kg before processing, was allowed a larger diameter and thickness than the final dimensions of the mirror. The blank had to be processed on all sides, a central hole of 500 mm diameter had to be bored out and to locate the support units 24 sockets 180 mm in diameter were drilled in the back.

More than 1700 kg of glass had to be removed and stringent requirements were specified in regard to the final quality of the mirror. Grinding would have taken too long with standard methods, therefore after some experiments, it was decided to process the blank by a faster method.

The polishing was done on a special grinding-polishing machine which was designed and manufactured at the Gorky Milling Machine Factory.

The machine (Figure 4-1) is 11.2 m long, 7 m wide and 6.5 m high, weighs 142 tons and is mounted on a foundation weighin gmore than 450 tons. The left and right carriages support a gantry which can move along the guide ways.

On the front of the gantry a lathe-like carriage, a drilling-boring and a retouching carriage are mounted. All three carriages have the necessary degrees of freedom and can be set on the blank as required.

Figure 4-1 Special grinding–polishing machine at the State Optical and Mechanical Plant (Leningrad) used in the manufacture of the 2.6-m primary mirror for the Crimean Astrophysical Observatory.

The preliminary operations (edging, boring) and surface generation from a master form are done at an auxiliary bench (see Figure 4-1).

The main bench (see Figure 4-2) is used for the optical work on the mirror. The bench and mirror can be turned 90° from the horizontal position by a special mechanism. Both benches are rotated independently by means of electric motors with continuous speed control.

Before optical processing begins a guiding rod is connected to the gantry and to a slider. A carriage with a pneumohydraulic cylinder to control the pressure of the tool on the mirror can be moved on rollers along tracks (see Figure 4-2). The carriage has a spindle to which grinding or polishing tools can be attached. Reciprocating motion of the carriage is produced by an electric motor through a speed reducer and a connecting-rod-crank mechanism.

The 500 mm hole in the center of the mirror was bored by a tubular diamond drill, and took 2 hours and 15 minutes. The 24 sockets in the back were drilled and bored by multiblade heads made of hard alloy.

The diameters of the sockets were calibrated by a honing head using an abrasive slurry.

Tiny cracks which might have appeared after the mechanical processing were removed by washing all the sockets with hydrofluoric acid, followed by neutralization.

All further work on the mirror was done on the main bench with a lever-axial support system. Fine grinding and polishing of the mirror to the radius of the nearest comparison sphere of 19,953 mm was done by a special grinding tool of an aluminium alloy (Figure 4-3).

The surface of the grinding tool was channeled into squares. A similar tool was used for the polisher, but with a smooth surface. Pitch was deposited on the heated surface of the tool and cut into squares. The polisher was then shaped by pressing into the blank sphere. Then, the spherical surface was polished.

After the sphere was polished, preliminary parabolization of the mirror was done by using the same polisher. A petal-shaped (Figure 4-4) lap was employed; this is more effective for turning up the edge. The center of the mirror was worked on using small polishing tools.

During processing, the pressure of the tool on the surface of the mirror was maintained throughout and regulated by the pneumohydraulic cylinder.

Narrow zones were smoothed out by a set of circular and elliptical polishing tools. A circular tool appropriately shaped to the zone width,

The manufacture for the Crimean Astrophysical Observatory

Figure 4-2 View of the main bench of the machine shown in Figure 4-1. Grinding and polishing is carried out on this bench.

Figure 4-3 Grinding tools made of an aluminium alloy used in the manufacture of the 2.6-m primary mirror.

The manufacture for the Crimean Astrophysical Observatory

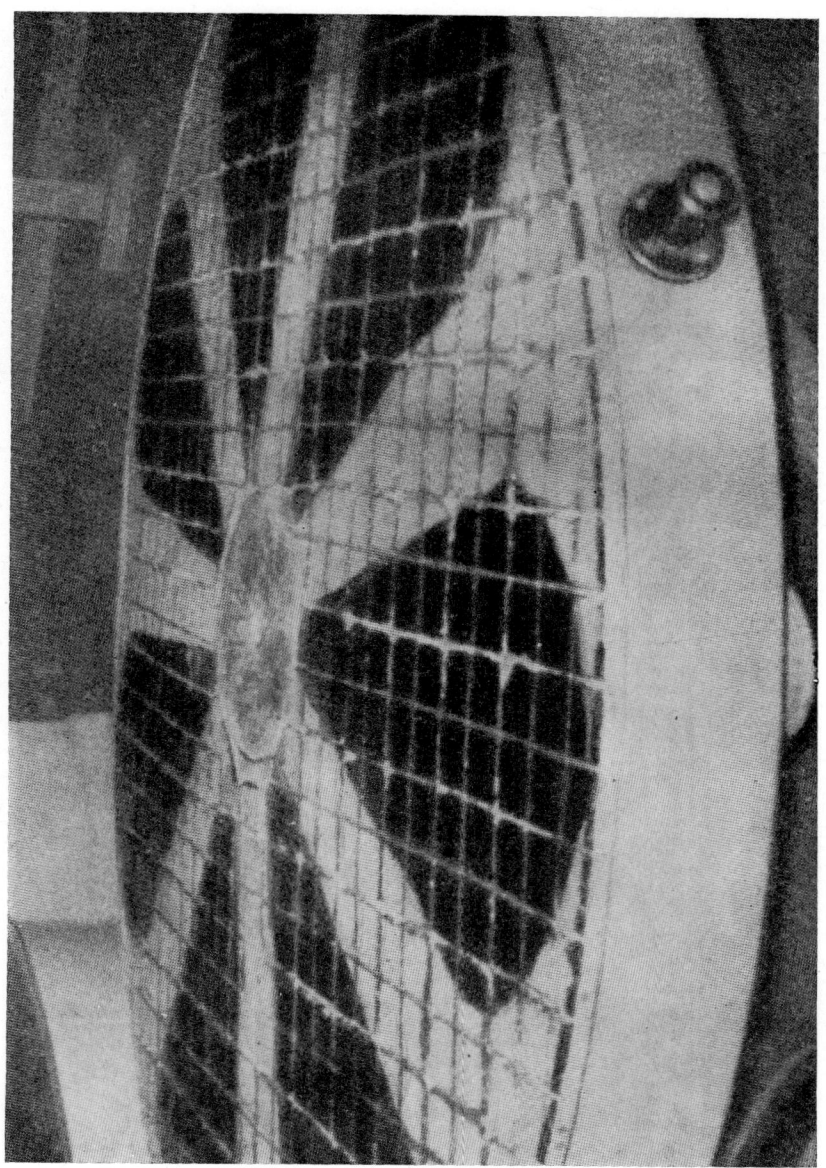

Figure 4-4 Petal-shaped lap used for parabolization of the 2.6-m primary mirror.

Figure 4-5 Retouching of the 2.6-m mirror. The tool is mounted in the retouching carriage of the grinding-polishing machine (see Figure 4-1).

was mounted in the frame of the retouching carriage (Figure 4-5). The elliptical polishers were used to retouch the narrow zones at the edge of the mirror by hand. In the retouching process the small polishing tools and the large ones were used alternately to obtain a smoother surface.

During parabolization and retouching of the main mirror, the image quality of an illuminated point source was frequently examined by the compensation shadow method devised by D. D. Maksutov.

After an aluminization test to determine the uniformity in thickness of the aluminium layer, the mirror was again examined under manufacturing conditions.

Examination of point source images by means of a microscope and photographs established that the diameter of the images were in the order of 10 microns; a pulsed light source was used.

The main mirror is now installed in the telescope and the Hartmann test is being carried out using stars as light sources.

CHAPTER **I-5**

The electronic digital control system of the PM-700 telescope*

YU. A. BELYAEV

Main Astronomical Observatory
Pulkovo

1 INTRODUCTION

The development of electronic computers offers new opportunities in the design of automatic instruments and control systems.

If automatic systems could be applied to the control of large telescopes to provide complete automation of the instruments and the operating processes, the accuracy and reliability of the observational results would be greatly increased, and unique instruments could be used more productively. Moreover, an electronic control system would simplify the mechanical aspect of the telescope drive since precision worm-gears could be replaced by coarser (or even friction) mechanical transmission.

A telescope controlled by computer may have any type of mount. The use of a computer has been considered, in particular, for telescopes with equatorial, altazimuth and horizontal mounts [1].

We first applied computer control to an equatorially mounted telescope, since this has a simpler drive mechanism.

The first stage was to test the principle of the control system for which we developed and studied a simplified model of an electronic computer to control an equatorially mounted telescope [2]. This was done between

---

* The contribution No. 5 "The control system of the PM-700 telescope" by Yu. A. Belyaev *et. al.* has been replaced by this contribution in the English version. This contribution appeared in the *Izv. GAOAN SSSR*, **22**, Issue 4, No. 169 (1961). The substitution was made in order to give a clearer description of the work presented in Contribution No. 5 of the Proceedings of the First Conference on *New Techniques in Astronomy*.

1954 and 1956, and the tests showed that further work in this direction would be worth while.

It was then decided to construct an electronic digital control system (ETsUM) which was capable of securing the required accuracy in the control of an actual telescope under real operating conditions. The ETsUM which was developed and constructed has been intended for the PM-700 (700 mm aperture) reflector at the Main Astronomical Observatory (Pulkovo).

2 BASIC CHARACTERISTICS AND BLOCK DIAGRAM OF THE ETSUM

The control system sets automatically the telescope on the celestial object and tracks it with a given accuracy [1]. With this control system the participation of the astronomer in setting the telescope is reduced only to set the coordinates of the given object on the console dials. From this moment on, the ETsUM will monitor and correct the motion of the telescope, computing all the necessary data for each instant.

To perform these functions, the ETsUM must carry out the following operations:

(1) Computation of sideral time S for each instant when corrections are made to the attitude of the telescope.

(2) Input from the control console of the values of right ascension α and declination δ.

(3) Computation of the value of the hour angle t.

(4) Computation of the actual value of the hour angle t_1, which the telescope takes up at the given time.

(5) Computation of the actual value of the declination δ_1.

(6) Computation of the correction components for mean atmospheric refraction in hour angle Δr_t and declination Δr_δ and allowance for future usage of these corrections in the ETsUM.

(7) Computation of the magnitude of the error in hour angle Δt and in declination $\Delta \delta$.

(8) Introduction of corrections corresponding to the errors Δt and $\Delta \delta$ to the telescope attitude.

The introduction of corrections for flexure and for the relationship between refraction, pressure and temperature was not planned at this stage.

In Figure 5a-1 we give the block diagram of the ETsUM, and in Figure 5a-2 the time diagram of one cycle of operation.

The ETsUM operates with 18 bits, which provide an accuracy of

$$\frac{360°60'60''}{2^{18}} \approx 4\rlap{.}''94. \tag{5a-1}$$

The nineteenth digit is a sign digit.

The repetition rate or cycling frequency f_u, i.e., the frequency with which corrections are introduced to the attitude of the instrument is given in sidereal time* by:

$$f_u = \frac{2^{18}}{24 \times 60 \times 60} = 3.034074074 \text{ (s sec)}^{-1}. \tag{5a-2}$$

This frequency is generated and regulated within the required limits by the velocity control mechanism MUS [3] and can be continuously adjusted from -2.5% to $+5\%$ for lunar and planetary observations.

The ETsUM includes a signal generator ZG, power amplifiers UM_1 and UM_2 and a frequency divider DCh_1 to which a frequency of 1000 (s sec)$^{-1}$ is fed from a time service crystal clock. The output from the UM_1 and UM_2 is fed to the MUS which gives a reference frequency of the order of 400 Hz and the modulated stable frequency of approximately 6.068 (s sec)$^{-1}$. For planetary observations, the modulated frequency may be changed by adjusting the frequency of the generator ZG. Using a detector in the frequency divider DCh_2, the frequency modulated signal is demodulated and divided by two. As a result of shaping in DCh_2 trigger pulses at a rate of $f_u \approx 3.034$ (s sec)$^{-1}$ are delivered at the output; these pulses determine the repetition frequency of the operating cycles of the ETsUM.

The DCh_2 triggers the synchronizing pulse generator GSI and the single delayed pulse generator GOZI with pulses at a rate of $f_t \approx 12.136$ (s sec)$^{-1}$ (see Figure 5a-2). These pulses drive the GSI which generates trains of 20 rectangular pulses 210 µsec wide at the rate of f_t. These pulses serve to synchronize all the arithmetic units of the ETsUM. The frequency of synchronization is 2,300 Hz and defines the bits frequency.

Simultaneously with the output of synchronization from the 19 output busses of the GOZI circuit, single-delayed pulses ZI shifted in relation to one another in such a way that each of them coincides in time with only

---

* In the English version sideral second is abbreviated s sec.

The electronic digital control system of the PM-700 telescope 35

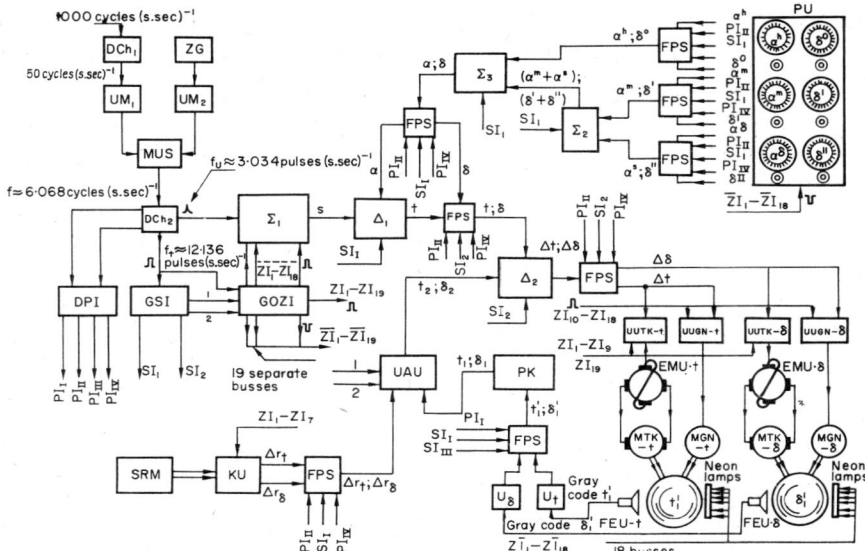

Figure 5a-1 Block diagram of the electronic control system (ETsUM) for the PM-700 reflector at the Main Astronomical Observatory, Pulkovo.

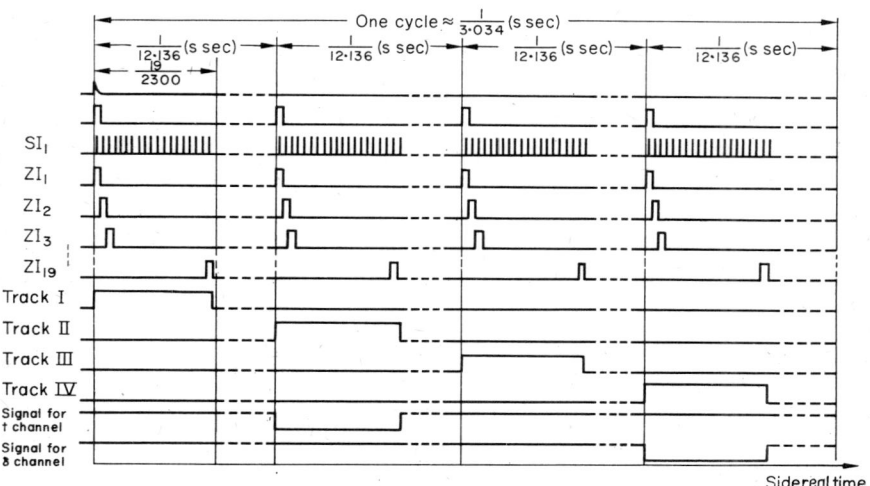

Figure 5a-2 Diagram of one cycle of operation (frequency with which corrections are introduced into the attitude of the instrument) of the ETsUM.

one bit of the number. The ZI pulses are used to transform the parallel numbers into sequential numbers, and vice versa, to form different numbers, to read the shaft angle encoders, and also in the output units of the ETsUM.

For sequential control of all the operations, a switching pulse transmitter DPI is used; these pulses are positive and rectangular in shape. These pulses divide each cycle of operation of the ETsUM into four tracks, during which the whole program of operations is performed:

Track I Readout of the shaft angle encoder in the polar axis of the telescope, and transformation of the readout into binary code, t_1.

Track II Input, computation and output of data for t-coordinate.

Track III Readout of the shaft angle encoder in the declination axis and transformation of the readout into binary code, δ_1.

Track IV Input, computation and output of data for the δ-coordinate.

The duration of each track is determined by the time needed to execute the arithmetic operations with 19-bit numbers, i.e., is equal to

$$\tau = \frac{19}{2300} \text{ s sec.} \qquad (5\text{-}3)$$

In the interval between two successive Tracks II, the error in the t-axis is generated by using the fine correction unit UUTK-t and the coarse setting unit UUGN-t; simultaneously, in the interval between two successive Tracks IV the units UUTK-δ and UUGN-δ are connected for the δ-axis.

The four-track system of operation enables the same units of the ETsUM to be used in the computation of the data for t and the δ coordinates.

In order that the initial or calculated data is fed only in the revelant track and only to the proper units, switching circuits FPS, controlled by pulses from the DPI are used.

Each cycle of operation of the ETsUM is triggered by a pulse from DCh$_2$ (repetition rate ≈ 3.034 s sec^{-1}) fed to the 18-bit adder Σ_1. A new number is setup and stored in the adder every $1/3.034$ s sec.

This number is the value of sidereal time S obtained by adding the number of pulses which arrive; the adder, in effect, counts time intervals, each $1/3.034$ s sec interval.

In Track II, the value of S is read in and simultaneously converted from parallel into sequential form, and fed to the subtractor Δ_1.

The value of right ascension α is fed into the ETsUM from the control console PU. The values of α in hours, minutes and seconds of time setup on the dials on the control console are transformed into sequential binary numbers using special coded punched tapes; these values are then transmitted to the adders Σ_2 and Σ_3. The first of them in Track II adds the seconds and minutes, and Σ_3 adds the hour to the previous sum.

Similarly, only in Track IV, the values of the seconds, minutes and degrees of δ are added using Σ_2 and Σ_3. The coded values for α and δ are given in the PU on punched tape.

The value of α obtained in this way in Track II is fed into the second input of the circuit Δ_1 which computes the hour angle t to which the telescope must be set at a given time.

The value of δ from the Σ_3 output is fed to the Track IV by the switching circuit FPS directly to the subtractor Δ_2.

The values of t_1' and δ_1' which give the actual equatorial coordinates of the telescope for a given attitude are obtained from shaft angle encoders in the Gray system [4]. The readout from the encoder is made by using a series of neon lamps, triggered sequentially by pulses $\overline{ZI}$* shifted by one digit. The signals from the neon lamps are fed to the photomultiplier tubes FEU-t and FEU-δ.

The resulting sequential Gray coded values are transformed in the Tracks I and III by the code coordinates converter PK into ordinary binary code t_1 and δ_1 values. In Tracks II and IV these values are applied to the universal arithmetic unit UAU.

In the UAU block to the values t_1 and δ_1 are added the component corrections for mean refraction Δr_t, and Δr_δ:

$$t_2 = t_1 \mp \Delta r_t,$$
$$\delta_2 = \delta_1 \mp \Delta r_\delta. \qquad (5a\text{-}4)$$

The UAU adds or subtracts depending on the sign of the correction.

The components Δr_t and Δr_δ are input into the UAU in the form of binary numbers by the coding unit KU. This unit is controlled by the analog computer SRM [5] which gives the mean refraction corrections.

The output values t_2 and δ_2 of the UAU are (in the appropriate track) fed to the subtractor Δ_2, where errors for hour angle $\Delta t = t - t_2$ and declination $\Delta \delta = \delta - \delta_2$, are calculated. The servomechanisms are controlled depending on the value and sign of Δt and $\Delta \delta$.

When the magnitude of the Δt or $\Delta \delta$ is greater than, or equal to, 2^9 ($4\rlap{.}''94$) either the coarse setting control unit UUGN-t or UUGN-δ is set into

---

* A line above the letters indicates negative pulses.

operation to control the servomotors MGN-t and MGN-δ respectively. The coarse setting rate is constant.

When the errors Δt or $\Delta \delta$ are less than or equal to $(2^9-1)\ 4\rlap{.}''94$ fine correction control units UUTK-t or UUTK-δ connect the servomotors MTK-t or MTK-δ. In this case, if the errors Δt and $\Delta \delta$ are less than or equal to $(2^8-1)\ 4\rlap{.}''94$ the generating is done at a rate proportional the magnitude of the error until this error becomes equal to zero. For this purpose, the UUTK-t and UUTK-δ convert the discrete values of the errors into a voltage which controls the amplidynes EMU-t and EMU-δ. The direction of rotation is determined by the sign of the errors. Figure 5a-3 gives the rate of the correction signal versus the error.

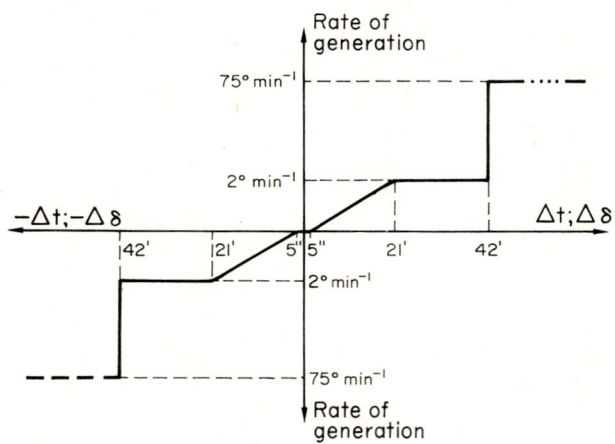

Figure 5a-3 Voltage control for the amplidynes EMU-t and EMU-δ versus errors Δt and $\Delta \delta$.

All the arithmetic units of the ETsUM with the exception of the adder Σ_1 operate sequentially. This principle fully secures the required speed of operation and yet enables the number of circuit elements to be considerably reduced.

There are no intermediate memory units in the machine, as their role and basic functions are performed by the input and output units, the adder Σ_1 and the coordinates converter PK, which are constructed on a combined parallel-sequential principle.

The ETsUM uses almost exclusively electron vacuum tubes (approximately three hundred).

3 DESCRIPTION OF THE ETsUM BLOCKS

3.1 CONTROL CIRCUITS

3.1.1 Switching pulse transmitter DPI

The block diagram of the DPI is shown in Figure 5a-4. The switching pulses PI_I to PI_{IV} are used to control the tracks of the ETsUM (see Section 2). By these pulses the computations of t and δ are separated in time, and so the same arithmetic units may be used for both computations.

Pulses from the DCh_2 which are rectangular in shape are used for switching commands (see Figure 5a-5; pulses a, b, c and d). The signals a and b are pulses at a rate of approximately 6.068 (s sec)$^{-1}$ and the signals c and d are pulses at a rate of approximately 3.034 (s sec)$^{-1}$ obtained by dividing by two the frequency 6.068 (s sec)$^{-1}$. The signals a and b by a differentiating network are transformed into pulses e and f which are fed to the inputs 1 (see Figure 5a-4) of the coincidence circuits SS_{P_1} to SS_{P_2}. The signals c and d

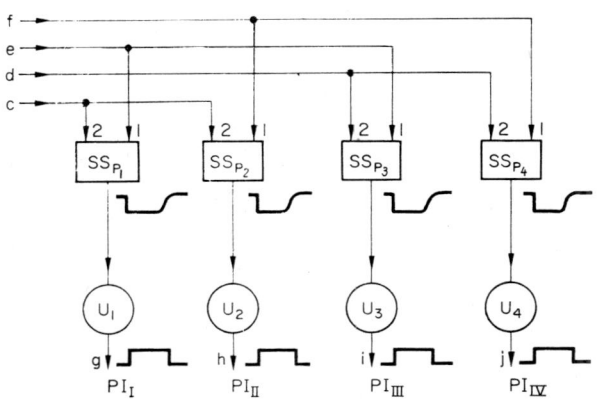

Figure 5a-4 Block diagram of the switching pulse transmitter DPI.

are fed to the inputs 2 of these circuits. With this combination of voltages at the inputs of the coincidence circuits, a switching pulse PI for the corresponding track is obtained at the output of each of the amplifiers U_1 to U_4 (pulses g, h, i, j in Figure 5a-5). The duration of these pulses is approximately 9 μ sec, which is just a little longer than required for a 19-bit sequential number to pass through the vacuum tube circuit controlled by these pulse. The oscillogram of a switching pulse PI is shown in Figure 5a-6.

Each of the circuits SS_{P_1} through SS_{P_4} of the DPI uses the vacuum tube 6Zh8, and the amplifiers U_1 through U_4 uses triodes 6N8.

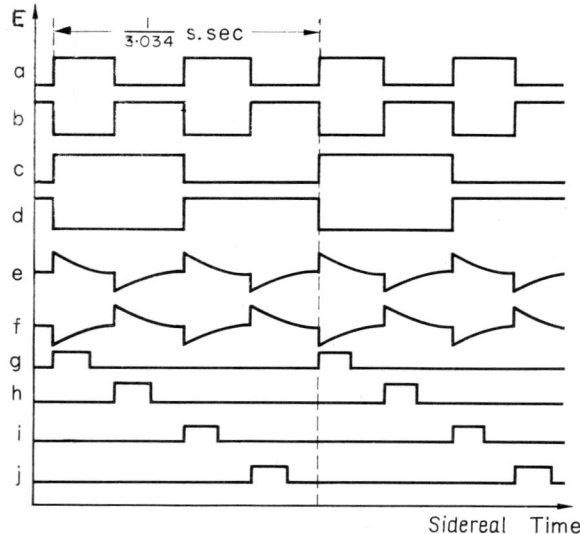

Figure 5a-5 Waveshapes of the signals from the frequency divider DCh_2 and the switching pulse transmitter DPI.

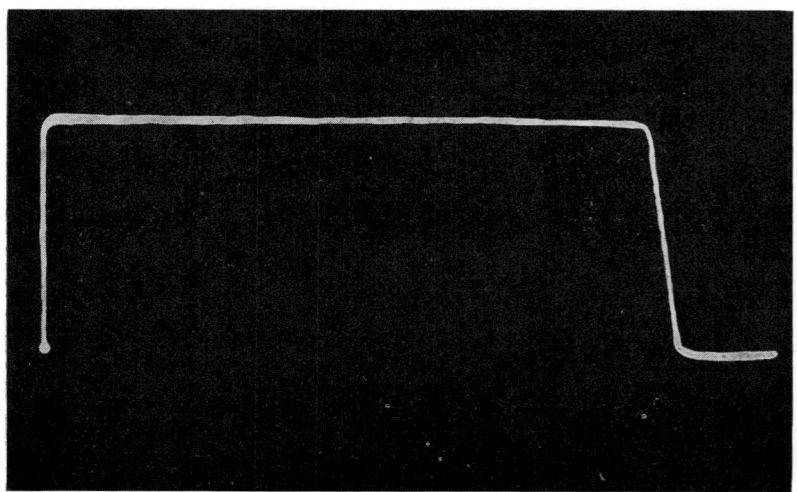

Figure 5a-6 Oscillogram of a switching pulse PI at the output of the switching transmitter DPI.

3.1.2 Synchronizing pulse generator GSI

For reliable operation of the sequential arithmetic units it is necessary that the pulses are synchronized and have identical width and shape.

In the ETsUM synchronization and shaping of the pulses is done using coincidence circuits FPS or FS which are controlled by rectangular synchron-

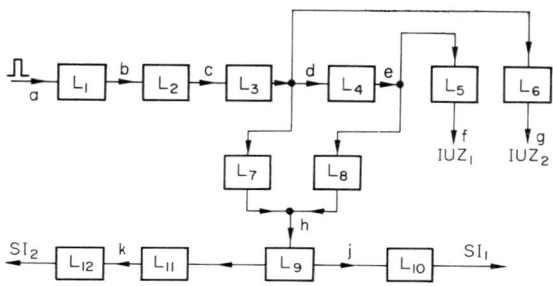

Figure 5a-7 Block diagram of the synchronizing pulse generator GSI.

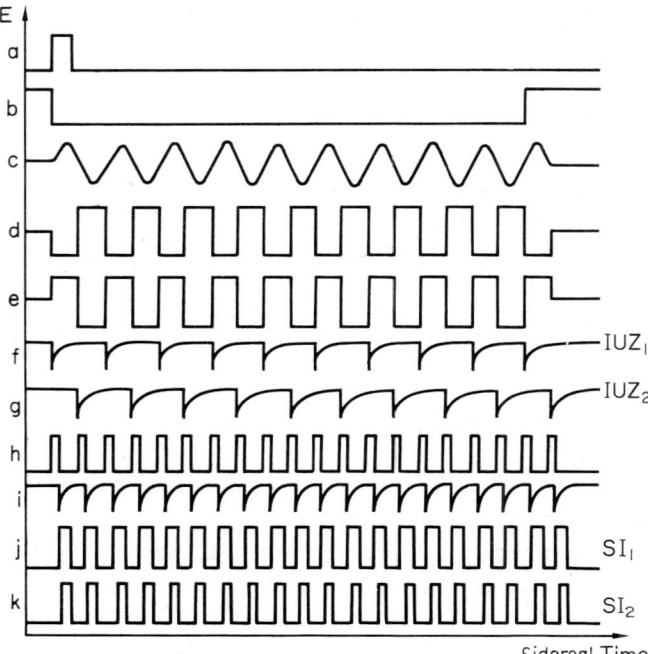

Figure 5a-8 Waveshape of the signals at different test points of the GSI (see Figure 5a-7).

izing pulses SI_1 and SI_2. The GSI is the synchronizing pulse generator which must in turn be synchronized, since the initial commands for the GSI are pulses at a rate of $f_t \approx 12.136$ pulses (s sec)$^{-1}$ and $f_a \approx 3.034$ pulses (s sec)$^{-1}$ obtained from the DCh_2. Therefore the GSI does not synchronize continuously but only in those intervals of time when the ETsUM is performing operations with numbers. Triggering of the GSI is done by pulses at a rate of $f_t \approx 12.136$ pulses (s sec)$^{-1}$ which arrive at the start of each track. This assures synchronization of the GSI and the operating tracks of the ETsUM. The pulse rate of the GSI generated does not depend on the triggering frequency, which may be changed for observation of the planets and the moon.

The block diagram of the GSI and the waveshape of the signals at different points of the GSI are given in Figures 5a-7 and 5a-8, respectively.

When a trigger pulse a is applied to the relaxation oscillator (vacuum tube L_1) it generates a negative pulse b. This pulse excites the "ringing" oscillator (vacuum tube L_2) which generates a 1150 Hz sinusoidal signal c. The number of periods (10) of this signal is determined by the width of the pulse b. The sinusoidal oscillations are converted by vacuum tubes L_3 and L_4 into rectangular pulses d and e, which after passing through the stages L_5 and L_6 are transformed into the train of pulses f and g or IUZ_1 and IUZ_2 (see also Figure 5a-1). These pulses latter are used for synchronization of the single delayed pulse generator GOZI. In this case the IUZ_1 pulses synchronize the ZI odd bit pulses output by the GOZI and the pulses IUZ_2 th even digits.

For convenience, in shaping the code pulses using coincidence circuits the synchronization pulses are delayed by 90 μ sec with respect to the ZI pulses. This delay is introduced by the circuitry associated to vacuum tubes L_7 and L_8, where the pulses h are shaped from the d and e pulses. Their frequency is double the frequency of the c oscillations, i.e. 2300 Hz. Using the pulses i at this frequency the relaxation oscillator (vacuum tube L_9) is triggered.

The rectangular pulses j generated by L_9 and after amplification by L_{10} are used as synchronzing pulses SI_1.

Since the code pulses after being processed by several arithmetic units are substantially affected by width changes, the synchronization of the pulses fed to the $\varDelta_2$ circuit is performed by the SI_2 pulses which are narrower; the stages L_{11} and L_{12} generates these pulses.

Oscillograms of pulses SI_1, SI_2, IUZ_1 and IUZ_2 are shown in Figure 5a-9.

a)
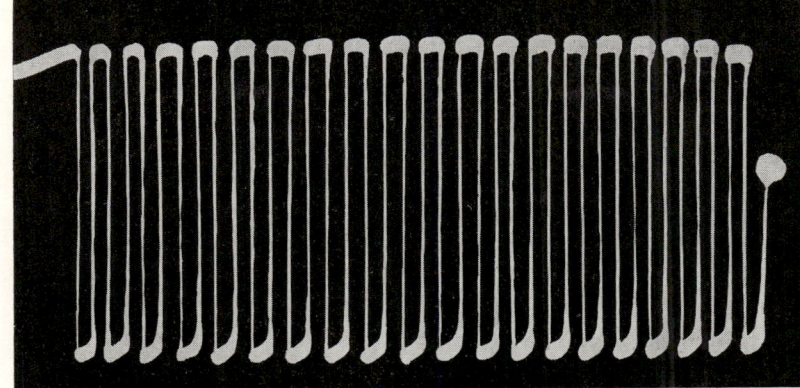

b)
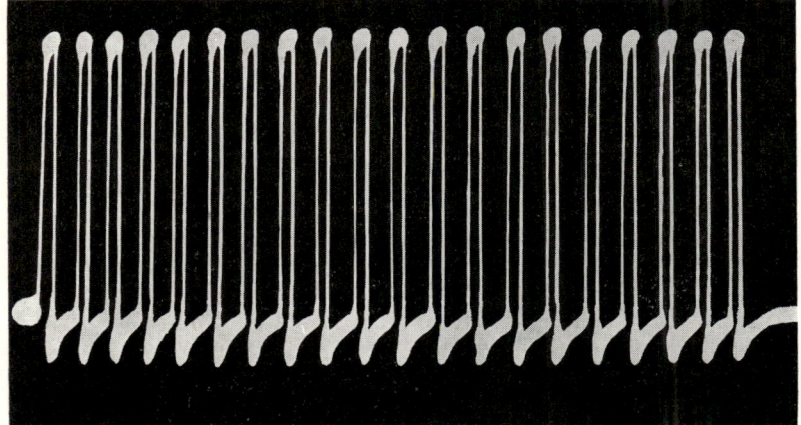

c)

d)

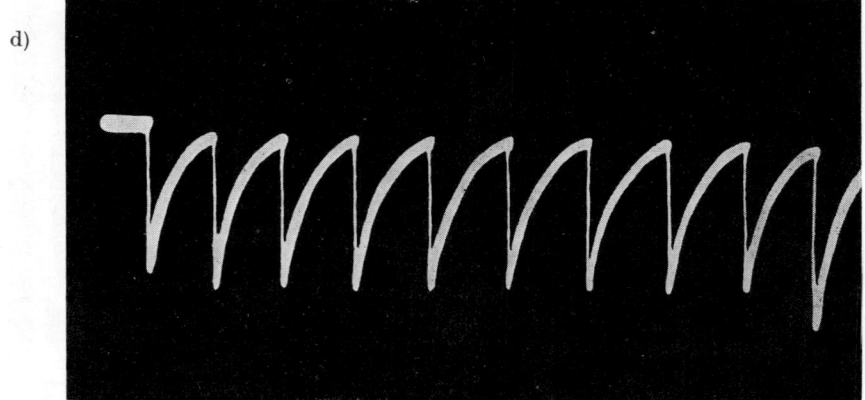

Figure 5a-9 Oscillograms of the output pulses of the synchronizing pulse generator GSI (see Figure 5a-7): a, SI_1; b, SI_2; c, IUZ_1; and d, IUZ_2.

3.1.3 Single delayed pulse generator GOZI

The circuit of the GOZI, shown in Figure 5a-10, consists of a chain of relaxation oscillators $R_1, R_2, \ldots, R_{18}$. The triggering of the GOZI as well as of the GSI, is done at the beginning of each track of the ETsUM operation by the c_1 pulses (see Figure 5a-11) which are fed by the DCh_2 at a rate of $f_t \approx 12.136$ pulses (s sec)$^{-1}$. In this case all the relaxation oscillators generate sequentially rectangular pulses $c_1, c_2, \ldots, c_{18}$, the trailing edges

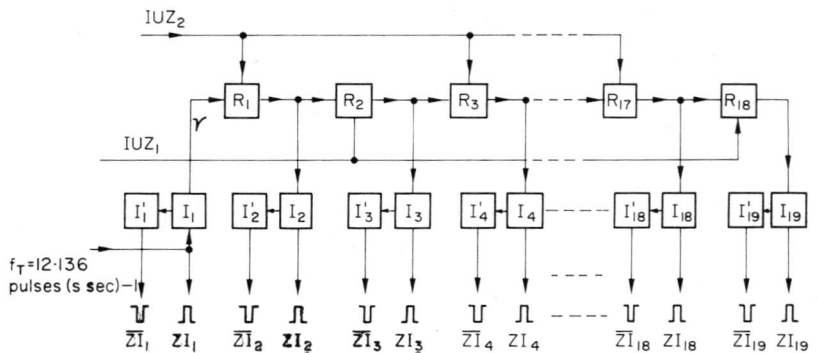

Figure 5a-10 Block diagram of the single delayed pulse generator GOZI.

of which are used to shape the delayed pulses ZI_2, ZI_3, ..., ZI_{19}, d_2, d_3, ..., d_{19}, by the inverters I_2, I_3, ..., I_{19}.

Each relaxation oscillator is triggered by the trailing edge of the pulse delivered by the preceding oscillator. The duration of the pulses c and therefore the delay of the pulses d are fixed by blocking the oscillators. This is achieved by the control pulses IUZ_2 and IUZ_1 (a, b) which are fed by the GSI (see description of GSI). The control pulses precedes by 90 μ sec the synchronization pulses SI_1 (e). Therefore, each of the 19d pulses generated by the GOZI, developed at the respective bus with a well-defined time relationship respect to the synchronization pulses.

The width of the d pulses is 400 μ sec and so the synchronization pulses SI_1 are centered in time with respect to the d pulses. This allows the synchronization and shaping of the pulses by the FPS and FS.

The fact is that all the initial sequential numbers in our ETsUM are obtained starting from the pulses delivered by GOZI. The methods of

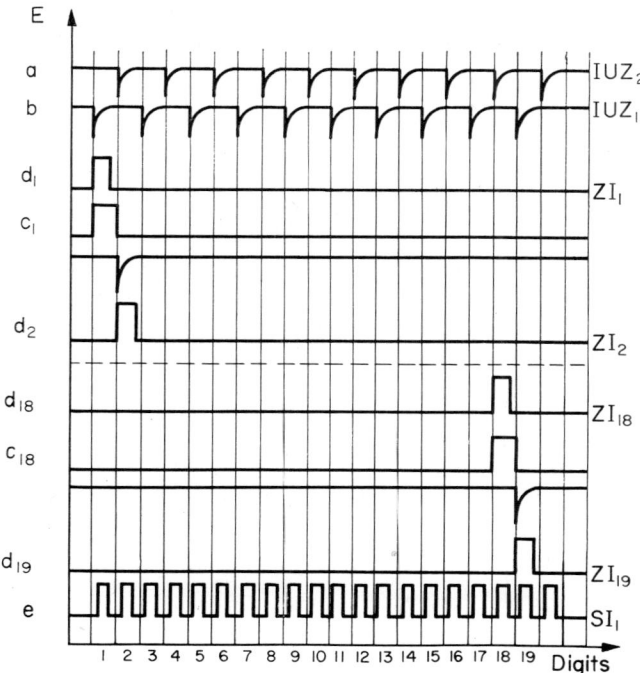

Figure 5a-11 Waveshape of the input and output signals of the single delayed pulse generator GOZI and the synchronizing pulses SI_1.

obtaining these numbers differ (see Sections 3.1.4, 3.2.1, 3.2.4 and 3.2.5), and so the pulses may differ considerably in shape from one another.

The FPS and FS output pulses, corresponding to any number, have a standard rectangular shape and determined by the characteristics of the synchronization pulses. This is illustrated by Figure 5a-12, where the following oscillograms are shown: (a) pulses of the 8th digit at the output

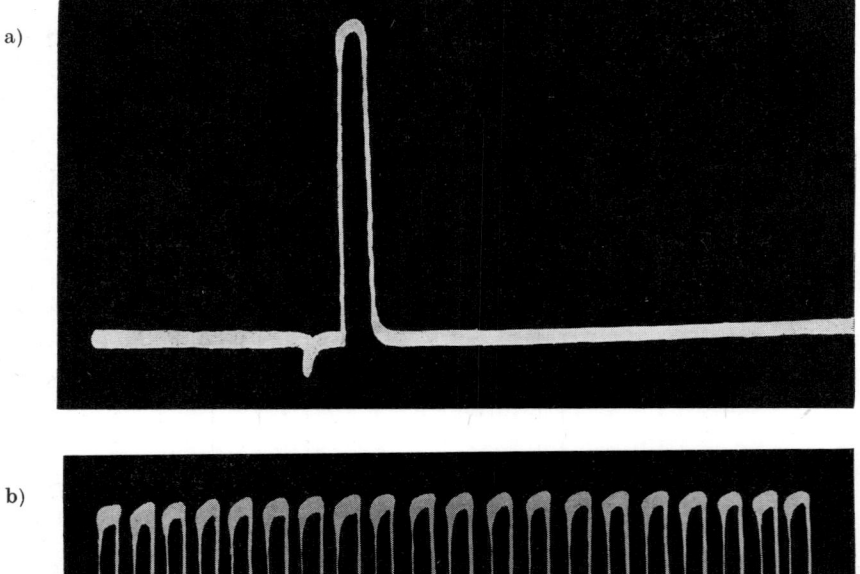

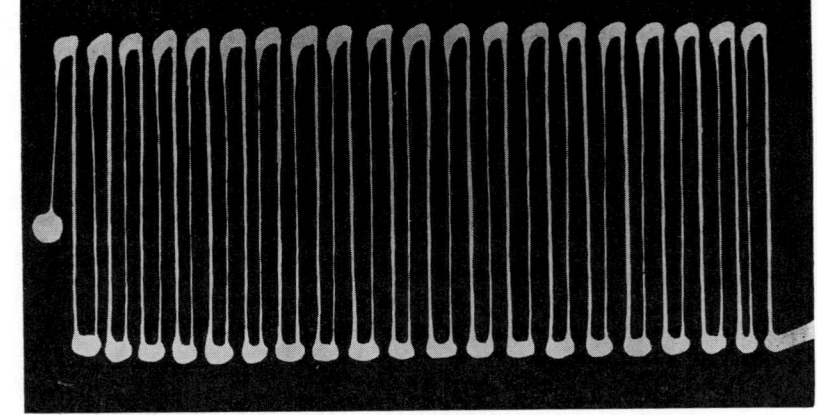

Figure 5a-12 Pulse shaping by the FPS and FS circuits; (a) pulse of the 8th digit at the GOZI output bus; (b) synchronization pulses; (c) pulses corresponding to a number at the input of the FPS; (d) pulses at the FPS output; (e) shape of the pulses (corresponding to a number) when they coincide with synchronization pulses.

c)
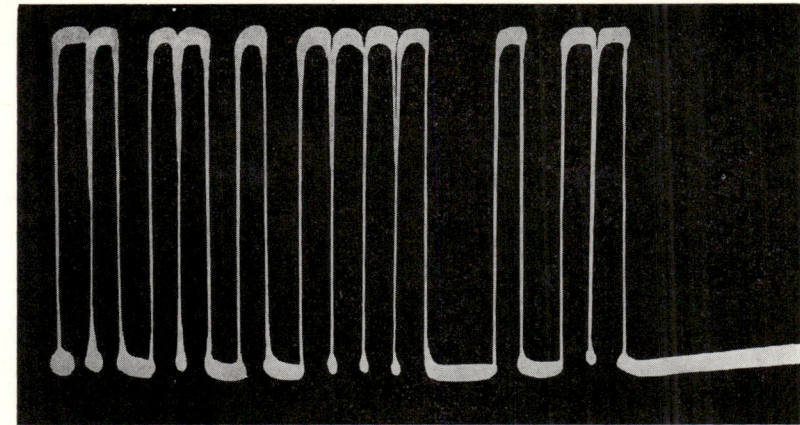

d)
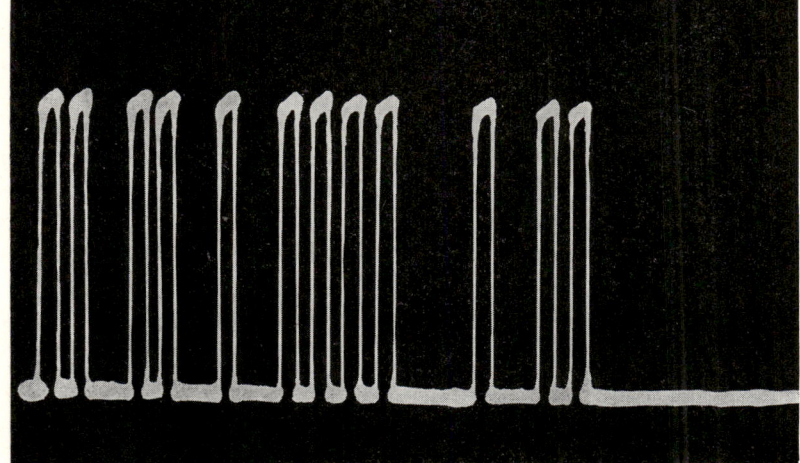

e)
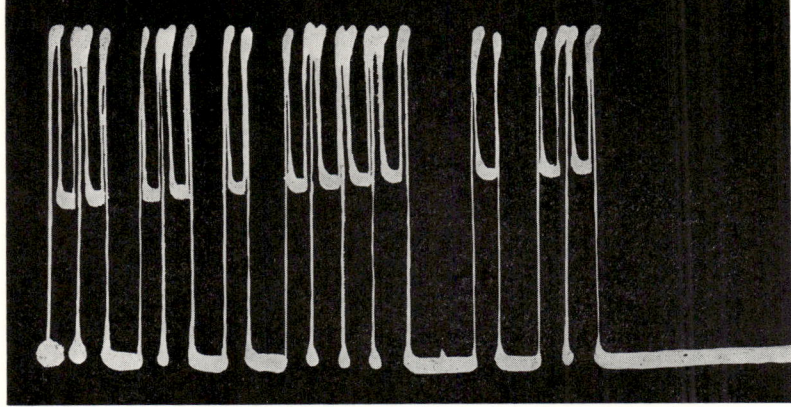

47

bus of the GOZI; (b) synchronization pulses; (c) pulses of the number at the input of the coincidence circuit; (d) pulses of the number at the output of the coincidence circuit; and (e) shape of the pulses (corresponding to a number) when they coincide with synchronization pulses.

Apart from the positive pulses ZI_1 to ZI_{19} the GOZI also generates the negative pulses $\overline{ZI}_1$ to $\overline{ZI}_{19}$ which coincide with the former in time.

3.1.4 Control console PU

To set the telescope on a given star, its coordinates are input into the ETsUM from the control console PU. These coordinates are the right ascension α and the declination δ. These quantities are represented by sequential binary numbers, and generated by a coded punched tape and ZI and $\overline{ZI}$ pulses. The principle of generation is illustrated by the circuitry given in Figure 5a-13.

The negative pulses $\overline{ZI}_1$ to $\overline{ZI}_{18}$ generated by GOZI and shifted in time by one digit arrive through the contacts Ks at the grid of the vacuum tube via the 18 separate buses. Each pulse coincides with the corresponding digit of the number. Depending on the combinations of closed and open contacts Ks, different binary numbers corresponding to these combinations will appear at R_g. For example, if the contacts 1, 2, 5, 8, 10, 14, 16 and 18 are closed, then at the anode of the tube the sequence of pulses corresponding to the binary number 101010001010010011 will be out.

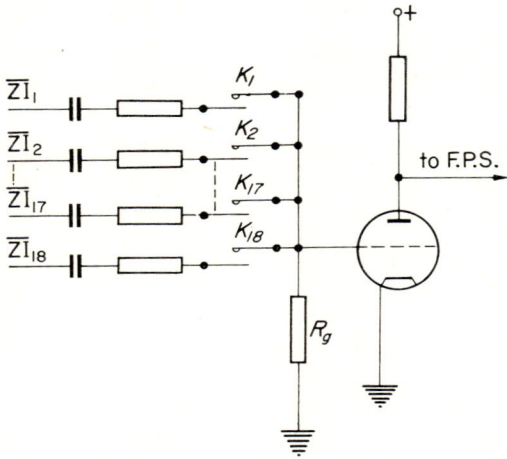

Figure 5a-13 Circuitry showing the generation principle of the coded values of α and δ at the control console PU.

The contacts are closed across holes in the coded punched tapes (Figure 5a-14) of the PU. The six binary numbers representing α in hours, minutes and seconds of time, and δ in degrees, minutes and seconds of arc respectively are generated by the circuit shown in Figure 5a-13. The punched tapes are set in the required position by dials on the PU. The quantity α is input with an accuracy of 1/3 sec and δ with an accuracy of 5″.

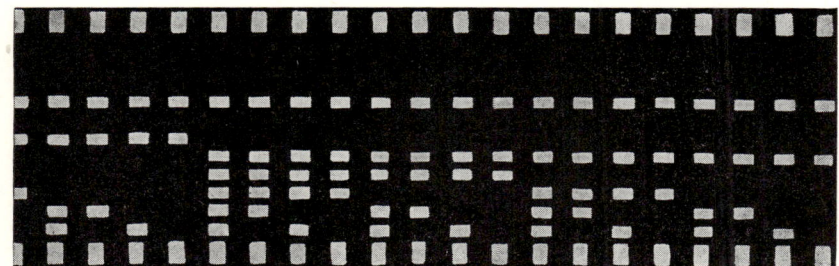

Figure 5 a-14 Sample of the coded punched tape in the control console PU.

In order to sum the hours, minutes and seconds of α, and also the degrees, minutes and seconds of δ, the adders Σ_2 and Σ_3 are used. The separate values of α and δ are input into the adders along common channels, but the addition takes place at different times. This is secured by the use of three identical switching units FPS; one of these is shown in Figure 5a-15. Each switch of the FPS consists of two vacuum tubes (V_α and V_δ) and an inverter I. The tubes used in the FPS are 6Zh8 pentodes.

To the first grid of V_α a switching pulse from the Track II (PI_{II}) is fed from the DPI, to the second grid, the signal α^h, and to the third grid, the synchronization pulses SI_1 from the GSI. Correspondingly, the pulse from the Track IV (PI_{IV}), the signal δ^o and the synchronization pulses SI_1 are fed to the tube V_δ.

The numbers α^h and δ^o generated from $\overline{ZI}$ pulses, like the synchronization pulses, are fed to the grids of the 6Zh8 tubes simultaneously in each of the four tracks, but the tubes are triggered only when there is coincidence between the signals on all three grids.

Therefore, the choice of the track in which the given numbers must be introduced into another circuit is determined by the pulses from the DPI which in the Track II unblocked the tube V_α and in Track IV the tube V_δ.

It is thus clear that the numbers α^h and δ^o arrive at the common output of the tubes. The width of the pulses for these numbers is determined by the synchronizing pulse width, since the latter are shorter. Thus, the switching circuit FPS, in addition, performs the role of synchronizer and pulse shaper (see waveshapes c, d and e in Figure 5a-12).

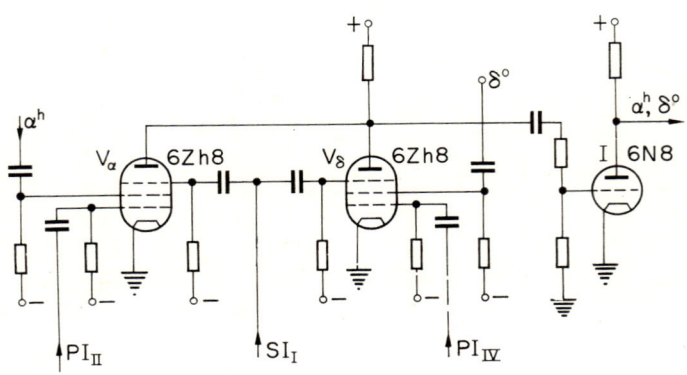

Figure 5a-15 Circuitry of the switching unit FPS.

3.2 ARITHMETIC AND LOGICAL UNITS

3.2.1 Adder Σ_1 and transformation of parallel numbers into sequential numbers

The adder Σ_1, is an 18-digit binary counter, consisting of multivibrators Tr_1 to Tr_{18} (see Figure 5a-16); this adder transmits the sidereal time S. Pulses from the DCh_2 are fed to the Σ_1 at a frequency $f_u \approx 3.034$ pulses (s sec)$^{-1}$. The value of the sidereal time (the number of periods in 1/3.034 s sec which has elapsed since the circuit was triggered, is given by the state of the multivibrators in Σ_1, i.e., by a parallel number.

For further processing, the parallel numbers obtained in Σ_1 are converted into sequential numbers. For this purpose, the digits of the number set up on Tr_1 to Tr_{18} are applied in the form of a voltage (high or low) to the input 1 of the corresponding vacuum tubes V_1 to V_{18}, which are blocked in the absence of a pulse at the inputs 2. During readout, a pulse from DCh_2 ($f_t \approx 12.136$ pulses (s sec)$^{-1}$) is fed to the input 2 of V_1 and this simultaneously triggers GOZI. The ZI pulses, time shifted by one bit, are applied to the inputs 2 of the corresponding tubes V_2 to V_{18}.

All the tubes will conduct in turn and will successively, digit by digit, output the number set up on the trigger register Tr_1 to Tr_{18}. The resulting sequential number, is fed to the inverse cascades I and through the shaping circuit FS to the subtractor Δ_1. An oscillogram taken at the output of the adder Σ_1 is shown in Figure 5a-17.

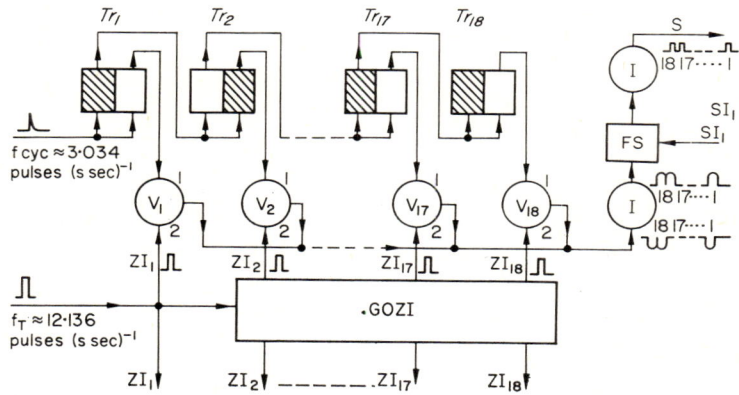

Figure 5a-16 Block diagram of the adder Σ_1 consisting of 18-digit binary counter.

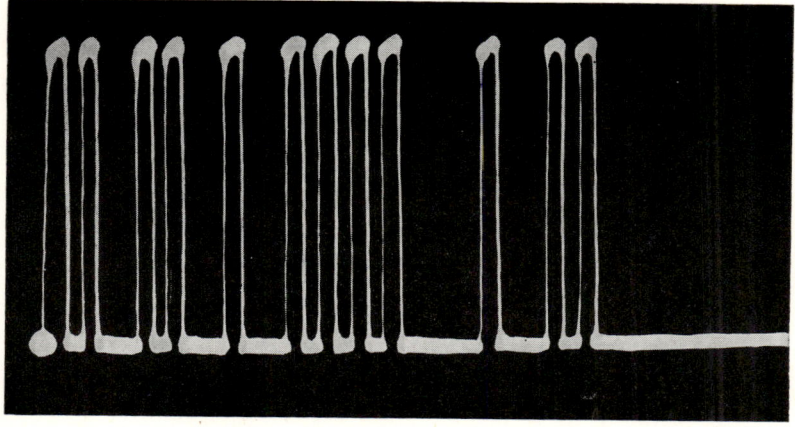

Figure 5a-17 Output pulses from the adder Σ_1.

3.2.2 Adders Σ_2 and Σ_3

The operation of the sequential arithmetic units of the ETsUM is based on the pulse amplitudes addition [6, 7].

When an adder is designed on this principle, the reliability of operation is reduced.

However, from our point of view, this drawback is compensated to a great extent by an increase in reliability due to the sharp reduction in the number of logical elements in the circuitry.

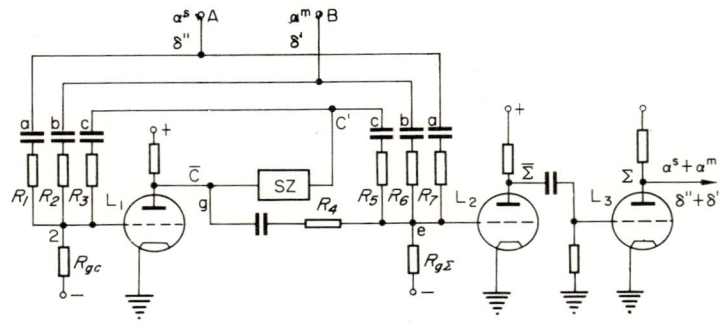

Figure 5a-18 Circuit of the adder Σ_2.

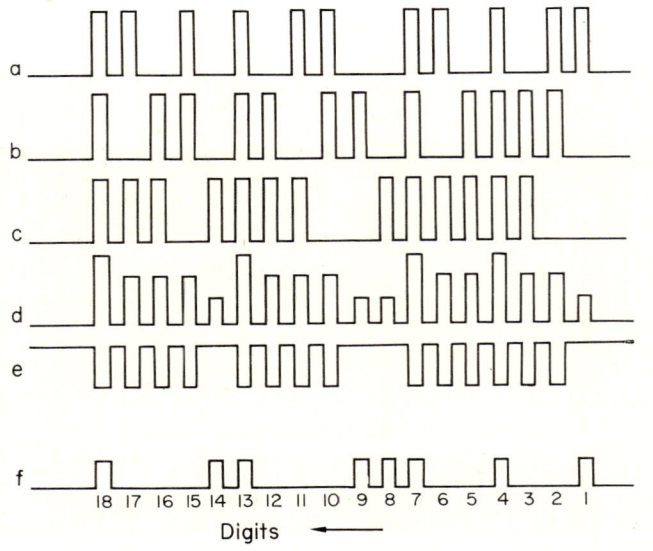

Figure 5a-19 Pulses at different steps of the binary addition process by the adder Σ_2.

In the circuit of the adder Σ_2 shown in Figure 5a-18, binary addition is executed as follows. Pulses corresponding to the codes of the first and second terms as well as the carry pulses (Figures 5a-19, a, b and c respectively) are fed at the inputs A, B and C' of the adder. The pulses are fed to the grids of the vacuum tubes L_1 and L_2 across the resistors R_1, R_2, R_3, R_5, R_6, R_7 (see Figure 5a-18); the added signal (Figure 5a-19, d) develops across R_{gc}. When L_1 is in a conductive state, due to two or three pulses fed simultaneously to the terminals A, B and C', if negatively bias the carry code (Figure 5a-19, e) is separated on the anode of (L_1) in the form of negative pulses.

In the delay circuit SZ, the code (Figure 5a-19, e) is delayed by one bit and applied to the bus C' (Figure 5a-19, c).

Apart from the positive pulses a, b, c, the negative pulses e, which are the carry signals before they are delayed, are fed across R_4. The amplitude of these pulses is such that it fully compensates the positive signal obtained in the presence of two pulses simultaneously at the outputs.

When pulses arrive simultaneously at all three terminals A, B, C' a negative pulse from L_1 compensates the positive pulse from the three signals by only 2/3 of its amplitude. Thus, in the addition of the signals a, b, c and e on the grid of L_2 a combination of pulses f, which are the sum of a and b, is separated out. The pulses of the sum Σ are amplified by the

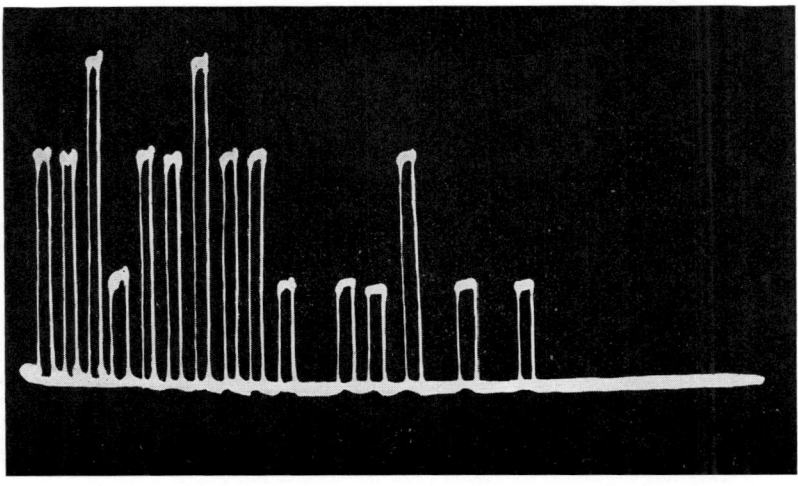

Figure 5a-20 Oscillograms at the input of an arithmetic circuit (adder) showing differences in pulse levels handled by the circuit.

triodes L_2 and L_3 which deliver positive pulses. These output pulses are fed from the adder Σ_2 to the adder Σ_3. The circuit of Σ_3 is similar to that of Σ_2 and differs only in the inversion stage for the codes of α and δ.

It has been shown in practice that arithmetic circuits based on the principle of voltage summation are very easily adjusted. The difference in amplitude levels for different combinations of signals at the inputs is about 30 V (Figure 5a-20). This is sufficient to secure reliable operation of the circuit.

3.2.3 Subtractors $\varDelta_1$ and $\varDelta_2$

The design principle of the ETsUM subtractor circuits is also based on the pulse amplitudes addition.

The logical operations of the substractors differ from those of the adders only in the output of the carry pulses to the next bit [6].

Therefore, the circuit of the subtractor $\varDelta_1$ (Figure 5a-21) which outputs the pulses $t = S - \alpha$ has been made similar to the one considered above, but the carry pulses are obtained by the additional tube L_1. The sidereal time $\bar{S}$, right ascension α and carry pulses C' delivered by this tube and delayed by the delay circuit SZ by one bit are fed in the form of sequential binary numbers to the grid of L_1 from the inputs A', B and C' across resistors. In this case a single pulse at the terminals B or C' or of two pulses simultaneously will turn L_1 conductive.

But in the presence of a positive pulse at one of the terminals B or C' and a negative pulse at the terminal A' the tube L_1 will remain blocked.

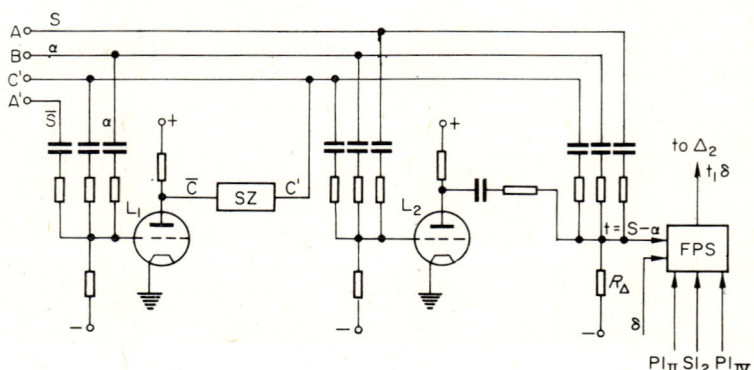

Figure 5a-21 Subtractor circuit $\varDelta_1$ of the ETsUM.

The electronic digital control system of the PM-700 telescope

When pulses are applied to all three terminals simultaneously, a negative pulse from A' compensates the total signal from the pulses at B and C' in accordance with the carry rules in the subtractor circuits.

The negative carry pulses $\bar{C}$ from L_1 are fed to the delay circuit SZ and then to the bus C' of the subtractor.

The circuit of $\varDelta_2$ does not differ in principle from that $\varDelta_1$.

3.2.4 The universal arithmetic unit UAU

The UAU is used to introduce the component corrections for mean refraction $\varDelta_{r_t}$ and $\varDelta_{r_\delta}$ in hour angle and declination. It may also be used to introduce any other corrections to the telescope position. The magnitude of the refraction components is calculated by the special computer SRM mounted on the telescope. To transform the analog quantities generated by the SRM into digital quantities, a shaft encoder unit KU [8] is used. The values of $\varDelta_{r_t}$ and $\varDelta_{r_\delta}$ are given in 6-digit binary code. The generation of the numbers in the KU is done with the use of the GOZI, as in the control consoles (Figure 5a-13).

From KU, the components $\varDelta_{r_t}$ and $\varDelta_{r_\delta}$ are fed to the switching circuit FPS, where they are divided in time and into the Tracks II and IV and sent to the UAU. The block diagram of the UAU is shown in Figure 5a-22.

The carry pulses $\bar{S}$ corresponding to the additions $t_1 + \varDelta_{r_t}$ and $\delta + \varDelta_{r_\delta}$ are delivered by the tube L_1 (6Zh8 type) and the carry pulses $\bar{S}$ corresponding to the subtraction $t_1 - \varDelta_{r_t}$ and $\delta_1 - \varDelta_{r_\delta}$ by L_2. The method of

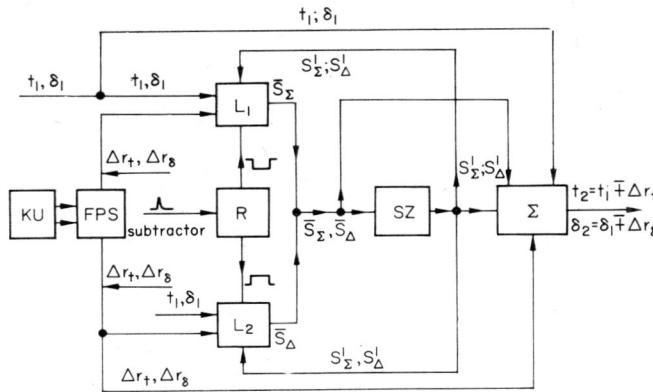

Figure 5a-22 Block diagram of the universal arithmetic unit UAU of the ETsUM.

selection of the signals $\bar{S}_\Sigma$ and $\bar{S}_\Delta$ is given in the descriptions of the adder Σ_2 and the subtractor Δ_1.

A relaxation oscillator R is used to control L_1 and L_2. When addition is being performed, L_1 is conductive and so the carry pulses S'_Σ arrive at the circuit Σ across the circuit SZ with a one-digit delay. In this case the carry pulses S'_Δ are applied to the circuit Σ.

The circuit Σ is an adder similar to the Σ_2 and Σ_3, and is used also in Δ_1 and Δ_2 to obtain the subtraction.

Depending on what carry pulses (S'_Σ or S'_Δ) are applied to Σ, we shall have the sum or the difference of the two numbers at its output.

3.2.5 Readout of the shaft angle encoders and code conversion

The reading of the telescope attitude is done at any time by means of encoders in the Gray system [4]. The values t'_1 and δ'_1 are readout by photomultiplier and neon tubes from the disk of the shaft angle encoders. The neon tubes are connected in parallel to the anode loads of the tubes in the GOZI which generate the negative pulses $\overline{ZI}_1$–$\overline{ZI}_{18}$.

The neon tubes are driven in the same sequence as the ZI pulses, i.e., with a shift in time by one bit. Therefore, each of them flashes exactly at the times when the given bit has to be readout.

As a result, the encoder readout signals at the anode of the photomultiplier is given in sequential Gray code. However, arithmetic operations in Gray code are complicated, and so it must be converted into binary code.

The transformation formula has the form

$$a_i = a_{i+1} \sim a'_i, \tag{5a-5}$$

using the notation of Hilbert and Ackermann [9]. Here a_i is the binary digit in the ith place; a'_i the representation of the ith place in Gray code.

In the notation of number theory this formula becomes

$$\sum_{j=i}^{n-1} a_j \equiv a_i \pmod{2}; \quad (i = 0, 1, 2, ..., n-1). \tag{5a-6}$$

The rule for sequential conversion from the Gray code into ordinary binary code using a single decoding trigger is simple if the Gray code is read starting from the higher digits.

Sequential reading of the Gray code starting from the lowest digit leads to indetermination. To determine a_0 we must know a'_{-1}, which is equal to zero if $a_0 = 0$ and to one if $a_0 = 1$. But since a_0 itself is unknown, the result is not uniquely defined.

The true result is always obtained in either one of the halves of the conversion trigger, but it is not known beforehand on which it will be. The result on the other half will be the inverse.

It is clear from the above that there are three methods for converting sequential Gray code into binary:

(1) Conversion starting from the highest digit. The resulting binary number must be stored (using a trigger register, for example) so that it can be read from the lowest digit.

(2) Conversion starting from the lowest digit. In this case the direct code of the two resulting codes (direct and inverse) is chosen also using a trigger register and a nineteenth pulse in the inverse code.

(3) A conversion trigger is set before the code is read [2].

In the ETsUM the first variant is used. The block diagram of the PK code converter is given in Figure 5a-23.

The quantities t_1' and δ_1' are read starting from the highest digit. The readout given in sequential Gray code is fed through the amplifiers U_t and U_δ to the vacuum tubes V_t and V_δ respectively.

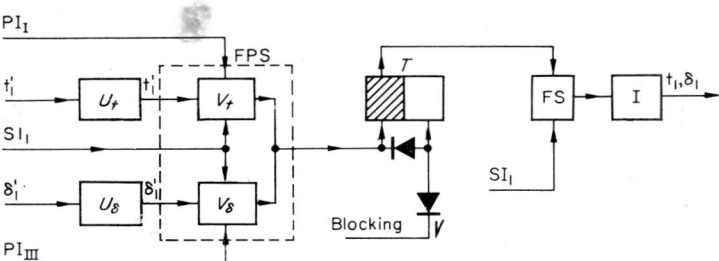

Figure 5a-23 Block diagram showing the principle of operation of the PK code converter.

Suppose that the shaft angle encoder output t_1' represented by the number a (see Figure 5a-24) is fed to one input of the V_t. The pulses b obtained by differentiating the synchronization pulses are fed to the second input of the V_t. The pulses c (PI_1 pulses from the DPI) which determines Track I, are fed to the third input of the V_t. The pulses d are obtained at the output of the V_t.

Every odd pulse in the series d, fed to both inputs of the trigger Tr, will trigger it, if T_r was first blocked by the pulse e, and every even pulse will block it. The signals f generated by the trigger are fed to one input of the shaping-coincidence circuit FPS and the synchronization pulses g to the FPS second input.

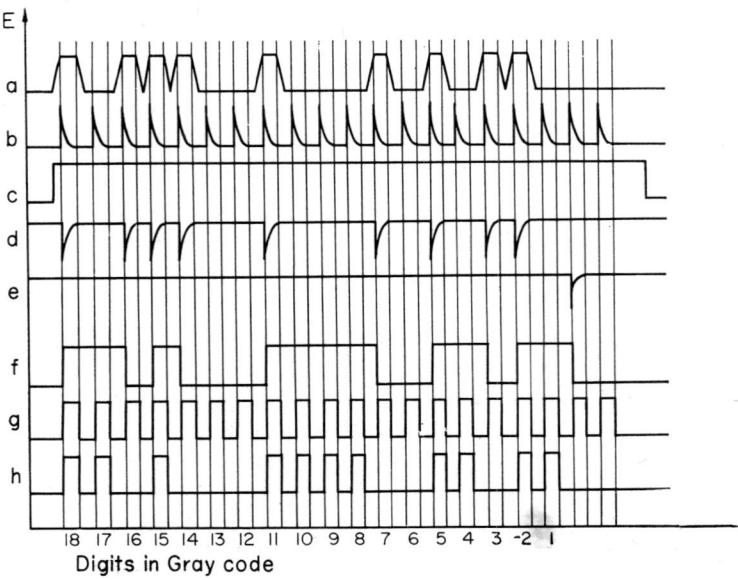

Figure 5a-24 Waveshapes of the signals during the conversion process of t_1' and δ_1 from sequential Gray code into binary code.

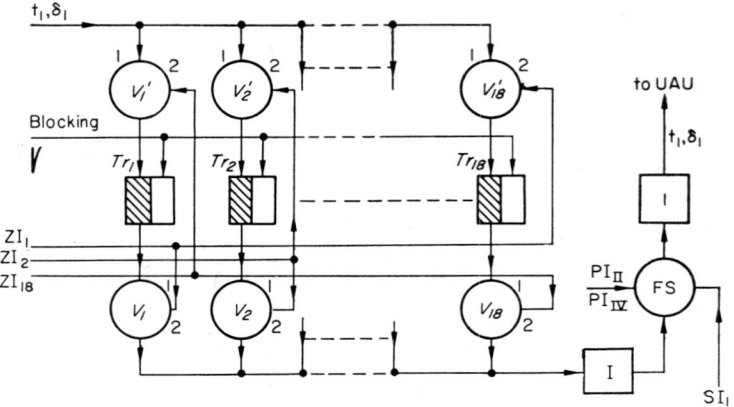

Figure 5a-25 Block diagram of the parallel trigger register in the PK code converter.

The electronic digital control system of the PM-700 telescope   59

As a result we obtain at the output the binary number h, corresponding to the Gray code number a.

Similarly, only in Track III will the readout of the shaft angle encoders for δ_1' be converted to the binary code.

The conversion method has as a main drawback that the resulting binary codes start from the higher digits. Therefore, before being fed into the arithmetic units, they are stored and readout again from the lower digits.

For this purpose a parallel trigger register (see Figure 5a-25) is provided in the PK.

The numbers t_1 and δ_1 are fed to the register in the Tracks I and III respectively, and are readout in the Tracks II and IV respectively. The sequential numbers are converted into parallel numbers and vice versa.

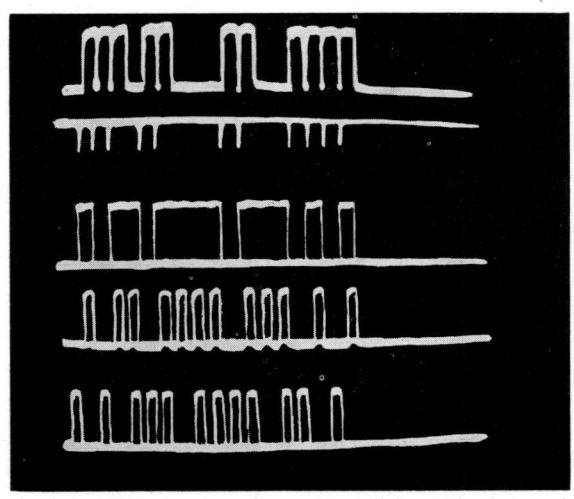

Figure 5a-26 Oscillograms showing the operation of the PK code converter (from sequential Gray code into binary).

At the start of Tracks I and III, i.e., before the numbers are fed into the register, the register is cleared by the pulse which is fed simultaneously to all triggers Tr_1 to Tr_{18}. The numbers t_1 and δ_1 are fed to the register by the pulses ZI_1 to ZI_{18} controlling the vacuum tubes V_1' to V_{18}'. To readout t_1 and δ_1 starting from the lower digits the vacuum tubes V_1 to V_{18} are used.

To illustrate the operation of the PK converter Figure 5a-26 shows the following oscillograms:

(a) Readout pulses in Gray code (starting from the upper digits).

(b) Triggering pulses in Gray code at the input of the FPS.

(c) Voltages at the anode of the conversion trigger.

(d) Binary code at the FS output.

(e) Readout of the PK register in binary code (starting from the lower digits).

3.3 OUTPUT UNITS OF THE ETsUM

3.3.1 Fine correction control units UUTK-t and UUTK-δ

The corrections of the telescope in the hour angle rate are made by the UUTK-t unit. This unit generates a rate proportional to the error signal.

The UUTK-t (Figure 5a-27) includes a parallel trigger register consisting of seven triggers Tr_1–Tr_7 in which the lowest digits of the quantity Δt are stored. The digits of the number set up in the register are sent in the form of voltages E_0, $2E_0$, $4E_0$, $8E_0$, $16E_0$, $32E_0$, $64E_0$ taken from the corresponding anodes of the Tr_1 to Tr_7 to the adding dc amplifier SU_{+t}.

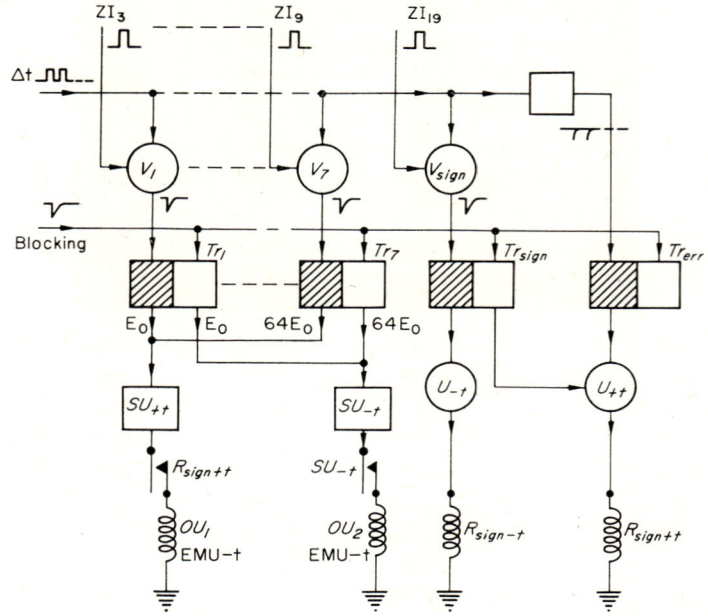

Figure 5a-27 Block diagram of the fine correction control unit UUTK-t.

Similarly, voltages are applied to the amplifier SU_{-t} from the other anodes of the triggers (negative pulses). The control voltages for the amplifiers allows a speed control with a dynamic range of 64 using the fine correction motor MTK-t. For errors in the limits $63E_0$–$127E_0$, the correction is done at a constant speed, corresponding to the value $63E_0$. The control voltage at the output of the SU_{+t} amplifier is used for positive values of the error, and the voltage from the SU_{-t} for negative values. The sign of the error Δt is stored in the sign trigger Tr_{sign}.

In the presence of a pulse in the 19th place of the number t (the error is then negative) Tr_{sign} is excited and energizes the relay R_{sign-t} via the amplifier U_{-t}. One of the contacts of this relay connects the control winding OU_2 of the EMU-t to the amplifier SU_{-t}. The fine correction motor operates to slow down the telescope motion.

In the absence of a pulse in the 19th place (the error is positive) the trigger Tr_{sign} is not excited and a high voltage in this case is applied to U_{+t}. But this condition alone is not sufficient to drive the amplifier U_{+t}. This amplifier is driven only when two high voltages are applied to it, from triggers Tr_{sign} and Tr_{err}. Trigger Tr_{err} is excited if at least one of the digits of Δt is one, i.e. it indicates the presence of an error.

Thus, the amplifier U_{+t} energizes the relay R_{sign+t} only in the presence of a positive error. In this case the control winding OU_1 of the EMU-t is fed from SU_{+t} and OU_2 is disconnected. The correction motor will now operate to speed up the telescope motion. In the absence of error both relays, R_{sign+t} and R_{sign-t} are de-energized, therefore, no correction is introduced. Disconnecting the control windings for an error equal to zero

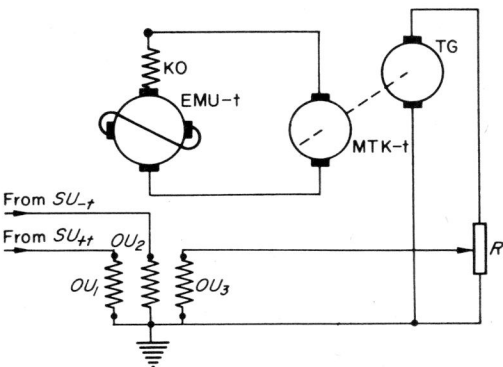

Figure 5a-28 Circuitry of the hour angle fine correction drive.

has greatly simplified the problem of keeping zero voltage in the adding amplifiers SU_{+t} and SU_{-t} and thus assuring the absence of sporious signals in the motors. The amplifiers SU use 6N2P triodes and have cathode followers at the output. The hour angle fine correction uses the amplidyne EMU-t as shown in Figure 5a-28; the amplidyne is of the EMU-3a type [10]. For fine correction a motor SL-369 is used, and for introducing negative velocity feedback a tachometer generator TG, SL-221 A type is used. The feedback voltage is applied to the third control winding OU_3 of the EMU-t.

The control system for δ is exactly similar to the t control system.

3.3.2 Coarse setting control units UUGN-t and UUGN-δ

Coarse setting of the telescope in hour angle and declination uses AOL-011/4 type motors (MGN-t and MGN-δ in the block diagram of Figure 5a-1). The rate of coarse setting is constant, and is 75° per minute.

Two similar units, the UUGN-t and UUGN-δ are provided in the ETsUM to control these motors.

The block diagram of the UUGN-t is shown in Figure 5a-29. The error Δt computed in sequential code in the circuit Δ_2 is fed to UUGN-t.

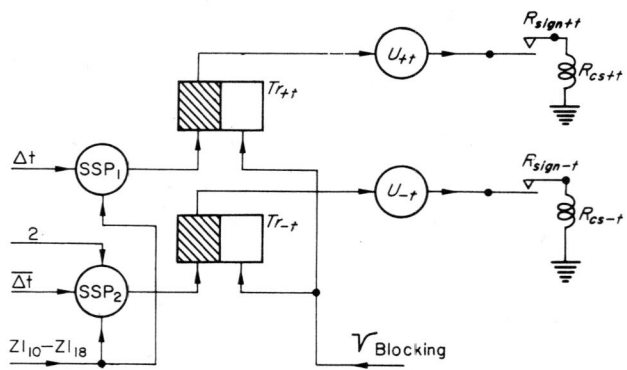

Figure 5a-29   Block diagram of the coarse setting control unit UUGN-t.

The coincidence circuits SSP_1 and SSP_2 are so designed that pulses starting from the 10th place of the number are omitted in the control triggers Tr_{+t} and Tr_{-t}. The positive pulses fed to Tr_{+t} and the negative to Tr_{-t}.

Pulses (ones) at the output excites the triggers Tr_{+t} and Tr_{-t} which in turn energizes the coarse setting relays R_{cs+t} and R_{cs-t} through the amplifiers U_{+t} and U_{-t}, reversing the sense of rotation of the servomotor.

The coils of the relays R_{cs+t} and R_{cs-t} are connected to the amplifiers across the contacts of the corresponding relays R_{sign+t} and R_{sign-t} used in the fine correction control unit UUTK-t; this allows for the sign of the error (Figure 5a-27). Direction of rotation of the motor, depends upon the closure of the given relay. Thus, two conditions are required to connect each coarse setting relay: the excited state of the trigger and the closure of the contact of the corresponding sign relay. In our circuit, these two conditions allow the magnitude of the error and its sign to be uniquely defined for all possible combinations of pulses in the number Δt.

4 CONCLUSION

All the basic blocks of the electronic digital control system ETsUM for a telescope were completed and had been debugged by the end of 1957. During 1958 the ETsUM was used for various auxiliary purposes.

At the same time, the operation of the blocks of the ETsUM itself were checked. Everyday use of the ETsUM showed that it operated sufficiently reliably and smoothly. A final estimate of its reliability, the accuracy which it obtains and the convenience for the astronomer can only be made after it has been used under operating conditions.

Results of the tests will be published at a later date.

The computer is of an experimental nature, and so the ETsUM was constructed mainly using electron tubes. In the future, it will be better to use more reliable circuits using semiconductors and ferrites. This will enable us to construct smaller, more reliable and economical instruments. The complete programming of the whole operation of the telescope justifies the use of electronic digital control not only for altazimuth and horizontally mounted telescopes, but also for telescopes using equatorial mounts.

4.1 ACKNOWLEDGEMENTS

The control system we have described was developed under the direct supervision of N. N. Mikhelson. A. V. Dravskikh, V. S. Sumin and A. V. Koroler assisted in the design and assembly of the separate blocks (power supplies, power amplifiers for the MUS and frequency dividers DCh_1 and DCh_2). To these, the author wishes to express his gratitude.

4.2 NOTE FROM THE EDITOR (ENGLISH VERSION)

Contribution No. 5 "The control system of the PM-700 telescope" by Yu. A. Belyaev *et al.* which appeared in the original Proceedings of the First Conference on *New Techniques in Astronomy* has been replaced in

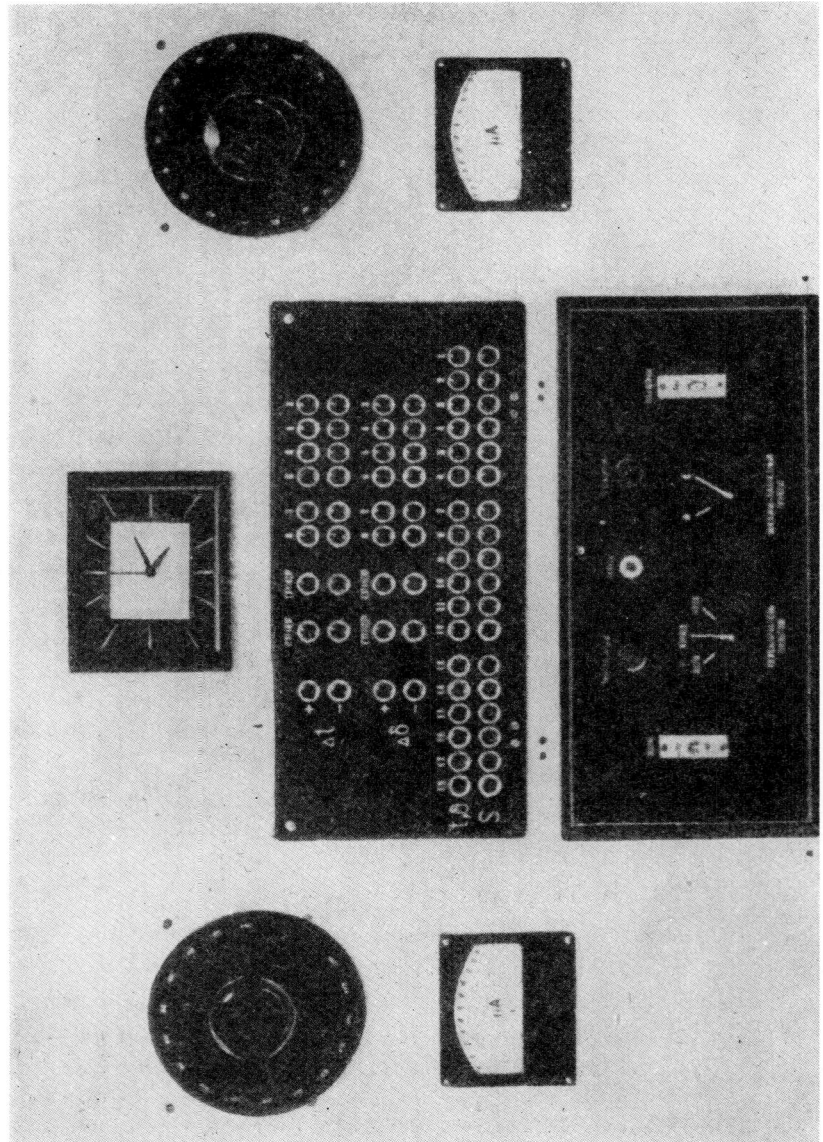

Figure 5a-30 General view of the electronic digital control system ETsUM console of the PM-700 telescope.

The electronic digital control system of the PM-700 telescope

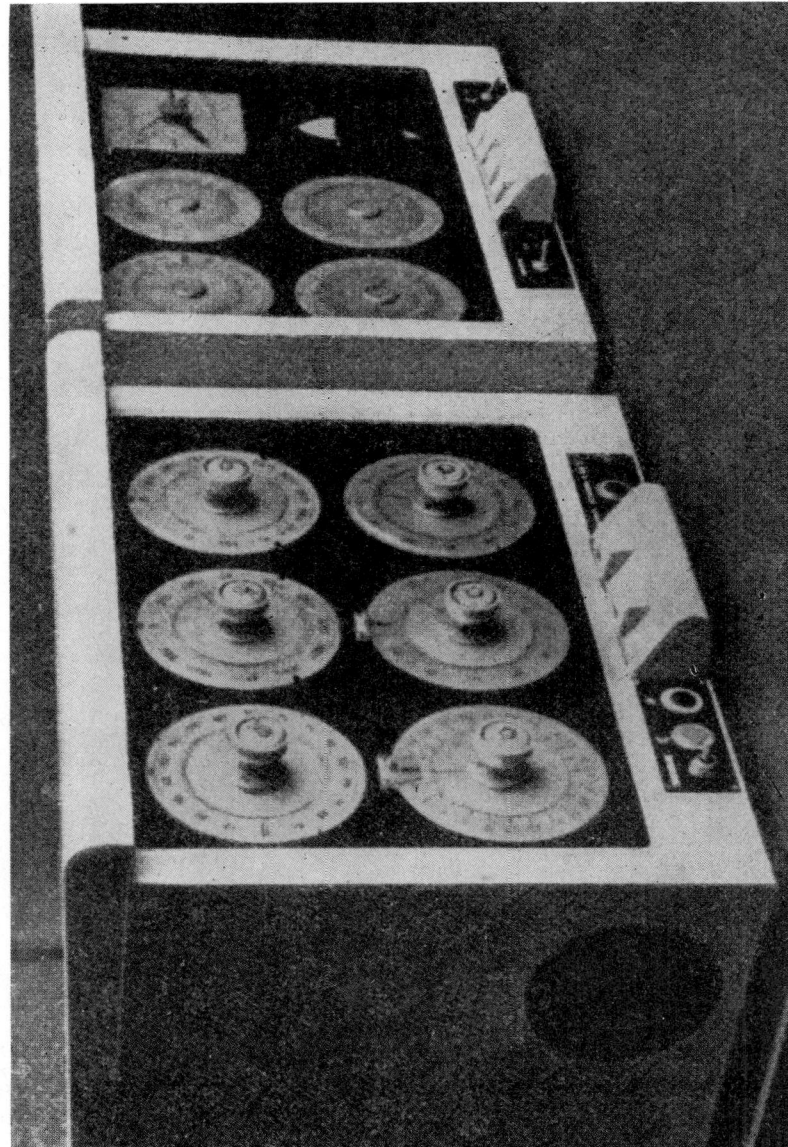

Figure 5a-31 Control of the PM-700 telescopes. Left—the automatic control console; right—the semiautomatic control console.

Figure 5a-32 Velocity control mechanism MUS for lunar–planetary RA drive mechanism with covers removed.

this English version by a comprehensive contribution, No. 5a. The substitution was made in order to give a clearer description of the work presented in Contribution No. 5. Nevertheless, in that contribution there appears a series of photographs (Figures 5a-30 to 5a-35) which are absent from Contribution No. 5a and which show different units and subassemblies of the ETsUM. The inclusion of these photographs into the present text will serve to give the reader an idea of the type of hardware used in the ETsUM. Figures 5a-36 and 5a-37 are photographs of the PM-700 telescope

Figure 5a-33 Shaft angle encoder for the declination axis with covers removed.

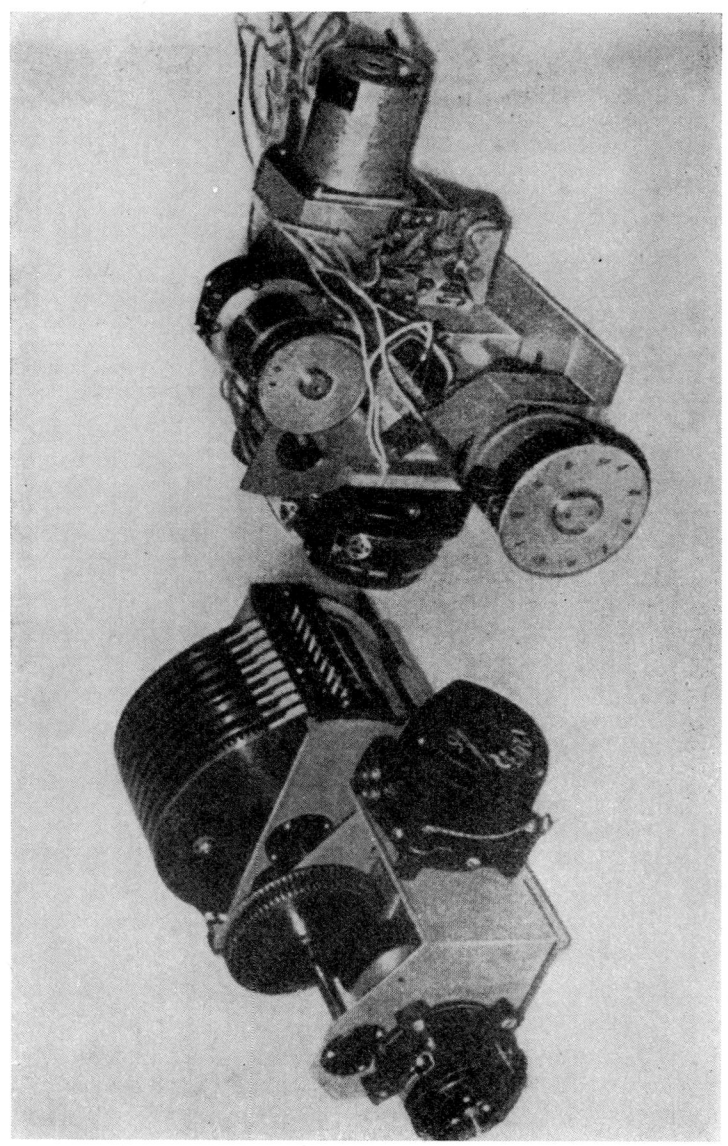

Figure 5a-34 Units for refraction correction with the declination lunar–planetary drive.

Figure 5 a-35 Analog computer SRM for mean refraction correction. The hemispherical protective housing has been removed.

Figure 5a-36 Photograph of the PM-700 telescope at the Main Astronomical Observatory, Pulkovo. At the central part of the mount (intersection of the polar and declination axes) the computer SRM for mean atmospheric refraction correction can be seen. At the right side of the photograph the automatic control console can be seen and next to it (right) the semiautomatic control console (photograph courtesy of the USSR Academy of Sciences).

Figure 5a-37 Photograph of the PM-700 telescope building at the Main Astronomical Observatory at Pulkovo (courtesy of the USSR Academy of Sciences).

and building, respectively, which were obtained directly by the Editor from the USSR Academy of Sciences and which are also included in the present text for the same purpose.

REFERENCES

1. N. N. Mikhelson, "A horizontal telescope mounting", *Izv. GAO AN SSSR*, **22**, Issue 4, No. 169, 131–140 (1961).
2. Yu. A. Belyaev and N. N. Mikhelson, *Report on the Development of a Method for Driving a Telescope Using an Electronic Computer* (Otchet o razrabotke maketa privoda teleskopa s primeneniyem elektronnoi vychislitel'noi mashiny) (in manuscript), Pulkovo (1957).
3. V. M. Konshin and N. N. Mikhelson, "A driving mechanism of a telescope for observations of the moon and planets", *Izv. GAO AN SSSR*, **22**, (1961).
4. J. J. J. Kernahan, "A digital wheel", *Bell Lab. Record*, **32**, 126 (1954).
5. N. N. Mikhelson, "A computer which corrects the position of a telescope for mean refraction", *Izv. GAO AN SSSR*, **21**, No. 162, 149 (1958).
6. D. Yu. Panov (Ed.), *High-Speed Computers* (Bystrodeistvuyushchiye vychislitel'nye mashiny) IL, Moscow (1952).
7. R. K. Richards, *Arithmetic Operations in Digital Computers*, Van Nostrand's University Series in Higher Mathematics (1955).
8. F. V. Maiorov, *Electronic Regulators* (Elektronnye regulyatory), Gostekhizdat, Moscow (1956).
9. D. Hilbert and V. Ackermann, *Principles of Mathematical Logic* (Grundzüge der Theoretischen Logik) (Trans. Hammond, Leckie and Steinhardt, Ed. Luce) Chelsea Publishing Co., New York (1950).
10. A. S. Belonovsky and B. M. Mensky, *Automated Drive Mechanisms Using Amplidynes* (Avtomatizirovannyi privod s elektromashinnym usilitelem) Voenizdat, Moscow (1956).

CHAPTER **I-6**

An automatic guiding and setting system for altazimuth telescopes

YU. A. SABININ and P. V. NIKOLAYEV

*Institute of Electromechanics on Automation and Machine Construction
Leningrad*

The stellar interferometer designed by V. P. Linnik is one of several astronomical instruments with an altazimuth mount. Its automatic guiding system has the basic elements and components required for the control of any altazimuth telescope.

The construction of giant telescopes has forced designers to consider altazimuth mount systems because of their advantages compared to the equatorial mounts. These advantages include simpler construction of the moving parts of the telescope, stiffness of the telescope mount kept mainly on the vertical plane (which allows weight reduction) and the suitability of hydrostatic bearings for both elevation and azimuth axes. Also this type of mount gives favorable load conditions of the mirrors since they rotate only (with respect to the vertical) around the elevation axis, making the design of a reliable support system simpler. Corrections for atmospheric refraction and flexure in an altazimuth telescope are needed only for one axis (elevation), which simplifies the introduction of automatic control.

Despite the obvious advantages of such a system there is not yet a single large optical astronomical instrument with altazimuth mount and automatic guiding. The first of such instruments, the Linnik stellar interferometer, is now nearing completion.

The sidereal drive mechanisms used in the equatorially mounted telescopes are unsuited for altazimuth telescopes, since in this case the apparent motion of a star must be compensated by driving the telescope around the two axes and at a variable speed rate. The rates of rotation in azimuth A

and in zenith distance z can be computed using the following expressions:

$$\frac{dA}{dt} = \frac{1}{\sin z}[\sin \varphi \sin z + \cos \varphi \cos z \cos A], \qquad (6\text{-}1)$$

$$\frac{dz}{dt} = \cos \varphi \sin A, \qquad (6\text{-}2)$$

where φ is the local latitude.

It follows from Eq. (6-1) that the rate of A, can take any value from 0 to ∞, depending upon the coordinates of the object, and from Eq. (6-2) that the rate for z can vary from 0 to $\cos \varphi$ rev day^{-1}.

With present-day techniques, precision drives with a wide range of velocity control cannot be constructed. Thus, the limit in the azimuth guiding rate for the interferometer has been taken as 8 rev day^{-1}. Equation (6-1) shows that, for $\varphi = 60°$ and the chosen rate limit, a "blind" zone exists at the zenith subtending a solid angle of 8° aperture. However, with the operating speeds from 0 to 8 rev day^{-1} around the azimuth axis and from 0 to 0.5 rev day^{-1} around the elevation axis ($\cos 60° = 0.5$), with the possibility of zero speed and reversing, a single servomechanism cannot provide the smooth guiding of the interferometer over the entire speed range and with the accuracy of sidereal drives, even with ideal conversion of the equatorial coordinates to A and z.

The guiding accuracy is improved by the use of two independent servomechanisms: (1) a coarse guiding servomechanism which operates in conjunction with a coordinates converter, and (2) a photoelectric automatic guiding servomechanism which "locks on" the star in the optical field of the guider. The photoelectric servomechanism, in this case, corrects the errors of the coarse guiding system. The equatorial coordinates are transformed into horizontal coordinates by means of an analog converter which uses rotary transformers.

In the horizontal system (X', Y', Z') and the equatorial (X, Y, Z), the Z-axis coincides; it lies on the plane of the horizon (Figure 6-1(a)) and points to the west. Therefore, to change from equatorial to horizontal coordinates, it is sufficient to rotate the first system around the Z-axis by the angle $\theta = 90° - \varphi$ where φ is the local latitude. This is due to the fact that the Y'-axis in the horizontal system is perpendicular to the plane of the horizon (Figure 6-1(b)), and the Y-axis in the equatorial system points to the pole of the celestial sphere at an angle φ with respect to the horizon. From Figure 6-1(b), the horizontal coordinates of the unit vector

$\bar{B}$ given as a function of the equatorial coordinates are:

$$x' = x \cos \theta - y \sin \theta = x \sin \varphi - y \cos \varphi,$$
$$y' = x \sin \theta + y \cos \theta = x \cos \varphi + y \sin \varphi, \qquad (6\text{-}3)$$
$$z' = z.$$

The position of an object on the celestial sphere is usually given by two angles: in the equatorial system by the hour angle t and the declination δ, and in the horizontal system by the azimuth A and the zenith distance z, or sometimes, instead of z, the elevation h (Figure 6-2(a), (b)).

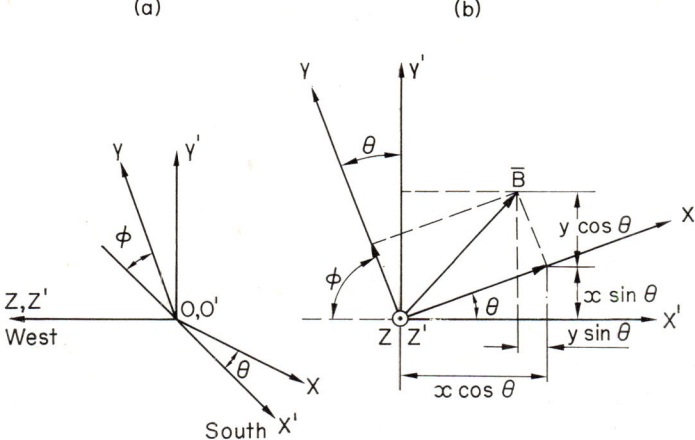

Figure 6-1 Relationship between the equatorial and horizontal coordinates systems.

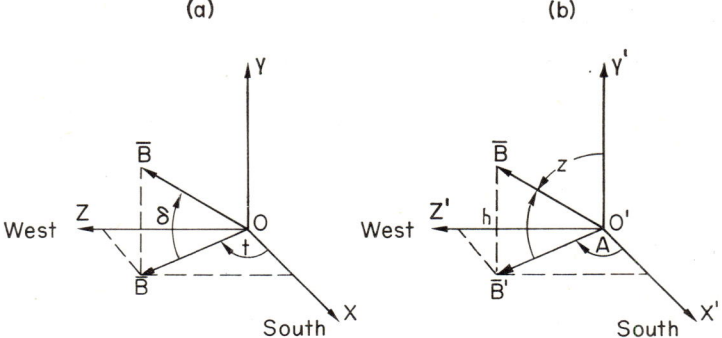

Figure 6-2 Coordinates of the unit vector in the equatorial (a) and horizontal (b) coordinates systems.

The actual values of A and z for given values of t and δ are obtained automatically by a coordinates converter developed by the Institute of Electromechanics. This converter has rotary transformers VT as the basic elements and performs successively the following operations:

(1) It computes the components x, y, z, of the unit vector $\bar{B}$ defined by the angles t and δ in the equatorial system;

(2) from x, y, z, it computes the projections of the unit vector in the horizontal coordinates system x', y', z';

(3) from x', y', z', the coarse guiding servomechanism, computes the angles A and H of the unit vector in the horizontal system.

A simplified diagram for a similar coordinates converter is shown in Figure 6-3. The simplification consists of removing the scaling from the circuit and taking all the transformation coefficients of the VTs as $K = 1$.

When the rotor of VT_δ turns an angle δ, voltages proportional to $\cos \delta$ and $\sin \delta = y$ (Figure 6-2(a)) appear at its secondary windings. The voltage proportional to $\cos \delta$ is fed to the primary winding of VT_t. When the rotor of VT_t turns an angle t, the signals

$$\cos \delta \sin t = z = z',$$
$$\cos \delta \cos t = x. \tag{6-4}$$

are induced in its secondary windings.

During adjustment, the rotor of VT is set to an angle φ. Thus when signals proportional to
$$\sin \delta = y$$
and
$$\cos \delta \cos t = x \tag{6-5}$$

are fed into its primary windings, the voltages

$$\cos \delta \cos t \sin \varphi + \sin \delta \cos \varphi = x \sin \varphi + y \cos \varphi = x'$$
$$\cos \delta \cos t \cos \varphi + \sin \delta \sin \varphi = x \cos \varphi + y \sin \varphi = y', \tag{6-6}$$

appear at the secondary. Then signals proportional to $-z'$ and x' are fed to the primary of VT_A, the rotor of which is coupled to the azimuthal axis of the telescope. In the secondary windings of VT_A the following signals are induced:
$$-x' \sin a + z' \cos a,$$
$$x' \cos a + z' \sin a, \tag{6-7}$$

where a is the angle which defines the position of the VT_A rotor or, in other words, the angle of rotation of the telescope around the A axis.

An automatic guiding and setting system for altazimuth telescopes 77

The voltage $-x' \sin a + z' \cos a$ is fed to the input of the electronic amplifier and then to the servomotor after power amplification. The servomotor drives the telescope and the rotor of VT_A in azimuth until the voltage at the amplifier input becomes equal to zero, i.e. until

$$-x' \sin a + z' \cos a = 0. \tag{6-8}$$

This condition will be satisfied when $\cot a = x'/z'$ and the angle a itself is equal to the required angle A, as shown in Figure 6-2(b).

The voltage at the secondary of the VT_A rotor

$$x' \cos a + z' \sin a \tag{6-9}$$

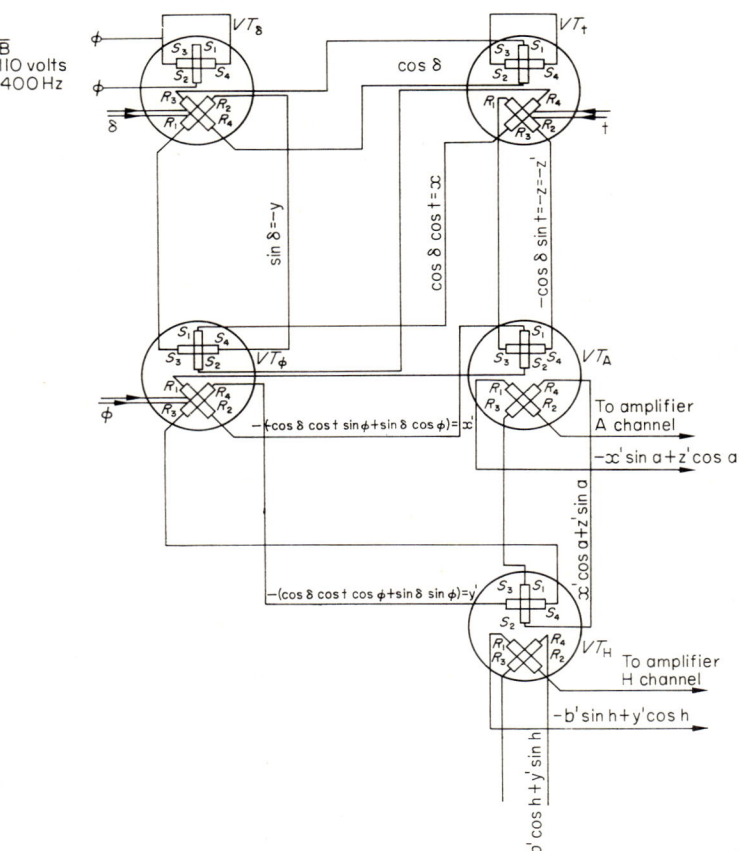

Figure 6-3 Simplified circuit diagram of the coordinates converter.

when $a = A$, i.e. when

$$\left.\begin{aligned} \sin a &= \frac{z'}{\sqrt{x'^2 + z'^2}} \\ \cos a &= \frac{x'}{\sqrt{x'^2 + z'^2}} \end{aligned}\right\} \quad (6\text{-}10)$$

expresses the projection $\bar{B}'$ of the vector $\bar{B}$ on the plane $X'O'Z'$ (Figure 6-2(b)), as can be seen by substituting the values of $\cos a$ and $\sin a$ in Eq. (6-9):

$$x' \cos a + z' \sin a = \frac{x'^2}{\sqrt{x'^2 + z'^2}} + \frac{z'^2}{\sqrt{x'^2 + z'^2}} = \sqrt{x'^2 + z'^2} = b', \quad (6\text{-}11)$$

where b' is the modules of $\bar{B}'$. Voltages proportional to b' and $-y'$ are fed at the primary of VT_H and the signals

$$\begin{aligned} -b' \sin h + y' \cos h, \\ b' \cos h + y' \sin h. \end{aligned} \quad (6\text{-}12)$$

are taken from the secondary, and h is the elevation of the optical axis of the instrument.

The voltage $-b' \sin h + y' \cos h$ is fed to the input of the amplifier of the channel H, which drives the motor of the H-axis. The motor rotates the H-axis of the telescope and the rotor of VT_H coupled to it until the following relationship is satisfied:

$$-b' \sin h + y' \cos h = 0$$

or

$$\cot h = b'/y'. \quad (6\text{-}13)$$

It is clear from Figure 6-2(b) that $\cot H = b'/y'$, and so Eq. (6-13) is satisfied only when $h = H$.

The output voltage $b' \cos h + y' \sin h$ at the secondary of the rotor of VT_H is not part of the solution of the problem, but physically when $h = H$, it is a quantity proportional to the modulus of $\bar{B}$. This can be seen by expressing $\cos h$ and $\sin h$ in in terms of $\tan h$ and putting $h = H$ in the second expression given in Eq. (6-12). We then have

$$b' \cos h + y' \sin h = \frac{b'^2}{\sqrt{b'^2 + y'^2}} + \frac{y'^2}{\sqrt{b'^2 + y'^2}} = \sqrt{b'^2 + y'^2} = B. \quad (6\text{-}14)$$

where b is the modulus of $\bar{B}$.

For normal operation of the servomechanism for the A axis and over the entire range, a unit is included to keep constant gain amplification (ARU). This is required since the sensitivity of the azimuth A error transmitter (VT_A), depends on the coordinates of the celestial object. When the error signal of the servomechanism for the A-axis satisfies the equation

$$-x' \sin a + z' \cos a = 0, \qquad (6\text{-}15)$$

the voltage at the secondary of the rotor of VT_A is proportional to the projection b' of the unit vector $\bar{B}$ on the plane $X'O'Z'$ and since the modulus of $\bar{B}$ is taken equal to unity, $b' = \cos H$. Thus, the voltage at the secondary of the VT_A, corresponding to an error of $90°$, does not remain constant during the operation of the coordinates converter, but changes as the height of the object changes as $\cos H$. Therefore for any other values of the error, the voltage U at the input of the amplifier for the A channel, will be a function of H and can be expressed by the relationship

$$U = \cos H \sin a_1, \qquad (6\text{-}16)$$

where a_1 is the error angle.

The ARU system must therefore change the gain of the amplifier so that for any changes in H the voltage at its output shall be a function of the error angle a_1 only. For the H axis an ARU is not required, since the maximum voltage is proportional to the modulus of $\bar{B}$ and does not depend on the coordinates of the object.

In practice, the transformation coefficients of a VT are different from 1 and when parallel circuits with a different number of VTs are used variation of the signal scale is unavoidable. In order to match the scales, the actual network of the coordinates converter has three scaling VTs to ensure that two signals transmitted to any of the decision VTs are of the same scale. In other respects coordinate conversion with scaling VTs is the same as described above. The complete block diagram of the coordinates converter and the coarse guiding servomechanisms are shown in Figure 6-4. The computer section of the diagram consists of five sine–cosine VTs and three MVTs. Of these, VT_A and VT_H are mounted in the telescope. It is planned for using readout units and handwheels to set the required values of $\varphi, \delta, s, \alpha$ and t. The sidereal time transmitter is a synchronous motor fed from a 220-V, 50-Hz supply. The value of the hour angle t is calculated from the formula $t = s - \alpha$ by algebraic summation using a mechanical differential. The servomechanism for each of the axes (A and H) consists of an electronic amplifier US, an amplidyne EMU, a servomotor ID, a tachometer generator TG, a reducer R and the corresponding error transmitters VT_A and VT_H. Strong negative feedback will be used in the servomechanisms to stabilize them and to increase smoothness of operation. It will be rigid* with respect to velocity and flexible† with respect to the voltage of the EMU. The coarse guiding servomechanisms of the interferometer together with the coordinates converter can also set the instrument automatically on a given object. For this purpose it is necessary only to have a unit which could change the transmission ratio from the servomotors to the drive of the A and H axes to increase the speeds. For the interferometer we chose a 200 rev day^{-1} setting rate; the change in the trans-

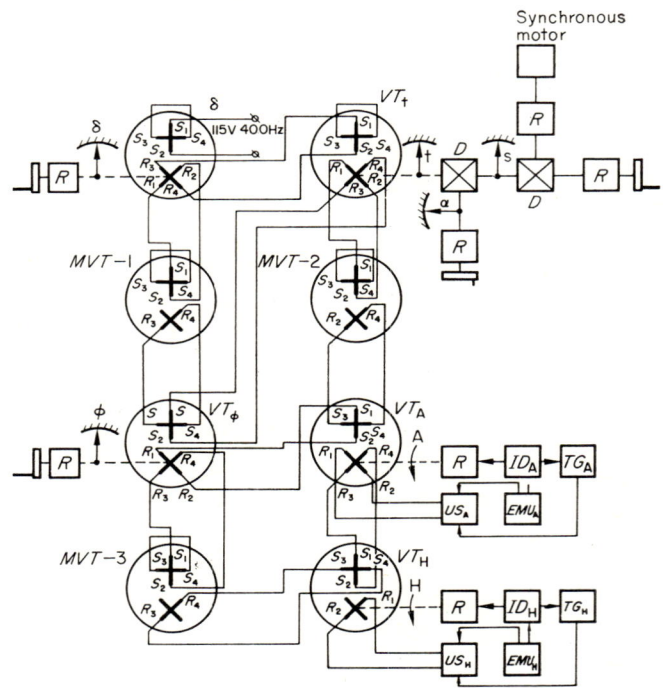

Figure 6-4 Complete block diagram of the coordinates converter and coarse guiding servomechanisms.

* In the Russian literature the term "rigid feedback" is used to designate negative feedback systems, when the amplitude of the feedback signal is proportional to the output signal. (Ed. English version.)

† The term "flexible feedback" is used to designate negative feedback systems, when the amplitude of the feedback signal is proportional to a given function of the output signal, e.g. its derivative or integral. (Ed. English version.)

mission ratios of the reducers are obtained by means of electromagnetic clutches. To set the interferometer on a star, the coordinates α and δ are fed into the coordinates converter, and then the servomechanism is switched on and at the rate of 200 rev day^{-1} the telescope is set at the selected attitude. Since the sidereal time transmitter operates continuously, the values of the hour angle are also fed continuously to the input of the coordinates converter. The actual values of A and H are generated by means of the coarse guiding servomechanisms, allowing continuous coarse guiding of the interferometer on the star. Transition from the setting mode to the guiding mode is made by switching the reducers back to guiding speeds. Laboratory tests of this coarse guiding and setting system for an altazimuth telescope showed that the error in the angle of rotation of the telescope axes does not exceed 7' over the entire operating range (from $z = 4°$ to $z = 90°$).

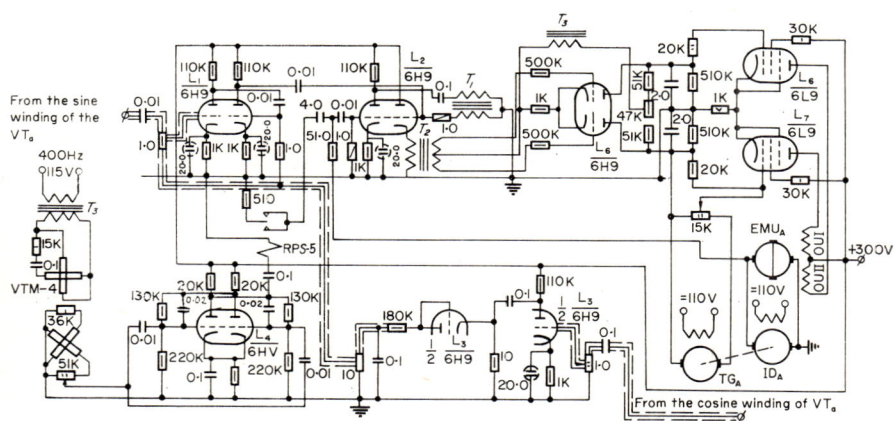

Figure 6-5 Circuit diagram of the coarse azimuth guiding servomechanism of the altazimuth mounted interferometer at the Main Astronomical Observatory (Pulkovo).

The main circuit for the coarse azimuth guiding servomechanism of the interferometer is shown in Figure 6-5. The error signal from the sine VT_A is amplified by a two-stage RC coupled amplifier using the vacuum tube L_1. The gain of these stages are regulated by the ARU system. The control signal for the ARU is the voltage taken from the cosine VT_A. This signal feeds the grid of the tube L_3, which is the amplifier and rectifier of ARU. The rectified signal of the ARU, smoothed by an RC filter, is fed to the grids of L_1 as negative feedback. The larger the signal of the ARU, the larger the negative bias at the grids of L_1 will be and, correspondingly, the smaller the gain. From L_1 the error signal is fed to the input of the right half of L_2 which is an adding cathode follower with transformer output. The voltage of the EMU (flexible negative feedback) amplified by the left half of L_2 is also fed to the input of the right half of L_2 through the transformer T_1. The flexible feedback signal is modulated by the polarized relay RPS-5, driven by a 200 Hz signal from a trigger excited by the tube L_4. A 400 Hz voltage taken from a phase shifter is used to drive the tube L_4, to synchronize the phases of the flexible feedback signal and the error voltage. For the phase shifter we use a rotary transformer VTM-4, aircraft type. A control signal is fed to the input of the phase discriminator (tube L_5) from the cathode transformer T_2' of tube L_2 and then to the grids of the dc balance amplifier (tubes L_6 and L_7), the anode circuits of which drive the control windings of the EMU. A feedback signal is fed from the tachometer generator TG_A to the input of the balance amplifier. All the high frequency circuits are fed from a single source ($U = 115$ V, 400 Hz). The system includes the amplidynes EMU-ZA, servomotors SL-361, and tachometer generators SL-161. The main circuit of the coarse guiding system of the interferometer for the H-axis differs from that of the A-axis only in the absence of the ARU.

    Fine guiding of the interferometer is done by a photoelectric automatic guiding servomechanism using dc servomotors. A detailed description of similar guiders has already been published [1–5]. A photomultiplier tube attached to the optics of the guider, which has a clear aperture of 250 mm, makes it possible to operate down to 6 th and 7 th magnitude stars if tubes of the FEU-17 type are used. To prevent the photoelectric guider from disturbing the coarse guiding system (whose error sensors are rigidly coupled to the axes of the interferometer), additional mechanical stages must be introduced into the kinematics of the instrument. With this mechanical arrangement, when the interferometer axes are driven by the photoelectric guider motors the rotors of VT_A and VT_H would turn by the same angle in the opposite direction.

    In practice it has been proven convenient to rotate the stator of the rotary transformers to avoid the use of mechanical differentials. Consequently, both systems (coarse guiding and photoguider), while acting on the same axes of the interferometer, are completely independent. A final evaluation of the overall dynamic and static accuracy of this arrangement can be made only after the performance tests in the Pulkovo interferometer are carried out. Earlier tests of the simultaneous operation of the drive mechanism and a photoelectric guider on an equatorial telescope were successful. The total automatic tracking error did not exceed 0.15 of the diameter of the stellar image on the focal plane of the photoguider.

REFERENCES

1. Yu. A. Sabinin, *Collected Works of Problems in Electromechanics* (Sb. rabot po voprosam elektromekhaniki), No. 4, Moscow and Leningrad (1960).
2. P. V. Nikolayev, *Collected Works of Problems in Electromechanics* (Sb. rabot po voprosam elektromekhaniki), No. 4, Moscow and Leningrad (1960).
3. Yu. A. Sabinin and P. V. Nikolayev, "A system with a semi-disk modulator of light beam for automatic guilding of telescopes", *Izv. KrAO AN SSSR*, **24**, 203 (1960).
4. S. O. Dobrogursky, V. A. Kazanov and V. K. Titov, *Computers* (Schetno-reshayushchiye ustroistva), Oborongiz (1959).
5. B. I. Stanislavsky, *Electric Computers* (Elekricheskiye schetno-reshayushchiye ustroistva), Sudpromgiz (1948).

CHAPTER **I-7**

Automatic control systems for astronomical instruments using a spherical coordinates converter

V. P. YEGOROV and YU. A. SABININ

*Institute of Electromechanics on Automation and Machine Construction
Leningrad*

1 INTRODUCTION

The progress of astronomy requires automation of astronomical instruments and of all accessory equipment. The automatic conversion of equatorial to horizontal coordinates of a celestial object and of the computation of the positional angle involves many complex problems.

Horizontal coordinates are needed as a final controlling parameter in automatic guiding systems for altazimuth telescopes. Also for automatic synchronization of the dome and the tube of equatorially mounted telescopes. The positional angle is used in automatic compensation for the rotation of the optical field, in observations at the Coudé focus of equatorial telescopes and at any focus of altazimuth telescopes. When a bellow is used to connect the telescope tube and the windscreen of the dome, it is necessary to rotate the coupling ring proportional to the change in the positional angle; this is to prevent it from twisting as the instrument changes attitude. For refraction corrections the positional angle is also needed as an intermediate controlling parameter in automatic and programmed control systems.

Several automatic control systems using coordinates converters based on rotary transformers (VTs) have been developed at the Institute of Electromechanics.

Prototype systems have been built and tested on models in the laboratory. In addition, two automatic dome guiding systems based on VTs have been installed in operating telescopes at the Crimean Astrophysical

Observatory for prolonged testing under observatory conditions. The results of both kinds of tests demonstrate that the systems are suitable for existing instruments and for those planned in the future.

2 AUTOMATIC CONVERSION TO HORIZONTAL COORDINATES AND DOME CONTROL SYSTEM

2.1 ASTRONOMICAL SPHERICAL COORDINATE SYSTEMS

The position of an object on the celestial sphere is uniquely defined by two spherical coordinates. Figure 7-1 shows the two systems of celestial spherical coordinates most used in astronomy, the horizontal and the equatorial (or parallactic). The horizontal coordinates of a celestial object σ are the height, h, or more often its complement zenith distance, z, and the azimuth A:

$$h + z = 90°. \tag{7-1}$$

The equatorial coordinates of the celestial object, σ, are the declination, δ, and the hour angle t (or right ascension α). The hour angle t, is defined as the difference between the sidereal time s and the right ascension, α:

$$t = s - \alpha. \tag{7-2}$$

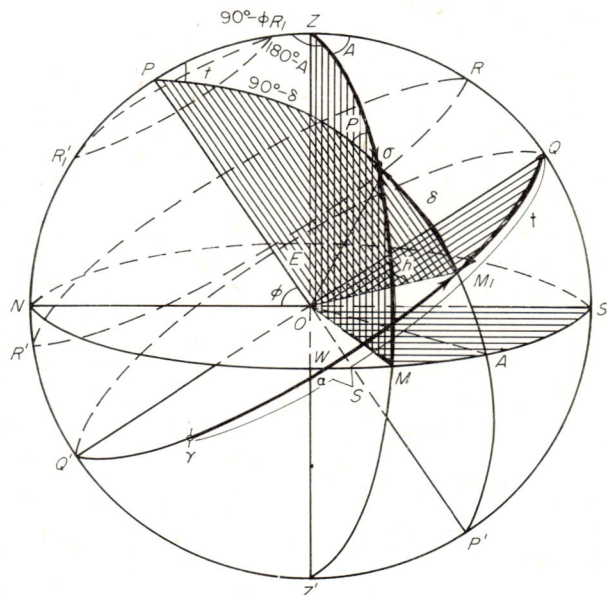

Figure 7-1 Horizontal and equatorial systems of coordinates.

The horizontal coordinates (A, z) can be calculated from (t, δ) using the following relationships:

$$A = \arctan \frac{\cos \delta \sin t}{\cos \delta \cos t \sin \varphi - \sin \delta \cos \varphi}, \quad (7\text{-}3)$$

$$z = \arctan \frac{\cos \delta \cos t \sin \varphi - \sin \delta \sin \varphi}{\cos A(\cos \delta \cos t \cos \varphi + \sin \delta \sin \varphi)}, \quad (7\text{-}4)$$

where φ is the local latitude.

2.2 AUTOMATIC SYNCHRONIZATION BETWEEN THE TELESCOPE ATTITUDE AND DOME

The positions of the telescope and the dome are given in different co-ordinates systems. Equatorially mounted telescopes are of either symmetrical or asymmetrical type. The coordinates of the window of the dome are given in the horizontal coordinates system. Synchronization of the dome and the instrument involves positioning the dome slit and the windscreen (which forms the window), so that the axis of the telescope tube passes through the center of the window. This is done by the spherical coordinates converter. The equatorial coordinates (t, δ) of the object under observation are taken from the telescope axes and fed into the converter to obtain (A, z). These values are then fed to the input of the system controlling the movements of the dome and screens.

The simplest case is the problem of automating the motion of the dome for a telescope with symmetrical mount. Figure 7-2 shows a schematic of the dome and a symmetrically-mounted telescope and three systems of coordinates with a common origin O: horizontal $OXYZ$, equatorial $OX_1Y_1Z_1$ and the coordinate system $OX_2Y_2Z_2$ rigidly connected to the telescope, in which the Z_2-axis points to the celestial pole P, the Y_2-axis coincides with the declination axis and the X_2-axis is perpendicular to it. When $t = 0$, the X_2-axis coincides with the line of intersection of the meridian and equatorial planes. When $t = 0$ the systems $OX_2Y_2Z_2$ and $OX_1Y_1Z_1$ coincide.

As the telescope moves with respect to t and δ, the point K lying on the tube axis describes a certain path on the hemispherical surface of the dome. The radius vector $\bar{r}$ drawn to K from the center of the dome coincides with the axis of the telescope tube. Simple trigonometric calculations show that for $\bar{r}$ in the various coordinate systems the relationships $A = f(t, \delta)$ and $z = f(t, \delta)$ maintaining $\varphi = $ const are exactly the same as Eqs. (7-3) and (7-4). Hence the coordinates A and z of the dome window coincide with the horizontal coordinates of the object.

When the telescope is asymmetrically mounted the problem is more complicated. Figure 7-3 gives a schematic showing the dome and a telescope with the usual off-axis mounting (two variants of displacement of the tube relative to the hour axis, t, are given) and the three coordinate systems used in Figure 7-2.

In constrast with the symmetrical mounting, the horizontal coordinates of the window cannot coincide with the horizontal coordinates of the

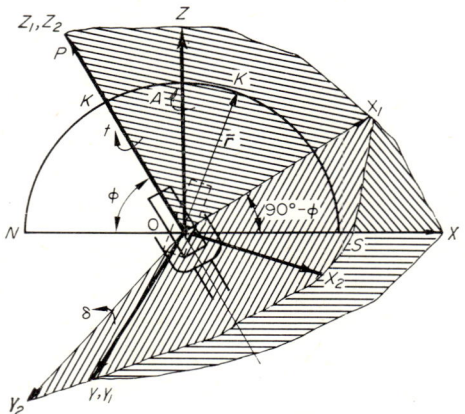

Figure 7-2 Schematic of the dome and a symmetrically mounted telescope, related to the reference frames used in the transformations of coordinates and computations.

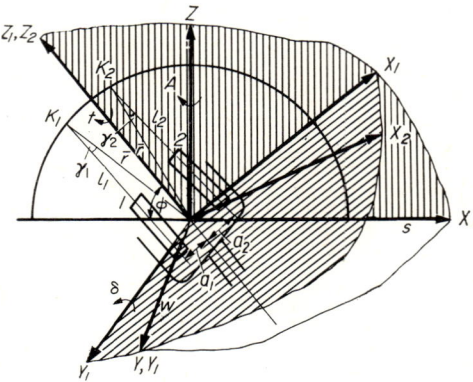

Figure 7-3 Schematic of the dome and an asymmetrically = mounted telescope related to the reference frames given in Figure 7-2.

object. To find the relationships $A = f(t, \delta)$ and $z = f(t, \delta)$, a correction for the magnitude and direction of the displacement of the tube (a_1 or a_2 in Figure 7-3) must be introduced.

Once the position of $\bar{r}$ in the different systems has been determined, we can find the relationships between the horizontal coordinates (A, z) of the window and the equatorial coordinates of the telescope, (t, δ):

$$A = \arctan \frac{\cos \delta \sin t \pm \dfrac{a}{l} \cos t}{\left(\cos \delta \cos t \mp \dfrac{a}{l} \sin t\right) \sin \varphi - \sin \delta \cos \varphi}, \qquad (7\text{-}5)$$

$$z = \arctan \frac{\left(\cos \delta \cos t \mp \dfrac{a}{l} \sin t\right) \sin \varphi - \sin \delta \cos \varphi}{\cos A \left[\left(\cos \delta \cos t \mp \dfrac{a}{l} \sin t\right) \cos \varphi + \sin \delta \sin \varphi\right]}, \qquad (7\text{-}6)$$

where a is the displacement of the tube axis relative to the hour axis, l is the distance measured along the tube from the declination axis to the dome. The value of a may be positive or negative depending on the direction of displacement of the tube in the off-axis mounting.

Telescopes occasionally have unusual off-axis mountings such as the one for the 50-inch telescope at the Crimean Astrophysical Observatory. In contrast to other off-axis mountings where the tube is relative to the hour axis. In this case the tube is extended off-center relative to the δ-axis.

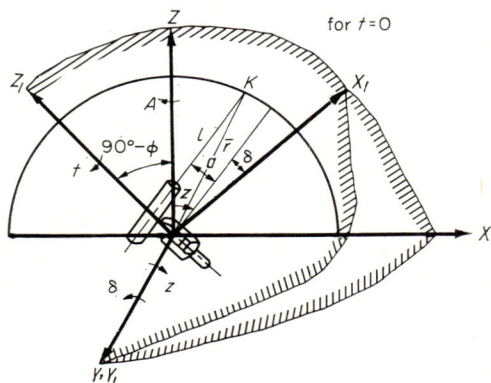

Figure 7-4 Schematic of the dome and the 50-inch telescope at the Crimean Astrophysical Observatory related to the reference frames used in the transformations of coordinates and computations.

Figure 7-4 gives a schematic of the 50-inch type telescope related to the reference frames used in the transformations of coordinates and computations. Using simple mathematical operations, we obtain the following relationships for this mounting:

$$A = \arctan \frac{\cos\left(\delta + \arctan \frac{a}{l}\right) \sin t}{\cos\left(\delta + \arctan \frac{a}{l}\right) \cos t \sin \varphi - \sin\left(\delta + \arctan \frac{a}{l}\right) \cos \varphi},$$
(7-7)

$$z = \arctan \frac{\cos\left(\delta + \arctan \frac{a}{l}\right) \cos t \sin \varphi - \sin\left(\delta + \arctan \frac{a}{l}\right) \cos \varphi}{\cos A \left[\cos\left(\delta + \arctan \frac{a}{l}\right) \cos t \cos \varphi + \sin\left(\delta + \arctan \frac{a}{l}\right) \sin \varphi\right]}.$$
(7-8)

A comparison of Eqs. (7-7) and (7-8) and Eqs. (7-3) and (7-4) shows that the relationships $A = f(t, \delta)$ and $z = f(t, \delta)$ for the symmetrical mount and the 50-inch telescope mount are identical. The sole difference lies in the fact that for the 50-inch telescope a constant correction factor, equal to $\arctan a/l$, is introduced into the value of δ.

In most instances domes for telescopes are not equipped with windscreens, so that only one horizontal coordinate, the azimuth of the dome slit, has to be computed by the automatic control system.

Two operating modes must be provided for by a system which matches the attitude of the telescope and the dome: setting and tracking. These two modes involve different rates for the motions of the instrument and the dome.

Operation in the near-zenith region creates considerable difficulties for the automation of the dome motion, as in this region small increments in the hour angle t correspond to large changes in A. When the observed object passes exactly across the zenith, i.e. when $\delta = \varphi$ (for a symmetrically mounted instrument) there is an instantaneous 180° change in the azimuth. When this happens in the setting mode, the dome must be rotated by 180° (the slit must move from east to west). In the guiding mode the telescope tube may move very differently, since in this case the instrument rotates about two axes, t and δ. It may therefore be necessary to rotate the dome from 0 to 180° depending on the angle between the axis of the telescope tube and the zenith. For an asymmetric mount the value of δ, for which an instantaneous change in the azimuth of the dome slit occurs, will be somewhat different from $\delta = \varphi$. Accordingly, the guiding system for the dome must be equipped with a unit which automatically moves the dome at the zenith point.

The basic component of the synchronization system is the coordinates converter PK, which has several variants. The first systems developed and tested at the Institute of Electromechanics for the dome control used electromechanical simulation as the systems type PKs (PK-I, PK-II, PK-III and PK-IV). The advantages of these PKs include reliable performance and their suitability to solve additional problems, such as protective halting of the telescope and the synchronization of its movements with those of the accessory equipment.

The circuits of the coordinates converters based on the use of VTs are the result of research on automatic dome control systems carried out at the Institute. These systems have a number of important advantages over the others, primarily the simplicity in the circuitry. By using a computer in the form of a VT cascaded circuit, we eliminate the need for supplementary remote transmission systems for the angles t, δ, A. This considerably simplifies the construction of the coordinates converter and consequently the circuits of the electronic blocks (dome drive, servomechanisms for t, δ, A, etc.). In addition, the solutions of $A = f(t, \delta)$, $z = f(t, \delta)$ are found with greater accuracy.

2.1.1 Automatic synchronization between the telescope attitude and the dome for a symmetrically mounted instrument (system using PK-V)

The first stage of the work at the Institute of Electromechanics was the development and testing of a dome control system with a PK based on VTs as applied to symmetrically-mounted instruments; this is the simplest automation problem. Figure 7-5 shows the block diagram of the spherical coordinates converter PK-V as applied to this system.

The PK-V converts the equatorial coordinates (t, δ) of the telescope into the horizontal coordinates (A, z) of the dome window as follows: the radius-vector $\bar{r}$ given in the form of a voltage U is fed into the input of a VT cascaded circuit. The projections of $\bar{r}$ in the equatorial system are obtained from the data t and δ. The projections of $\bar{r}$ in the horizontal system can be obtained by rotating the coordinate system around one of the axes by the angle φ or the complementary angle $\theta = 90° - \varphi$. Finally, these projections can be summed to obtain $\bar{r}$ and to find the angles A and z (or h).

In the PK-V circuit the rotary transformers SKVT-δ and SKVT-t receives the coordinate information, SKVT-φ receives the latitude information, and SKVT-A and SKVT-Z are used as synthetizers. MVT-1, MVT-2 and MVT-3 are introduced into the circuit to match the scales of the electrical signals and to balance the elements of the VT cascaded circuit.

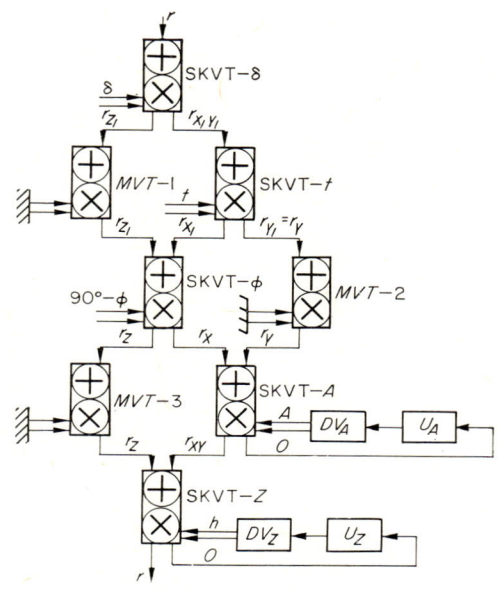

Figure 7-5 Block diagram of the PK-V type coordinate converter.

The rotations around A and z are produced by the servodrives (amplifier U_A, U_Z and servomotors DV_A, DV_Z).

Figure 7-6 shows the vector diagrams representing the functions performed by the different elements in the PK-V coordinates converter.

Voltage U, proportional to the radius-vector $\bar{r}$ is applied to the excitation winding of SKVT-δ. Since the absolute magnitude of $\bar{r}$ in the PK-V circuit is unimportant, the voltage $U \equiv r$ is applied to the input of the cascaded circuit directly from an ac network, and not through a linear rotary transformer (LVT). When the rotor of SKVT-δ rotates by the angle δ, we obtain at the output the projections of $\bar{r}$ in the plane of the declination circle $Z_1OX_2(Z_2OX_2) - r_{Z_1}$ and $r_{X_1Y_1}$. The quantity $r_{X_1Y_1}$ is fed to the

Automatic control systems for astronomical instruments 93

input of the SKVT-t. When its rotor turns by the angle t, we obtain at the output voltages which determine the projections of $r_{X_1Y_1}$ in the plane of the celestial equator, $X_1OY_1(X_1OY) - r_{X_1}$ and $r_{Y_1} = r_Y$. Voltages proportional to r_{X_1} and r_{Z_1} are fed to the input of the SKVT-φ. By setting the value of the rotation angle of the coordinates system in the meridian plane $XOZ(X_1OZ_1)$ in SKVT-φ, $\theta = 90° - \varphi$ (where φ is the local latitude), we obtain at the output the projections r_Z and r_X. The sum r_{XY} in the plane of the horizon $XOY(XOY_1)$ is obtained from the data r_X and $r_Y = r_{Y_1}$ applied to the input of SKVT-A. The azimuth value A is obtained as the angle of rotation of the SKVT-A rotor when the voltage on the sine VT is equal to zero. The vector $\bar{r}$ is obtained by adding r_Z and r_{XY}, on the vertical plane of the object ($r_{XY}OZ$). The voltages proportional to r_Z and r_{XY} are fed to the input of SKVT-Z and the values of $\bar{r}$ and the height, $h(z = 90° - h)$ are obtained. We can use SKVT-Z to generate both h

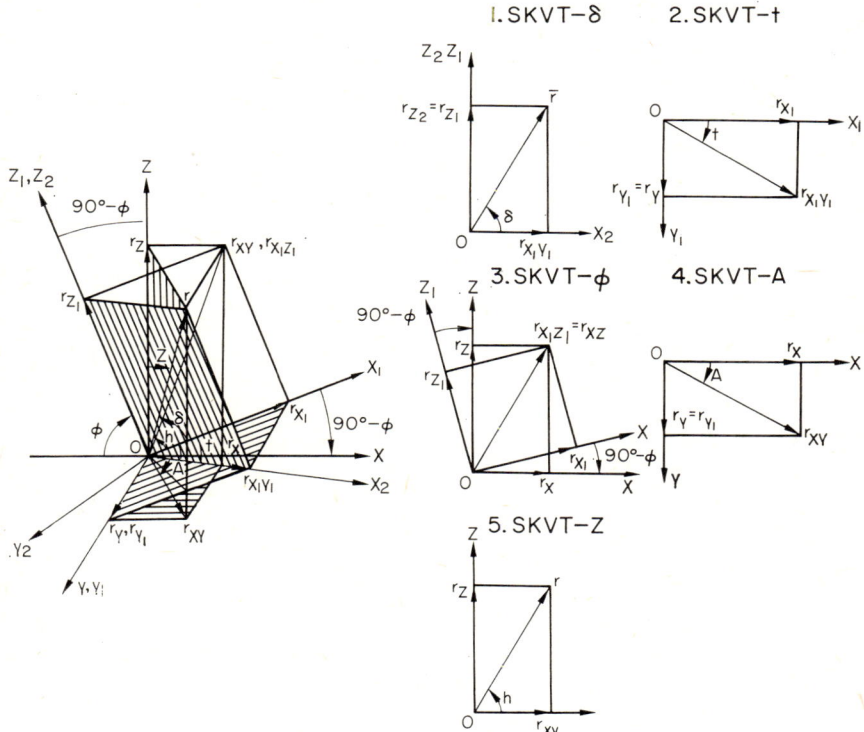

Figure 7-6 Vector diagrams representing the functions performed by the different components of the PK-V coordinates converter.

and z just as in SKVT-φ, by replacing the angle $90° - \varphi$ by φ. It depends on how the windings of SKVT-Z and SKVT-φ are connected to the windings of the other VTs and on the initial setting of the PK-V cascaded circuit.

The main circuit of the system using PK-V type coordinates converter to synchronize the motion of a symmetrically-mounted telescope and the dome, is given in Figure 7-7. The relationships between voltages in the VT windings of the PK-V are given by the set of relationships (7-9), (7-10), (7-11), (7-12) and (7-13).

The VT windings are labeled as follows: excitation winding $S_1 - S_2$, quadrature winding $S_3 - S_4$, sine winding $R_1 - R_2$, cosine winding $R_3 - R_4$.

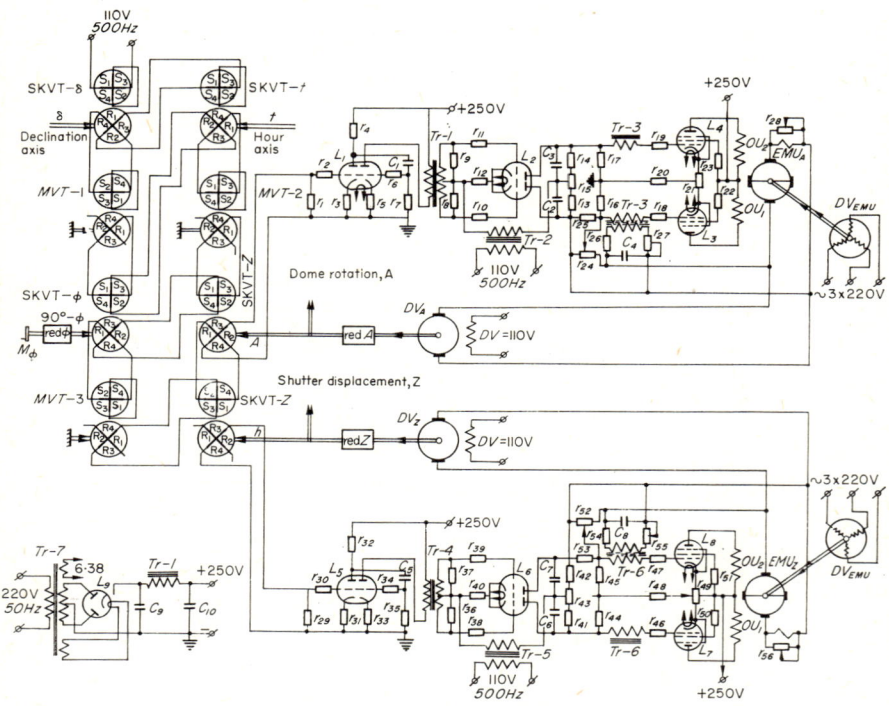

Figure 7-7 Main circuit of the system using PK-V type coordinates converter to synchronize the motion of the dome window and the attitude of a symmetrically-mounted telescope. L_1 and L_5 (6 N 8); L_2 and L_6 (6 N 9); L_3, L_4, L_7 and L_8 (6 P 9); L_9 (5Ts 4 S); SKVT-δ (I 6,713,014); SKVT-t (I 6,713,010); SKVT-φ (I 6,713,020); SKVT-A (I 6,713,050); SKVT-Z (I 6,713,049); MVT-1 and MVT-2 (I 6,713,153); MVT-3 (I 6,713,036); the types of DV_A, DV_Z, EMU_A and EMU_Z are selected on the basis of the static moments of the dome and windscreen.

We take the transformation coefficients of all the VTs equal to one in order to simplify the mathematical expressions.

1. SKVT-δ (coordinates mode)

$$U_{S_1-S_2} = U \equiv r, \quad (U = 110 \text{ V}, f = 500 \text{ cm})$$
$$U_{R_1-R_2} = U \sin \delta \equiv r_{Z_1}, \qquad (7\text{-}9)$$
$$U_{R_3-R_4} = U \cos \delta \equiv r_{X_1 Y_1}.$$

2. SKVT-t (coordinates mode)

$$U_{S_1-S_2} = U \cos \delta \equiv r_{X_1 Y_1},$$
$$U_{R_1-R_2} = U \cos \delta \sin t \equiv r_{Y_1} = r_Y, \qquad (7\text{-}10)$$
$$U_{R_3-R_4} = U \cos \delta \cos t \equiv r_{X_1}.$$

3. SKVT-φ (coordinates conversion mode)

$$U_{S_1-S_2} = U \sin \delta \equiv r_{Z_1},$$
$$U_{S_3-S_4} = U \cos \delta \cos t \equiv r_{X_1},$$
$$U_{R_1-R_2} = U[\cos \delta \cos t \cos (90° - \varphi) - \sin \delta \sin (90° - \varphi)]$$
$$= U(\cos \delta \cos t \sin \varphi - \sin \delta \cos \varphi) \equiv r_X,$$
$$U_{R_3-R_4} = U[\cos \delta \cos t \sin (90° - \varphi) + \sin \delta \cos (90° - \varphi)] \qquad (7\text{-}11)$$
$$= U(\cos \delta \cos t \cos \varphi + \sin \delta \sin \varphi) \equiv r_Z.$$

4. SKVT-A (synthesis mode)

$$U_{S_1-S_2} = U \cos \delta \sin t \equiv r_Y,$$
$$U_{S_3-S_4} = U(\cos \delta \cos t \sin \varphi - \sin \delta \cos \varphi) \equiv r_X,$$
$$U_{R_1-R_2} = U[(\cos \delta \sin t \cos A - (\cos \delta \cos t \sin \varphi \qquad (7\text{-}12)$$
$$\quad - \sin \delta \cos \varphi) \sin A] = 0,$$
$$U_{R_3-R_4} = U[\cos \delta \sin t \sin A + (\cos \delta \cos t \sin \varphi$$
$$\quad - \sin \delta \cos \varphi) \cos A] \equiv r_{XY}.$$

5. SKVT-Z (synthesis mode)

$$U_{S_1-S_2} = U[\cos \delta \sin t \sin A + (\cos \delta \cos t \sin \varphi$$
$$\quad - \sin \delta \cos \varphi) \cos A] \equiv r_{XY},$$
$$U_{S_3-S_4} = U(\cos \delta \cos t \cos \varphi + \sin \delta \sin \varphi) \equiv r_Z,$$

$$U_{R_1-R_1} = U\{(\cos \delta \cos t \cos \varphi + \sin \delta \sin \varphi) \cos h$$
$$- [\cos \delta \sin t \sin A + (\cos \delta \cos t \sin \varphi - \sin \delta \cos \varphi) \cos A] \sin h\}$$
$$= U\{(\cos \delta \cos t \cos \varphi + \sin \delta \sin \varphi) \sin z - [\cos \delta \sin t \sin A \quad (7\text{-}13)$$
$$+ (\cos \delta \cos t \sin \varphi - \sin \delta \cos \varphi) \cos A] \cos z\} = 0,$$

$$U_{R_3-R_4} = U\{(\cos \delta \cos t \cos \varphi + \sin \delta \sin \varphi) \sin h$$
$$+ [\cos \delta \sin t \sin A + (\cos \delta \cos t \sin \varphi - \sin \delta \cos \varphi) \cos A] \cos h\}$$
$$= U\{(\cos \delta \cos t \cos \varphi + \sin \delta \sin \varphi) \cos z + [\cos \delta \sin t \sin A$$
$$+ (\cos \delta \cos t \sin \varphi - \sin \delta \cos \varphi) \cos A] \sin z\} \equiv r.$$

We do not give similar equations for the windings of the scaling VTs, since when the transformation coefficients of all VTs are equal to one, the voltages at the input and output of the MVTs are equal. The voltages on the sine windings of SKVT-A and SKVT-Z are reduced to zero by means of servomechanisms.

The values of A and z generated by the systems in this case satisfy the set of Eqs. (7-3) and (7-4).

The input data t, and δ are fed to the system from the SKVT-t and SKVT-δ which are coupled to the hour and the declination axes of the telescope (Figure 7-7). The latitude is set in the SKVT-φ. The output data A and z (or h) are generated in SKVT-A and SKVT-Z, coupled to the corresponding axes of the dome and screen drives. The angular range of SKVT-t, SKVT-δ, SKVT-A and SKVT-Z is equal to 360°. A and z are generated by dc servomechanisms.

The systems for A and z are identical. The servomotors used are dc motors DV_A and DV_Z, the power amplifiers are amplidynes EMU_A and EMU_Z controlled by four-stage electronic amplifiers. The first and second stages L_1 (L_5) are voltage amplifiers, the third L_2 (L_6) is used as a discriminator. The fourth stage L_3, L_4 (L_7, L_8) is a push–pull dc power amplifier.

The load of the amplifiers are the control windings OUs of the corresponding EMUs. To stabilize the operation of the servomechanism, rigid negative voltage feedback and flexible negative feedback using the first derivative of the voltage output of the EMUs is fed in the last stage of the electronic amplifier through r_{24}, r_{25} (r_{52}, r_{53}) and Tr-3 (Tr-6).

The circuitry of the systems which generates A and z using dual-winding EMUs are shown in Figure 7-7.

As we have indicated, certain difficulties arise in the near-zenith region. In PK modeling-type circuits a special computer was used to control the dome when the telescope pointed at the zenith. The PK-V (Figures 7-5, 7-7) has the property that in the zenith region the error signal per unit of error angle falls sharply. While it reduces accuracy in the near-zenith region, the VT in the PK-V at the same time allows us to eliminate the need for a special dome driving system at the zenith by switching off the system which computes the dome azimuth. When the telescope tube has passed the zenith, the PK-V starts again to supply azimuth values.

The automatic dome control system using PK-V has undergone laboratory tests and operating trials. In 1958 a laboratory model of the system was constructed in which the telescope and dome were electromechanically simulated. The error in solving $A = f(t, \delta)$ and $z = f(f, \delta)$ for $\varphi = 45°$ in general did not exceed $\Delta A = \pm 10'$ and $\Delta z = \pm 10'$. Only in the close vicinity of the zenith did the azimuth error increase somewhat, reaching $\Delta A = \pm 30'$, with the error in z not exeeding $\Delta z = \pm 15'$.

In 1959 an automatic synchronization system for the telescope and dome with PK-V was attached to the 50-inch telescope at the Crimean Astrophysical Observatory. The dome of this telescope has no windscreen and so only one horizontal coordinate, the azimuth A of the dome slit, had to be computed. Accordingly the MVT-3, SKVT-Z and the channel for generating z are absent from the circuit.

As we have indicated, the 50-inch telescope has an unusual off-axis mount. The values of $A = f(t, \delta)$ and $z = f(t, \delta)$ for the dome are obtained from Eqs. (7-7) and (7-8), very similar to Eqs. (7-3) and (7-4). The similarity between the equations make it possible to use the PK-V system for the off-axis mounting of the 50-inch telescope without any basic change in the VT cascaded circuit with the exception of the adjusting of the zero position in the SKVT-δ.

In order for Eqs. (7-7), (7-8) and (7-3), (7-4) to be identical the following condition is necessary: $\delta = 0$ for the symmetric mount and $\delta = -\arctan a/l$ for the 50-inch telescope. The PKV- circuit for this telescope is adjusted by adjusting the zero values of δ of the instrument and of SKVT-δ.

The circuit of the PK-V system used in the 50-inch telescope is basically similar to that shown in Figure 7-7. A motor MI-42 T and an amplidyne EMU-50 with four control windings are used to drive the dome. Due to the large mechanical inertia of the dome the feedbacks in the system were greatly amplified using all four control windings of the EMU. In the experimental assembly the maximum azimuth speed of the dome was equal to 0.5 to 0.75 rev. min^{-1}.

In the circuit of the dome control, in addition to the automatic control, it is planned to have pushbutton operation of the dome drive motor with a wide range in speed regulation.

The results of experiments *in situ* have shown that the PK-V system is efficient and gives sufficiently high accuracy in the solution of the problem with operation in any mode (guiding and setting) and in any observable region of the sky, including the zenith. It was found to be feasible to operate the system with a single-motor drive and still retain sufficient accuracy without any need for a special dome drive system at the zenith.

Modification in the system can be made in the future; for example, a magnetic amplifier was tested at the same time as the electronic amplifier in the circuit for computing A in the 50-inch telescope.

The PK-V dome control system is now in experimental use in the 50-inch telescope for extensive tests under actual observing conditions. It should be indicated that this system is suitable for any latitude (readjustment is made by setting the corresponding angle in SKVT-φ).

2.1.2 Automatic synchronization between the telescope and the dome for an asymmetrically-mounted instrument (system using PK-VI)

As we have indicated, the off-axis mounting presents a complicated problem for automation of the dome compared with the system used for symmetrically-mounted telescopes. The basic circuit of the coordinates converter PK-VI developed at the Institute of Electromechanics is given in Figure 7-8.

In the PK-VI, the SKVT-δ operates in the coordinates mode, the SKVT-ε as a sine VT, the SKVT-t and SKVT-φ in the coordinate conversion, and the SKVT-A and SKVT-Z in the synthesis mode. The MVT-1, MVT-2 and MVT-3 are scaling transformers.

The functional relationship of the voltage on the different VT windings of the PK-VI are shown in Figure 7-8 (as in PK-V the transformation coefficients of all the VTs are taken equal to one for simplicity).

The voltages on the sine SKVT-A and SKVT-Z are made zero by the servomechanism. We can solve the equations for the voltages on these windings for A and z and after several transformations, using $\sin \varepsilon = \pm a/l$ and $h = 90° - z$ we obtain expressions for $A = f(t, \delta)$, $z = f(t, \delta)$ when $\varphi = $ const, which are identical with Eqs. (7-5) and (7-6).

When the PK-VI system is used the SKVT-t and SKVT-δ are coupled to the corresponding axes of the telescope, the SKVT-A and SKVT-Z to the dome and windscreen drive; the SKVT-ε and SKVT-φ may be placed anywhere. Circuits similar to those used in the PK-V system can be used for A and z.

The PK-VI system is universal and is suitable for observatories located anywhere (adjustment to the required latitude is made simply by turning the rotor of SKVT-φ). The circuit is applicable to any off-axis or sym-

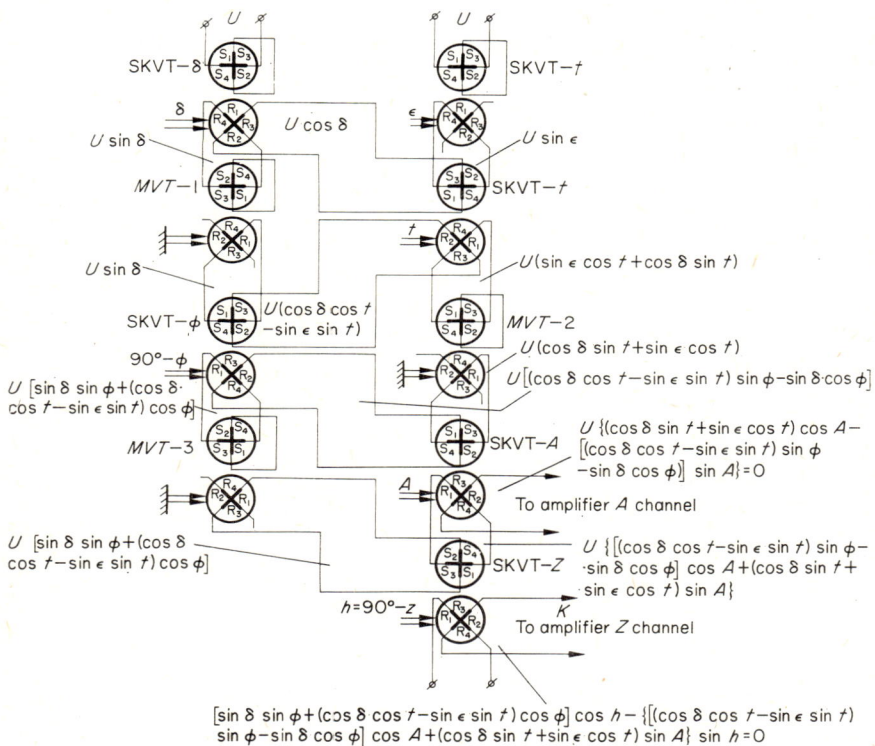

Figure 7-8 Basis circuit diagram of the PK-VI type coordinates converter.

metrically-mounted telescope ($\varepsilon = 0$). The nature and magnitude of displacement of the telescope tube are allowed for by a corresponding rotation of the rotor of SKVT-ε (the angle of rotation of SKVT-ε is uniquely defined by the design of the telescope and dome).

The magnitude of the tube displacement corresponds to the angle of rotation of the rotor, equal to $\varepsilon = \arcsin a/l$, and the direction of the tube displacement determines the phase of the SKVT-ε voltage output which is fed to the SKVT-t.

The system passed laboratory tests and in 1960 was mounted for trials on the double 406-mm astrograph at the Crimean Astrophysical Observatory. The circuit of the system tested was basically similar to that of Figure 7-7, except that only the azimuth A of the dome slit was calculated. Accordingly, the MT-3 and SKVT-Z were not included in the circuit. The motor MI-32 and the amplidyne EMU-5A with two control windings were used to drive the dome. The electronic amplifier circuit is similar to that used in the PK-V system.

Experimental use of automatic dome control systems using PKs based on an electromechanical modeling unit and with the use of VTs at the Crimean Astrophysical Observatory showed that the circuits based on VTs have a number of important advantages over the modeling unit type of PK. First of all, they are simpler in design and give a more accurate solution. Furthermore, if the PK is used in the form of a VT cascaded circuit there is no need to have additional systems for remote transmission of t, δ, A and z.

The circuits with PK-V and PK-VI do not require a special system to drive the dome at the zenith and the circuit as a whole is thus considerably simplified due to the absence of the halting system.

The PK-V and PK-VI systems can be used at any latitude by adjusting the angle in SKVT-φ; in this respect a system based on VTs is universal, and this is a great advantage over PK-type modeling units. In addition, the PK-VI system is applicable to any equatorial mount, whether symmetrical or not.

The error of the systems under normal conditions does not exceed $\pm 10'$, thus satisfying the requirements for the dome and windscreen drives. The PK-V and PK-VI systems are efficient in any guiding or setting mode, and for the observation of objects in any part of the sky. They can be recommended for use in existing instruments and in those planned in the future.

The Institute of Electromechanics has recommended that PK-V and PK-VI systems to be used with the 320-mm meniscus telescope, the 1-m telescope at the Byurakan Astrophysical Observatory and the 700/980-mm telescope at the Abastumani Astrophysical Observatory.

The PK systems based on VTs can also be used for automatic setting and guiding of optical instruments and the antennae of altazimuth radiotelescopes and for other problems associated with the conversion of spherical coordinates.

3 AUTOMATIC COMPUTATION OF THE POSITIONAL ANGLE (p)

3.1 INTRODUCTION

It is often necessary to compute the positional (parallactic) angle p as a final or intermediate regulating parameter (as in compensation for rotation of the optical field, automatic rotation of the dome bellows, etc.).

The relationship between the angle p and the equatorial and horizontal coordinates can be established by considering the spherical triangle $PZ\sigma$ associated with the object σ (Figure 7-1). The angle p can be determined in two ways, depending on whether the initial data is (t, δ) or (A, z). After simple trigonometric computation, we obtain the following equations for $p = f(t, \delta)$ and $p = f(A, z)$ with φ constant:

$$p = \arctan \frac{\cos \varphi \sin t}{\sin \varphi \cos \delta - \cos \varphi \cos t \sin \delta}, \qquad (7\text{-}14)$$

$$p = \arctan \frac{\cos \varphi \sin A}{\sin \varphi \sin z + \cos \varphi \cos A \cos z}. \qquad (7\text{-}15)$$

In either case a computer is needed to convert the coordinates. The input data for this computer will be $(t$ and $\delta)$ in one case and $(A$ and $z)$ in the other, and the output will be the value of p.

Accordingly, two variants of a positional angle computer based on VTs were developed; the PK-VII-A and the PK-VII-B.

3.2 COMPUTER FOR THE POSITIONAL ANGLE (SYSTEM USING PK-VII)

The basic circuits of the PK-VII (A, B) converters are given in Figures 7-9 and 7-10.

In the PK-VII-A the SKVT-φ and SKVT-t operate in the coordinates mode, the SKVT-δ in the coordinate conversion and SKVT-p in the synthesis mode.

In the PK-VII-B the SKVT-φ and the SKVT-A operate in the coordinates mode, the SKVT-Z in the coordinate conversion and SKVT-p in the synthesis mode. In both cases MVT-1 and MVT-2 are introduced to match the scales and to make the elements of the VT circuit symmetrical. The motion in p in the SKVT-p is generated by a servodrive similar to that in PK-V.

The functional relationships between the voltages at the separate VT windings of the PK-VII cascaded circuit are given in Figures 7-9 and 7-10 (as before, the coefficients of all the VTs are taken equal to one). The

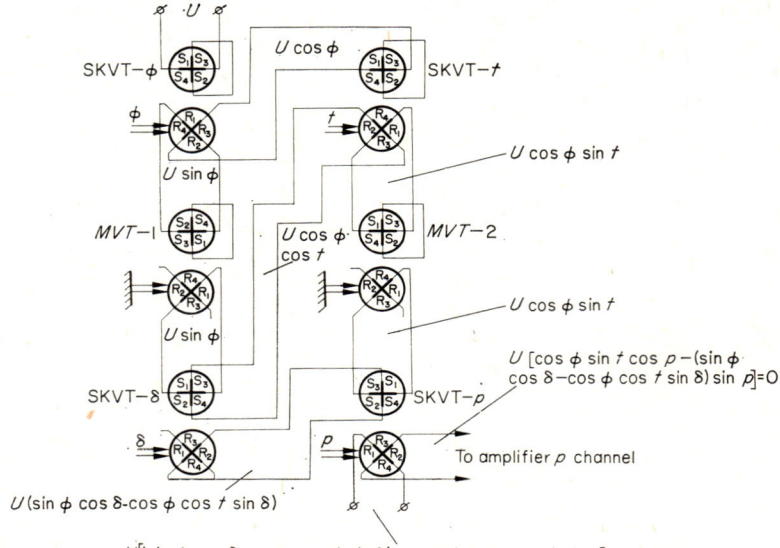

Figure 7-9 Basic circuit diagram of the PK-VII-A type coordinates converter.

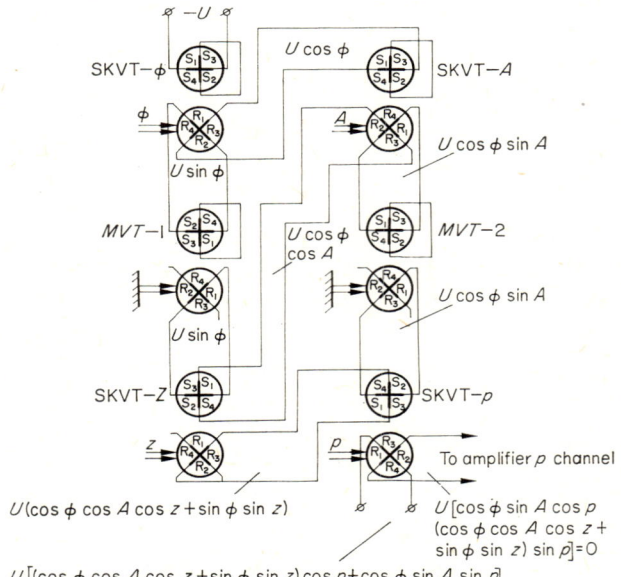

Figure 7-10 Basic circuit diagram of the PK-VII-B type coordinates converter.

voltages on the sine SKVT-p (Figures 7-9, 7-10) are reduced to zero by means of servomechanisms. We solve the equations for p and obtain the expressions $p = f(t, \delta)$ and $p = f(A, z)$ which are identical to Eqs. (7-14) and (7-15).

The angles t and δ are obtained from the SKVT-t and SKVT-δ which are coupled to the corresponding axes of the telescope in the equatorial mount. For an altazimuth instrument they are coupled to the corresponding axes of a computing unit which gives the current values of t and δ.

The values of A and z from the SKVT-A and SKVT-Z respectively are fed to units mounted on the corresponding axes of:

(1) the altazimuth telescope,

(2) the dome and windscreen drive, and

(3) the PK-V system.

In both variants the latitude φ is set in SKVT-p, which generates the p values.

In both cases we use a servomechanism consisting of an amplifier U_p and a servomotor DV_p. This servomotor feeds the values of p to the SKVT-p through a reduction gear mounted on the corresponding axis of the drive system (to drive the plateholder, support ring of the bellows, etc.). The angular range of the SKVT-p is 360°.

Circuits similar to those in the PK-V and PK-VI systems can be used as systems for computing p. The PK-VII underwent laboratory testing in 1960. A model of the PK-VII-A and PK-VII-B systems based on the PK-V systems was constructed, and the operation of both variations were studied. The error in solving $p = f(t, \delta)$ and $p = f(A, z)$ for $\varphi = 45°$ did not exceed $\Delta p = \pm 10 - 15'$. The accuracy in the solution of problems using the PK-VII can be improved by introducing a special feedback to increase the operating range of the system.

As in the PK-V circuit, the PK-VII system has a "blind zone" where the voltage of the error signal referred to the unit of error angle on the sine SKVT-p falls sharply. This blind zone is the zenith region for PK-VII-A and the circumpolar zone for PK-VII-B. At the zenith and the pole the value of p in the respective PK-VII systems is undefined, and the error sharply increases.

The voltage on the sine SKVT-p during mismatch condition can be estimated from the voltage on the cosine winding in the match condition. In Figure 7-11 the values of the maximum voltage on the cosine SKVT-p are given in the matched condition.

It has been established that in PK-VII-A this voltage is proportional to $\cos h = \sin z$ and in PK-VII-B to $\cos \delta$, and becomes equal to zero at the zenith (PK-VII-A) and at the pole (PK-VII-B). It is therefore advisable to use a special feedback circuit in the PK-VII system to increase the accuracy of operation.

One possible feedback system is illustrated in Figure 7-12. After amplification and rectification the voltage of the cosine winding of the SKVT-p is fed to the bias of the first stage of the electronic amplifier of the channel

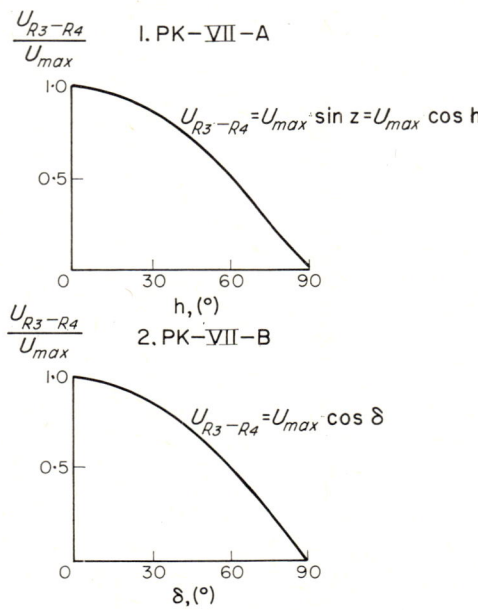

Figure 7-11 Voltage output at the cosine winding of the SKVT-p versus h and δ (matched condition).

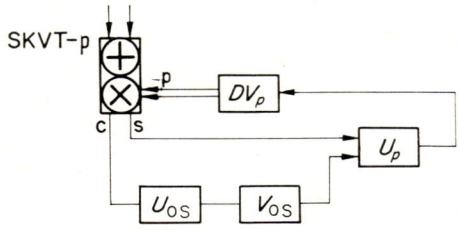

Figure 7-12 Block diagram of the feedback circuit in the PK-VII type coordinates converter.

for generating p. The circuit is adjusted so that when the voltage at the cosine winding is a minimum there is maximum amplification. The feedback voltage can also be taken from SKVT-δ (PK-VII-A) or SKVT-Z (PK-VII-B).

A similar feedback circuit can be used to increase the useful range and accuracy in the PK-V system for the azimuth. The maximum voltage in the cosine SKVT-A in the matched position in PK-V, changes as

$$U_{R_3-R_4} = U \sin z = U \cos h.$$

The feedback circuit in this case will be similar to that of PK-VII-A.

4 CONCLUSIONS

Laboratory tests of the system to solve $p = f(t, \delta)$, $p = f(A, z)$ for $\varphi = 45°$ showed that the error did not exceed $\Delta p = \pm 10$–$15'$.

The PK-VII systems have blind zones in which the error signal falls sharply and consequently the error in solving the equations increases. The voltage in the cosine SKVT-p has a simple functional relationship, so that a feedback circuit can be used to improve the accuracy in the blind zones (near-zenith zone for PK-VII-A and circumpolar zone for PK-VII-B) and to widen the operating range of the system.

The PK-VII system for generating the positional angle can be recommended for use in existing and planned altazimuth instruments and symmetrical equatorially-mounted telescopes.

When the PK-VII circuit is used for automatic rotation of the bellows that couple the telescope tube and the dome window together with the PK-V dome control system, it is better to use the PK-VII-B variant. Since the blind zones of the PK-V and PK-VII-B systems are different (zenith and pole), in this case, it is better to match the dynamic characteristics of the systems during operation in these zones. In addition, emergencies are less likely to arise in the general circuit, since the initial data for the PK-VII-B system depend on the output values of the PK-V, which provides more reliable synchronization in the operation of both systems as compared with the PK-VII-A. When PK-VII systems are used for other problems where the values of p are needed, the variant to be used must be decided on the basis of an individual analysis, taking into account the operating conditions of the overall control system.

REFERENCES

1. Ya. V. Novoseltsev and A. N. Lebedev, *Compuiers* (Schetno-reshayushchiye ustroistva), Mashgiz (1954).
2. S. O. Dobrogursky, V. A. Kazakov and V. K. Titov, *Computers* (Schetno-reshayushchiye ustroistva), Oborongiz (1959).
3. Yu. A. Sabirin and V. P. Yegorov, *A Tracking Drive with an Electromechanical Coordinate Converter for Automatic Synchronization of the Positioning of the Dome and Equatorial Telescope* (Sledyashchii privod s elektromekhanicheskim preobrazovatelem koordinat dlya avtomaticheskogo soglasovaniya polozhenii kupola i teleskopa na ekvatorialnoi montirovke), Filiate of VINITTI (1958).
4. Yu. A. Sabirin and V. P. Yegorov, "A system for the automatic coordination of the rotation of the dome with the movement of an equatorially mounted telescope", *Izv. KrAO AN SSSR*, **22**, 275 (1960).
5. Yu. A. Sabirin and V. P. Yegorov, "Systems of automatic operation of a telescope dome with transformation of spherical coordinates using rotating transformers", *Izv. KrAO AN SSSR*, **26**, 395 (1961).

CHAPTER **I-8**

Altazimuthal and positional drive mechanisms in a combined control system for altazimuth instruments

V. P. YEGOROV and YU. A. SABININ

Institute of Electromechanics on Automation and Machine Construction
Leningrad

The use of altazimuth instruments (optical and radio telescopes) is becoming much more common; but, while they have the advantage of being simpler in design, they require a more complex control system than the ones for equatorial instruments.

The equatorial coordinates of a celestial object are the declination δ and the hour angle t (or right ascension α). The hour angle is defined as the difference between sidereal time s and the right ascension α ($t = s - \alpha$). Both s and t are linear functions of time, whereas α and δ do not depend on time or on the position of the observer.

Accordingly, when an equatorial telescope is guided on an object, one coordinate (t) changes monotonically at the rate of 1 revolution per day, and the other (δ) remains constant, so that in this case guiding can be accomplished simply by using drives with uniform rate of 1 revolution per day; positioning adjustments can be made by use of a visual guider or an automatic photoguiding system.

The horizontal coordinates of an object are the height h, or the more commonly used zenith distance z ($h + z = 90°$) and the azimuth A. Both coordinates depend on time and on the position of the observer and change in accordance with a complex law, which causes certain problems in automatic control. The coordinates (A, z) can be found from (t, δ) using the equations

$$A = \arctan \frac{\cos \delta \sin t}{\cos \delta \sin \varphi \cos t - \sin \delta \cos \varphi}, \quad (8\text{-}1)$$

$$z = \arctan \frac{\cos \delta \sin \varphi \cos t - \sin \delta \cos \varphi}{\cos A (\cos \varphi \cos \delta \cos t + \sin \varphi \sin \delta)}, \quad (8\text{-}2)$$

where φ is the local latitude.

Setting and guiding systems can use the PK-V coordinates converter, based on rotary transformers, which was developed at the Institute of Electromechanics. The input data to the system will be taken from the corresponding axes of the computing unit which calculates the current values of t and δ. Then the system will calculate A and z for the corresponding axes of the altazimuth telescope.

However, coordinates converters should provide a non-uniform rate for the instrument, since the rates of change of horizontal coordinates are essentially variable. Since there is no speed control in the guiding unit of the telescope, and guiding is carried out only by position (angle), $A = f(t, \delta)$ and $z = f(t, \delta)$ cannot be solved with sufficient accuracy.

Instead of relying solely on a coordinates converter, another method which can be used is to control by velocity, as is usually done in equatorial telescopes. Tracking can then be done using the coordinates converter and a photoelectric servomechanism for automatic guiding on a star, which is required to attain the desired accuracy. For this purpose "altazimuthal" drive mechanisms are used, using computers which calculate the actual values of the velocities $\dot{A} = \left(\dfrac{dA}{dt}\right)$ and $\dot{z} = \left(\dfrac{dz}{dt}\right)$ for each value of t and δ. In contrast with equatorial drive mechanisms, the altazimuthal drive mechanisms, will have variable speed.

The dynamic conditions of operation of the coordinates converter will then be greatly improved, since it need only perform as a correcting unit.

Another important automation problem for altazimuth telescopes is to obtain the values of the positional angle p, which is also a complex function both of t, δ and of A, z:

$$p = \arctan \frac{\cos \varphi \sin t}{\sin \varphi \cos \delta - \cos \varphi \sin \delta \cos t}, \quad (8\text{-}3)$$

$$p = \arctan \frac{\cos \varphi \sin A}{\sin \varphi \sin z + \cos \varphi \cos A \cos z}. \quad (8\text{-}4)$$

In photographic and spectrographic work, the positional angle must be automatically computed to compensate for rotation of the optical field. Like A and z, the values of p can be obtained from a coordinate converter system based on rotary transformers, the PK-VII. From the same considerations as before "positional" drive mechanisms are needed to improve the accuracy in calculating p by obtaining the values of $\dot{p} = \left(\dfrac{dp}{dt}\right)$. The coordinates converter for p, performs the fine tracking; that means it is used only as a correcting unit.

Using spherical trigonometry, we can write the expressions for $\dot{A}$, $\dot{z}$ and $\dot{p}$ in the form

$$\dot{A} = \frac{\sin \varphi \sin z + \cos z \cos A \cos \varphi}{\sin z} ; \qquad (8\text{-}5)$$

$$\dot{A} = \frac{\cos \delta \cos p}{\sin z} ; \qquad (8\text{-}6)$$

$$\dot{z} = \frac{\cos \varphi \cos \delta \sin t}{\sin z} ; \qquad (8\text{-}7)$$

$$\dot{z} = \cos \varphi \sin A ; \qquad (8\text{-}8)$$

$$\dot{z} = \sin p \cos \delta ; \qquad (8\text{-}9)$$

$$\dot{p} = \frac{\cos z \cos \delta \cos p - \sin z \sin \delta}{\sin z} ; \qquad (8\text{-}10)$$

$$\dot{p} = \frac{\cos A \sin p}{\sin t} ; \qquad (8\text{-}11)$$

$$\dot{p} = \frac{\cos \varphi \cos A}{\sin z} . \qquad (8\text{-}12)$$

The rate of change of A, z and p depends essentially on the distance from the "danger zone", which is a region of the zenith basically determined by the declination. In the zenith region, the velocities $\dot{A}$ and $\dot{p}$ increase sharply. Thus, in guiding a velocity of 1 revolution per day (for the t-axis) even for $z = 2.5\text{--}3°$ the values of $\dot{A}$ and $\dot{p}$ become equal to ± 15 rev day^{-1}. For smaller zenith distances $\dot{A}$ and $\dot{p}$ increase even more and are theoretically infinite at the zenith. The maximum rate of change of z during guiding does not exceed ± 0.75 rev day^{-1}.

It is worth noting that the accelerations for the A, z and p axes are also complex functions, satisfying the relationships

$$\frac{d^2A}{dt^2} = -\frac{\cos \varphi \sin A}{\sin^2 z} [\sin z \cos z \sin \varphi + \cos \varphi \cos A(1 + \cos^2 z)], \quad (8\text{-}13)$$

$$\frac{d^2z}{dt^2} = \cos \varphi \cos A \left[\sin \varphi + \frac{\cos \varphi \cos A}{\tan z} \right], \quad (8\text{-}14)$$

$$\frac{d^2p}{dt^2} = -\frac{\cos \varphi \sin A}{\sin^2 z} [\sin z \sin \varphi + 2 \cos \varphi \cos A \cos z]. \quad (8\text{-}15)$$

The accelerations for the respective axes, depend greatly on the distance from the zenith region (such as the velocities), increasing rapidly for smaller zenith distances. In guiding, even for $z = 2.5$–$3°$ the accelerations $\ddot{A}$ and $\ddot{p}$ are equal to ± 200 rev. day^{-2}. For the z-axis, the value of $\ddot{z}$ is then equal to ± 15 rev. day^{-2}.

Clearly, the guiding system for altazimuth instruments must operate under very difficult dynamic conditions.

Computer circuits using rotary transformers have been developed at the Institute for altazimuthal and positional drive mechanisms.

Several variants are possible, since Eqs. (8-5) to (8-12) can be replaced by other equations. The variant chosen will depend on the specific problems, conditions of use and input data.

The block diagrams of one variant of altazimuthal and positional drive mechanisms are given in Figures 8-1 to 8-3. It consists of a computing unit and a generating system for the actual values of the velocity (the circuits given are those which calculate the values of $\dot{A}$, $\dot{z}$, $\dot{p}$ from Eqs. (8-5), (8-8) and (8-12).

The system for generating $\dot{A}$ is a computer circuit based on VTs with a divider unit (Figure 8-1). The VTs in the circuit are used as follows: SKVT-Z_1 in the coordinates mode; SKVT-A as a cosine VT; SKVT-Z_2 as a sine VT; SKVT-φ operates in the coordinate conversion and the LVT-$\dot{A}_1$ and LVT-$\dot{A}_2$ are used as linear VTs. The MVT-1, MVT-2 and MVT-3 are used to match the scales of the separate units of the system and for symmetry of the VTs.

The computer operates as follows (for simplicity we have taken the transformation coefficients of all VTs equal to one). The voltage U (110 V at 500 Hz) is fed to the input of the SKVT-Z_1 which delivers the output voltages $U \sin z$ and $U \cos z$; the last voltage is fed to SKVT-A, which delivers as an output $U \cos z \cos A$. Voltages proportional to $U \sin z$ and

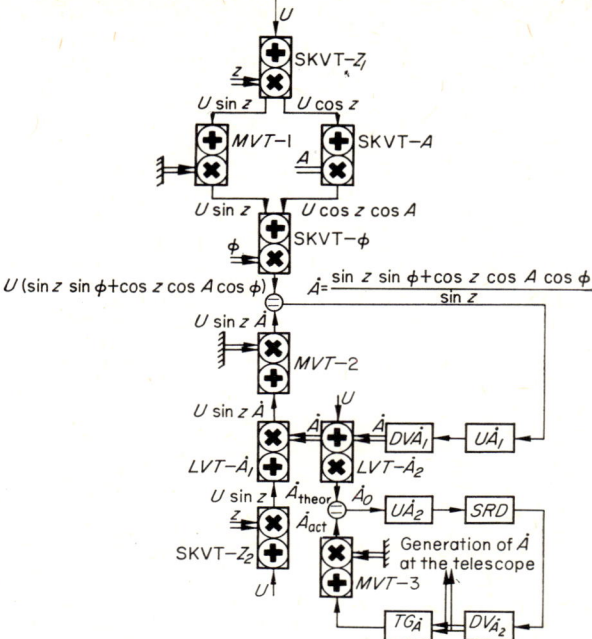

Figure 8-1 Block diagram of $\dot{A}$ channel for the altazimuthal drive mechanism.

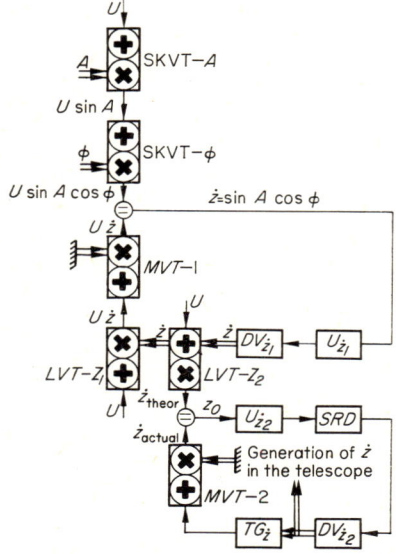

Figure 8-2 Block diagram of the $\dot{z}$ channel for the altazimuthal drive.

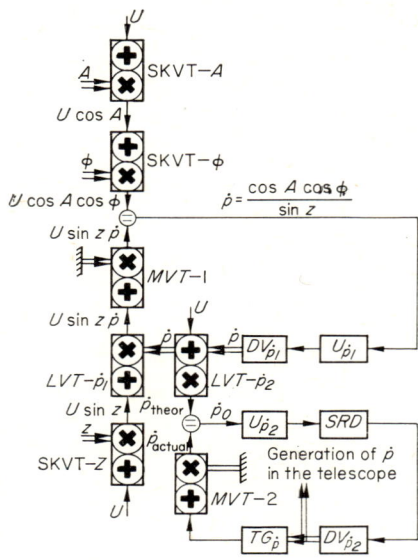

Figure 8-3 Block diagram of the $\dot{p}$ channel for the altazimuthal drive.

$U \cos z \cos A$ are fed to the input of the SKVT-φ. The voltage at the output of the cosine winding of the SKVT-φ is proportional to $U(\sin z \sin \varphi + \cos z \cos A \cos \varphi)$ and it is fed to the comparison unit. On the other hand, a voltage $U \sin z \dot{A}$ is also applied to the comparison unit (U at the input of SKVT-Z_2 gives $U \sin z$ at the output, and $U \sin z \dot{A}$ at the output of LVT-$\dot{A}_1$, and MVT-2. When the generated voltage reaches zero a "decoding" servomechanism is used so that the equation $U : (\sin z \sin \varphi + \cos z \cos A \cos \varphi) = U \sin z \dot{A}$ is satisfied, and so the value of $\dot{A}$ generated by the system is equal to

$$\dot{A} = \frac{\sin z \sin \varphi + \cos z \cos A \cos \varphi}{\sin z}. \tag{8-16}$$

The circuit of the computer can be somewhat simplified by removing SKVT-Z_2 from the system, with the voltage taken from the sine winding of SKVT-Z_1 and equal to $U \sin z$ applied to the input of LVT-$\dot{A}_1$.

The decoder consists of a comparison unit, the amplifier $U_{\dot{A}_1}$ and the generating motor $DV_{\dot{A}_1}$. The values of $\dot{A}$ are generated in LVT-$\dot{A}_1$ and LVT-$\dot{A}_2$.

LVT-$\dot{A}_2$ is the transmitter of the system which generates the values of $\dot{A}$. These values are generated by means of a motor $DV_{\dot{A}_2}$ which determines the rate of the altazimuthal telescope drive. The shaft of the motor Dv_{A_2} is coupled to the tachometer generator TG_A which produces a voltage proportional to the actual velocity $\dot{A}_{act}$ of the altazimuthal drive mechanism. After the signal has been scaled by MVT-3, the theoretical and actual values of $\dot{A}$ are fed to the comparison unit, which is the input to the servomechanism for $\dot{A}$ on the telescope. The circuit includes the amplifier $U_{\dot{A}_2}$ and the SRD system for motor velocity regulation. When the voltage proportional to $\dot{A}_0$ approaches zero the actual value $\dot{A}_{act}$ approaches the theoretical value $\dot{A}_{theo}$, supplied by the computer.

The system for $\dot{z}$ is a computer based on VTs with a multiplier unit (Figure 8-2). The VTs in the circuit fulfill the following functions: the SKVT-A, sine VT; the SKVT-φ, cosine VT; the LVT-$\dot{Z}_1$, and LVT-$\dot{Z}_2$ are used as linear VTs; the MVT-1, MVT-2 as scaling VTs. The computer operates as follows: the voltage U (110 V at 500 Hz) is fed to the input of the SKVT-A. The voltage proportional to $U\sin A$ at the output is fed to the SKVT-φ. From the SKVT-φ we obtained as an output, a voltage proportional to $U \sin A \cos \varphi$ which is applied to the comparison unit. Also, the voltage $U\dot{z}$ is fed to the comparison unit. The voltage $U\dot{z}$ is delivered at the output of MVT-1 and LVT-$\dot{Z}_1$, by feeding U to the LVT-$\dot{Z}_1$. When the generated voltage approaches zero, so that the equation $U\sin A \cos \varphi = U\dot{z}$ is satisfied, a decoding servomechanism is used and the value generated is equal to

$$\dot{z} = \sin A \cos \varphi. \qquad (8\text{-}17)$$

The decoder, as in the $\dot{A}$ system, consists of a comparison unit, the amplifier $U_{\dot{Z}_1}$ and the generator motor $DV_{\dot{Z}_1}$. The values of $\dot{z}$ are generated in LVT-$\dot{Z}_1$, and LVT-$\dot{Z}_2$.

The LVT-$\dot{Z}_2$ is the transmitter to the $\dot{z}$ generator system. The values of $\dot{z}$ are generated by means of a special motor $Dv_{\dot{z}_2}$ which determines the rate of the telescope drive. The circuit of the generating system for $\dot{z}_{act}$ and its principle of operation are exactly analogous to those of the $\dot{A}$ system.

The $\dot{p}$ system is in the form of a computer based on VTs with a division unit (Figure 8-3). The functions performed in the computer by the individual components are as follows: SKVT-A and SKVT-φ, cosine VTs; SKVT-Z, sine VT; LVT-$\dot{p}_1$, LVT-$\dot{p}_2$, linear VTs; the MVT-1 and MVT-2, scaling VTs.

The principle of operation and the components of the circuits of the decoding servomechanism and the p system on the telescope are exactly analogous to the $\dot{A}$ and $\dot{z}$ systems. The value of $\dot{p}$ generated by the decoding servomechanism is equal to

$$\dot{p} = \frac{\cos A \, \cos \varphi}{\sin z} \, . \tag{8-18}$$

The current values of $\dot{p}$ are generated by means of a special motor $Dv_{\dot{p}_2}$ which determines the rate of the positional drive mechanism.

All three circuits (Figures 8-1, 8-2, 8-3) can be combined in a single circuit with a reduction in the number of VTs. For example, we could combine SKVT-Z_1 of the $\dot{A}$ system and SKVT-Z of the $\dot{p}$ system.

As we have indicated, for each of the Eqs. (8-5) to (8-12), there is a corresponding circuit for the horizontal coordinates and positional drive mechanisms. As an example, Figures 8-4, 8-5, 8-6 show the basic circuits to solve Eqs. (8-6), (8-9) and (8-11); the values of p are used as input data. The circuit design principle is similar to that of the systems considered above. Each consists of the following components: a computer with decoding servomechanism and a servomechanism for generating actual values of $\dot{A}$, $\dot{z}$ and $\dot{p}$ of the telescope. The circuits of the servomechanisms for $\dot{A}$, $\dot{z}$, $\dot{p}$ are identical with those in Figures 8-1, 8-2, 8-3 and are omitted from Figures 8-4, 8-5, 8-6 (only the transmitters LVT-$\dot{A}_2$, LVT-$\dot{Z}_2$ and LVT-$\dot{p}_2$ are shown). As before all these circuits can be combined into one.

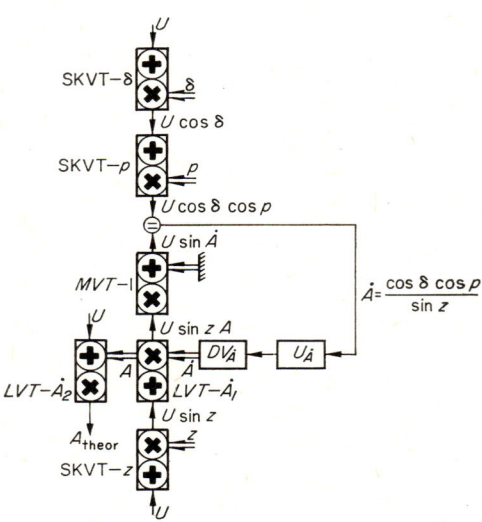

Figure 8-4 Block diagram of the $\dot{A}$ computing unit.

Altazimuthal and positional drive mechanisms 115

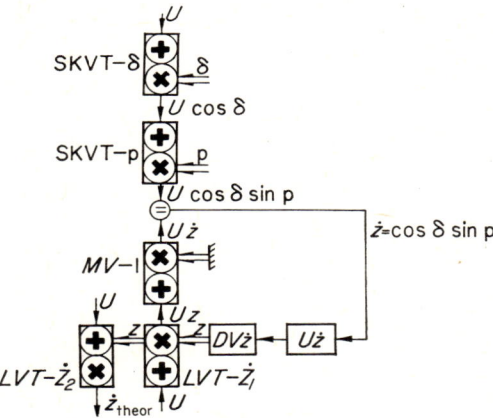

Figure 8-5 Block diagram of the $\dot{z}$ computing unit.

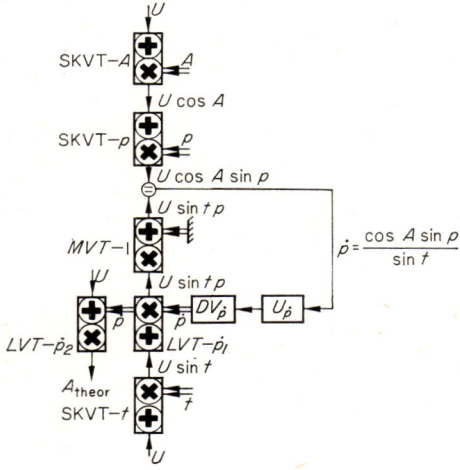

Figure 8-6 Block diagram of the $\dot{p}$ computing unit.

The application of a horizontal and positional drive mechanism does not exclude the use of systems with coordinate converters (PK-V and PK-VII) to generate A, z and p from the telescope.

The superposition principle can be applied to systems which compute the values of the horizontal coordinates and the rate of change of these coordinates. Neither system (guiding by position and velocity) excludes the other, but merely supplements it. The final purpose of regulation of both systems are the respective A, z and p axes of the altazimuthal instrument.

The use of a combined control system considerably simplifies the operating conditions of the system which generates the values of the actual coordinates of the celestial object. In this case the guiding system by position and using a PK operates in reverse near the zero position. Then the function of the guiding system by angle is changed in principle; its purpose is only to supply small angular corrections to the respective telescope axis. The velocity range in a guiding system by angle is greatly reduced owing to the fact that the guiding of the altazimuthal instrument is determined by the velocity system. The speed at which the values of A, z and p are generated can be greatly reduced, and the sensitivity of the circuit increased. A consequent increase in the accuracy of the solution of A, z and p may be expected, as well as a larger operating range due to the smaller danger zone near the zenith. Combined control for altazimuth instruments should give a smoother continuous generation of the coordinates values of the celestial object and a considerable improvement in the dynamic guidance performance.

The block diagram of the combined control of an altazimuth telescope by angle and velocity is shown in Figure 8-7.

The coordinates of the observed object are given in the equatorial system (right ascension α and declination δ). Using an adder for sidereal the time s and the right ascension α the current values of the equatorial coordinates t and δ and the latitude φ are input data to the coordinate conversion systems PK-V and PK-VII, which generate the values of the

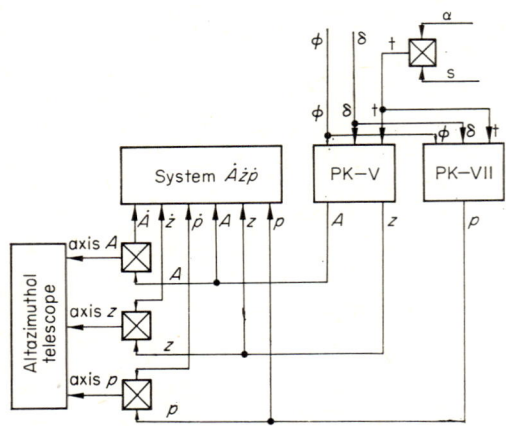

Figure 8-7 Block diagram of the combined control by position (angle) and velocity for an altazimuth instrument.

horizontal coordinates (A and z) and the positional angle p. These values and the $\dot{A}$, $\dot{Z}$ and $\dot{p}$ values (computed from A, z and p) are used to drive the corresponding axes. The values of the angles and velocities can be transmitted to the corresponding axes of the telescope using mechanical differentials which perform the function of adders. As a result, the combined control both by angle (A, z, p) and by velocity ($\dot{A}$, $\dot{z}$, $\dot{p}$) is performed for each of the axes of the altazimuth telescope.

In principle it is also possible to control an altazimuth instrument by acceleration, by using circuits which compute the values of $\ddot{A}$, $\ddot{z}$ and $\ddot{p}$.

Experiments show that the error in the solution of the equations $A = f(t, \delta)$, $z = f(t, \delta)$ and $p = f(t, \delta)$ (for systems which guide by angle with PKs using VTs) do not exceed 10–15′. Considering the relative simplicity of the VT cascaded circuits of the horizontal and positional drive mechanisms, and the fact that the improved dynamic operating conditions will improve the accuracy of generation of A, z and p, one would expect the error in the solution of the problem by a combined control systems to be of the same order.

CONCLUSIONS

Automatic guiding of altazimuthal astronomical instruments can be performed by means of combined control by angle and velocity. For this purpose altazimuthal and positional drive mechanisms with variable speed drives have been devised in the form of a special computer based on rotary transformers. Guiding by velocity with correction by angle greatly improves the control of the telescope when tracking, because the values of the horizontal coordinates and positional angle are generated continuously.

REFERENCES

1. S. O. Dobrogursky, V. A. Kazakov and V. K. Titov, *Computers* (Schetno-reshayushchiye ustroistva), Oborongiz (1959).
2. T. Ya. Khodorov, *Electromechanical Induction Computers* (Elektromekhanicheskiye induktsionny schetno-reshayushchiye ustroistva), Sudpromgiz (1960).
3. Yu. A. Sabinin and V. P. Yegorov, "Systems of automatic operation of a telescope dome with transformation of spherical coordinates using rotating transformers", *Izv. KrAO AN SSSR*, **26**, 395 (1961).

CHAPTER I-9

Some principles for the design of digital control systems for altazimuth instruments

S. V. KOROTKOV, V. A. MYASNIKOV and YU. A. SABININ

*Institute of Electromechanics on Automation and Machine Construction
Leningrad*

1 OPERATIONAL FEATURES OF ALTAZIMUTH INSTRUMENTS

The majority of telescopes today have equatorial mounts. In this type of mount, to compensate for the diurnal rotation of the earth, the telescope tube rotates around the polar axis of the mount at a uniform rate of one revolution per day. The control system for tracking of a star is then extremely simple.

However, for large equatorial telescopes the compensation for the shift of the optical axes is an extremely difficult problem. Consequently, for large telescopes it is better to use altazimuth mounts. For telescopes with main mirrors of several meters in diameter, the altazimuth mount has a number of important advantages over the equatorial (parallactic) mount. Some of the major advantages are simple mechanical design, favorable operating conditions for the bearings, a simpler support system for the mirror, and comparatively simpler compensation for flexure. These advantages make it possible to obtain sharp images without drift even though the moving components of the telescope weigh several tons.

However, the control system is much more complicated than for equatorial instruments, especially in the setting and guiding. This is due to the fact that the position of a celestial object is given in equatorial coordinates and since the instrument has to be driven by a precision drive mechanism, equatorial coordinates must be converted with high accuracy into horizontal coordinates.

In addition, servomechanisms must be used to compensate for the non-uniform rotation of the field in the focal plane.

A detailed analysis of mounting systems for astronomical instruments can be found in the specialized literature. We shall therefore discuss only the formulas which relate the equatorial and horizontal system of coordinates, since they will be required later in the choice of algorithms. Figure 9-1 gives a schematic showing the relationship between the two systems of coordinates.

In the equatorial system the basic parameters are the pole of the system P, hour angle t, declination δ and right ascension α. The pole lies on the meridian at an angle with respect to the horizon equal to the latitude of the observer φ. The hour angle is the dihedral angle between the plane of the celestial meridian and the plane of the declination circle of the object, and it is measured by the arc QM of the celestial equator from the upper point Q. The hour angle is calculated from the upper point of the equator in a clock-wise direction, i.e. in accordance with the diurnal rotation of the celestial sphere from 0 to 360°. The right ascension α is the arc of the equator from the vernal equinox γ to the base of the arc of declination. The vernal equinox γ is the point at which the sun intersects the equator in the spring (March 21), when it passes from the southern hemisphere of the celestial sphere into the northern hemisphere. The right ascension α is

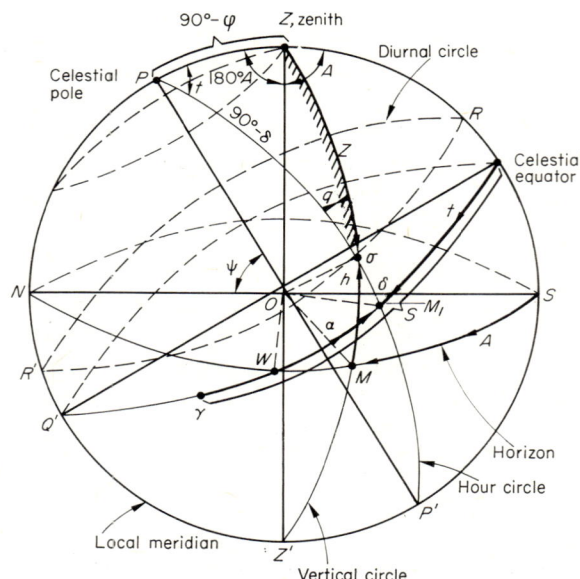

Figure 9-1 Horizontal and equatorial systems of coordinates.

expressed in hours, minutes and seconds, and is calculated from γ in a direction opposite to the diurnal motion of the objects and the hour angles. For each object, α remains constant despite the rotation of the celestial sphere because the vernal equinox itself has a diurnal motion (its apparent motion due to the rotation of the earth).

Therefore, the hour angle may be expressed as $t = s - \alpha$, where s is the sideral time, and a linear function of time. Sideral days are equivalent to 23 hours, 56 min and 4.08 sec mean solar time, and are defined as the interval of time between two successive upper culminations of the vernal equinox in the meridian of the observer. The declination δ is the distance from the object σ to the equator, measured along the arc $M_1\sigma$, and is positive in the direction from the equator toward P and negative toward P'.

Coordinates α and δ do not depend either on the time or the local latitude, so that only they are given in the star catalogs. The limits of variation of the parameters in the equatorial system are

$$\delta = 0° \quad \text{to} \quad \pm 90°,$$

$$t = 0 \quad \text{to} \quad 23^\text{h}56^\text{m}04\overset{\text{s}}{.}08.$$

In the horizontal system, Z is the pole of the system and the height h and azimuth A are the coordinates. The zenith distance z is the distance from Z to the celestial object σ; it is related to h by $z = 90° - h$. The azimuth A is the dihedral angle between the meridian plane ZSZ' and the vertical plane $Z\sigma MZ'$. It is calculated from the southern part of the meridian clock-wise (south, west, north, east). The range of values of the spherical coordinates in the horizontal system are

$$h = 0° \quad \text{to} \quad \pm 90°,$$

$$A = 0° \quad \text{to} \quad 360°.$$

By using the rules of spherical trigonometry it is possible to establish a relationship between the coordinates of the two systems.

For the sake of brevity, we use the following notation:

$$a_1 = \cos \delta; \quad a_2 = -b_1 \sin \delta; \quad a_3 = b_2 \cos \delta;$$
$$a_4 = b_2 \sin \delta; \quad a_5 = b_1 \cos \delta; \quad a_6 = b_1/a_1; \quad a_7 = \sin \delta; \quad (9\text{-}1)$$
$$b_1 = \cos \varphi; \quad b_2 = \sin \varphi.$$

Then the relationships taken from

$$\cos z = a_4 + a_5 \cos t, \tag{9-2}$$

$$\tan A = \frac{a_1 \sin t}{a_2 + a_3 \cos t}, \tag{9-3}$$

$$\sin z \sin A = a_1 \sin t, \tag{9-4}$$

$$\frac{dz}{dt} = \frac{a_5 \sin t}{\sin z}, \tag{9-5}$$

$$\frac{dz}{dt} = b_1 \sin A, \tag{9-6}$$

$$\frac{dA}{dt} = \frac{b_2 \sin z + b_1 \cos z \cos A}{\sin z}, \tag{9-7}$$

$$\frac{dA}{dt} = b_2 + b_1 \frac{\cos A}{\tan z}, \tag{9-8}$$

$$\frac{dA}{dt} = \frac{a_1 \cos q}{\sin z}, \tag{9-9}$$

$$\frac{d^2z}{dt^2} = \frac{a_5 \cos A \cos q}{\sin z}, \tag{9-10}$$

$$\frac{d^2z}{dt^2} = b_1 \cos A \left[b_2 + \frac{b_1 \cos A}{\tan z} \right], \tag{9-11}$$

$$\frac{d^2z}{dt^2} = b_1 \cos A \frac{dA}{dt}, \tag{9-12}$$

$$\frac{d^2A}{dt^2} = -\frac{b_1}{\sin z} \left[\frac{b_1 \sin 2A}{\sin z} + a_7 \sin A \right], \tag{9-13}$$

$$\frac{d^2A}{dt^2} = -\frac{B_1 \sin A}{\sin z} \left[a_7 + \frac{2B_1 \cos A}{\sin z} \right], \tag{9-14}$$

$$\sin q = \frac{B_1 \sin t}{\sin z}. \tag{9-15}$$

The control system for an altazimuth telescope should control the drive over a wide range of speeds, in two axes, including the reversal of the motion. While the speed of the telescopes in zenith distance does not exceed one revolution per day it becomes infinite in azimuth at $z = 0°$.

For a fuller analysis of the control system for an altazimuth telescope, let us consider the basic requirements. It must ensure guiding with an error not exceeding $\pm 0''\!.2$ relative to the line of sight to the object. To achieve such high accuracy two systems are used: a coarse guiding system (analogous to the drive mechanism in equatorial telescopes) and a fine guiding system (photoelectric guider). The coarse guiding system must include a coordinate converter with a maximum error not exceeding $\pm 5''$ at each point to allow for actual refraction and flexure.

Accelerations in the coarse guiding system must not exceed ten times the maximum calculated value of the acceleration at $z = 5°$. The control system cannot be used in the near-zenith region of solid angle $10°$ because of the sharp increase in speed and acceleration in azimuth.

The system must include a unit for automatic corrections of atmospheric refraction and flexure. The actual refraction is calculated from the formula

$$\Delta r_z = 60''\!.2 \frac{B}{760} \frac{273}{273 + T} \tan z, \qquad (9\text{-}16)$$

where B is the atmospheric pressure in millimeters of mercury, T is the temperature in °C and z is the zenith distance.

Note that this formula is approximate and will not give exact results, since the computational error amounts to about 10% (several seconds of arc).

The exact formula for the actual refraction correction is

$$\Delta r_z = 60''\!.2 \frac{B}{760} \frac{273}{273 + T} \tan z f(z), \qquad (9\text{-}17)$$

where $f(z)$ is a function which includes the odd power terms in the series expansion of $\tan z$.

Correct operation of the whole control system requires that the photoelectric guiding system should not disturb the coarse guiding system. A unit which compensates for the rotation of the field in the focal plane of the main focus and the fixed focus must be included.

The control system must also provide for a lunar-planetary drive mechanism and a device for tracking artificial satellites. Any velocities added to the basic motion must be able to be regulated within $\pm 5\%$ of the apparent diurnal motion of the celestial sphere.

The coarse guiding system must set the telescope on a given object. The maximum error, i.e. the maximum deviation of the image of the star from the line of sight of the telescope after setting, must not exceed 5″ in any direction, and to attain such high accuracy, corrections for actual refraction and flexure must be included in the system. The maximum setting rate in each of the horizontal coordinates must be $1°\ \text{sec}^{-1}$. The maximum value of the acceleration in either coordinate during setting must not exceed $0.^{\circ}1\ \text{sec}^{-2}$.

The transition from setting to guiding conditions must be accomplished automatically, without interruption. Adjustment to the position of a star after setting is done by fine-adjustment drives with push-button control, but care must be taken that the action on these drives does not disturb the guiding and setting system. These are, in brief, the main technical requirements for the control system.

Thus, it is clear that precision servomechanisms must be used in the setting and guiding systems. It has been shown that servo-mechanisms using analog circuits could calculate the require values with an accuracy of less than 0.01%, which is totally unacceptable in this case. With an error of 5″ in the operation of the setting and coarse guiding system, the accuracy of the servomechanism is 0.0005% or 20 times greater than that provided by the most accurate analog servomechanism. We must therefore use a digital tracking system (TsSS), the operating principle of which is identical to that of analog devices, but in which all operations are performed using digital techniques.

It is then possible to achieve any desired accuracy within the limits of accuracy of the actual angular measurement by the positional feedback transmitter.

Smooth motion of the telescope around the two axes, which is especially important when the velocity passes through zero values, can be secured by introducing velocity feedback circuits into the TsSS.

2 ANALYSIS AND CHOICE OF THE TsSS BLOCK DIAGRAM ON THE BASES OF SIMILARITIES WITH ANALOG SYSTEMS

Analog coordinates converters from equatorial to horizontal system (using rotary transformers) are available at present [1]. Let us consider the possibility of constructing a digital coordinate converter on the basis of analog computer systems. The simplest and most reliable solutions of trigonometric equations can be obtained in digital form by using digital integrating units or digital differential analyzers TsDA [2, 3].

A digital differential analyzer is an analog-digital machine which successfully combines the analog principle of solution with digital techniques. All mathematical operations are reduced to the single operation of integration, which is performed by the addition and subtraction of numbers. This is done by a digital integrator. The TsDA combines the advantages of analog and digital machines: reliability of operation (due to the relatively small number of electronic components and automatic operation); high accuracy; high speed of operation; continuity of solution; and simplicity of adjustment.

Unlike universal computers, TsDAs do not require complicated programming, have a convenient form of input, can generate various functions and are simple to use.

In a universal computer to obtain trigonometric functions such as those expressed by Eqs. (9-1) to (9-15), one would have to write a program to operate at given intervals of time. Two methods could be used:

(1) The program could use tabular values of the trigonometric functions stored in an external memory, and could calculate intermediate values by interpolation;

(2) The function could be expanded in series, the number of terms used depending on the accuracy required.

Despite the relatively high speed of universal computers such as the BESM or the Strela, which performs from 5,000 to 13,000 operations per second, this particular operation would require a machine a hundred times faster. Moreover, the specific problem of tracking celestial objects does not require a universal machine since fast compact TsDAs permit the program to be changed easily to compute a broad class of funtions.

The available literature [2, 3] hardly discusses the problems of digital differential analyzers, but the design of the main circuits for digital integrators described in Soviet and foreign works make it possible to write a number of algorithms for the solution of the relationships given by Eqs. (9-1) to (9-15).

Let us consider the main block diagrams for the conversion from equatorial to horizontal coordinates, using digital integrators.*

(1) Let us use Eq. (9-2), as the one most suitable for calculating z. A block diagram of a system using digital integrators that solves Eq. (9-2)

---

* For the interpretation of the block diagrams given in this contribution, the reader is referred to the English translation of [2] published by The MacMillan Company, 1964. (Ed. English version.)

is shown in Figure (9-2). Using integrators 1 and 2 we obtain $a_1 + a_5 \cos t$ i.e. cos z. Integrators 3, 4, 5 and 6 compute the function arc cos z.

(2) Another system to solve Eq. (9-2) and that uses digital integrators is shown in the block diagram given in Figure (9-3).

Unlike the first system (Figure 9-2), this one needs only three integrators to obtain arc cos z, by using the servointegrator 3.

(3) To compute the inverse trigonometric function

$$A = \text{arc tan} \frac{a_1 \sin t}{a_2 + a_3 \cos t}$$

from Eq. (9-3) the system shown schematically in Figure 9-4 may be used.

The integrators 3, 4 and 5 compute tan A, while 6, 7, 8 and 9 compute the inverse function arc tan $[a_1 \sin t/(a_2 + a_3 \cos t)]$.

(4) The block diagram of a second system to compute $A = \text{arc tan} \times [a_1 \sin t/(a_2 + a_3 \cos t)]$ is shown in Figure 9-5. It differs from the one given in Figure 9-4 in that a servointegrator is used in the inverse tangent circuit, thus reducing the number of integrators by one.

The algorithms to compute z and A are given in the block diagrams of Figures 9-2, 9-3, 9-4 and 9-5; these algorithms are fairly cumbersome. We could attempt to construct a TsSS to generate z and A in which functions of z and A rather than the values themselves are compared,

(5) To compute the zenith distance z and azimuth A, a system represented by the block diagram shown in Figure 9-6 can be used. The integrators 1 and 2 computes cos z_comp for given values of α and δ. Integrators 4 and 5 gives

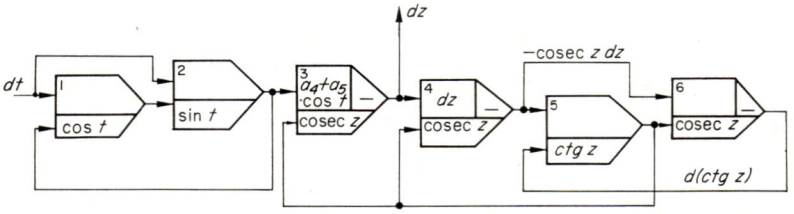

Figure 9-2 Block diagram of a system to compute arc cos z using digital integrators.

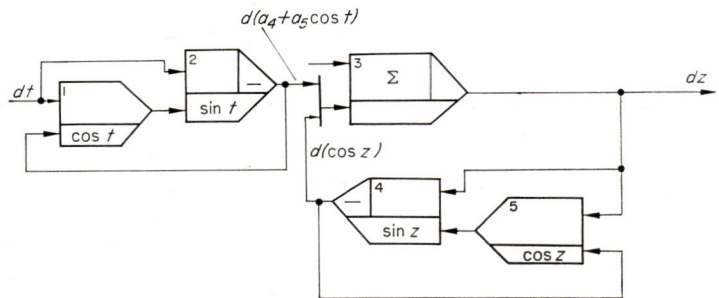

Figure 9-3 Block diagram of another system to compute arc cos z using digital integrators.

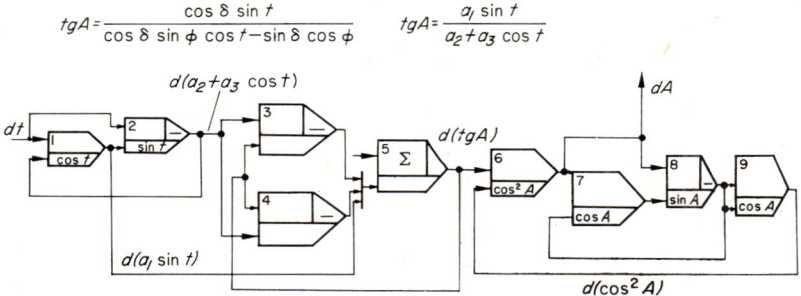

Figure 9-4 Block diagram of a system to compute arc tg A using digital integrators.

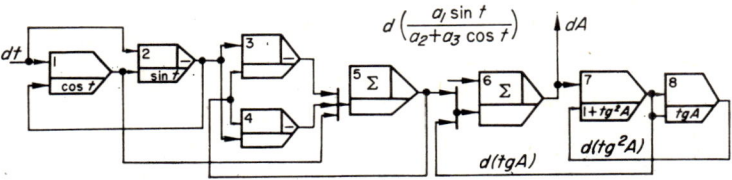

Figure 9-5 Block diagram of another system to compute arc tg A using digital integrators.

the actual values of $\cos z_{\text{act}}$. The servointegrator 3 continues to feed error pulses to the servomechanism IM until $\cos z_{\text{act}} = \cos z_{\text{comp}}$.

(6) The system indicated in Figure 9-6 also generates the azimuth values by solving Eq. (9-4); in this system the products $\sin z_{\text{act}} \sin A_{\text{act}}$ and $\cos \delta \sin t_{\text{comp}}$ are compared.

(7) A TsSS system can be designed to satisfy Eq. (9-3). For this purpose we rewrite Eq. (9-3) in the following form:

$$\sin A (a_2 + a_3 \cos t) = \cos A \, a_1 \sin t.$$

This system (Figure 9-7) compares the products $\sin A_{act}(a_2 + a_3 \cos t_{comp})$ and $\cos A_{act} \, a_1 \sin t_{comp}$; their difference is fed to the servomechanism IM.

The systems discussed in 5, 6 and 7 with block diagrams given in Figures 9-6 and 9-7 are used also in analog systems; for example the system

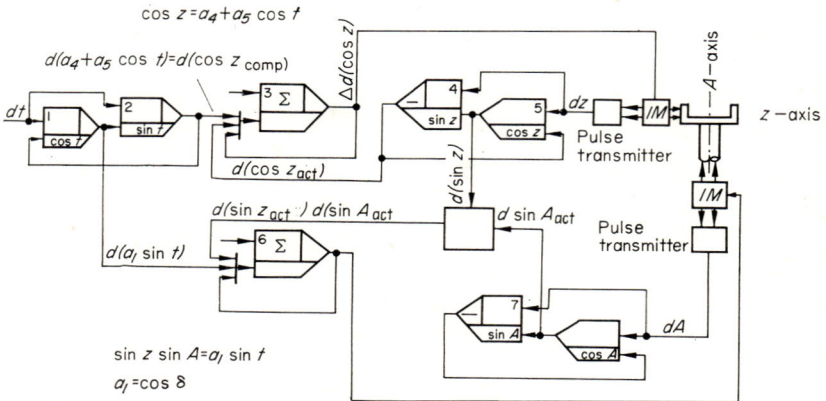

Figure 9-6 Block diagram of the digital system to compute zenith distance z and azimuth A.

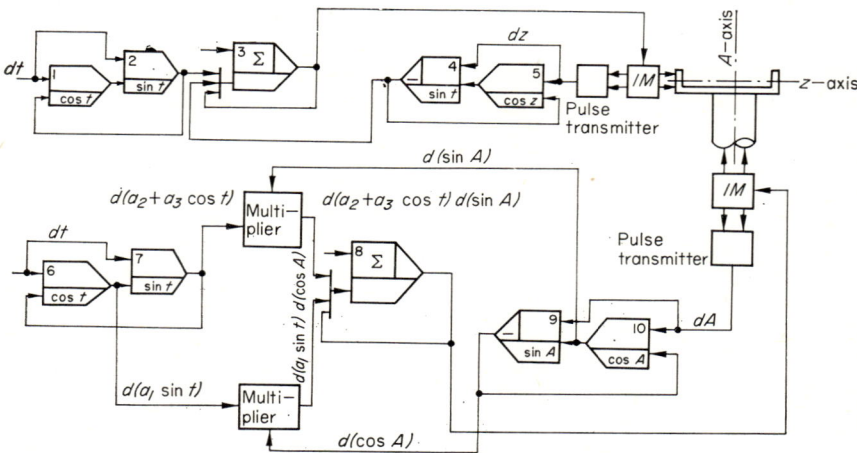

Figure 9-7 Block diagram of another system to compute zenith distance z and azimuth A using digital integrators.

discussed in 7 (Figure 9-7) has been used in a coordinates converter using rotary transformers. This converter developed at the Institute of Electromechanics is used in an automatic setting and coarse guiding system of an altazimuth telescope.

Consideration of the preceeding algorithms shows that the simplest of them are obtained by comparing functions of z and A rather than the values themselves. However, an analysis of the relationship between the output signal of the comparison circuit and the error in the coordinate for the schemes discussed in 5, 6 and 7 gives the following results.

For the system given in Figure 9-6 and the z channel, the output error signal $\varDelta$ of the comparison circuit is given by

$$\varDelta = \cos(z \pm \varDelta z) - (\sin\varphi \sin\delta + \cos\varphi \cos\delta \cos t) = \pm\sin\varDelta z \sin z,$$

where $\varDelta z$ is the error in z; φ, δ, and t are computed values. For sufficiently small $\varDelta z$, $\sin\varDelta z \approx \varDelta z$ and then $\varDelta = \pm\varDelta z \sin z$.

For the same system given in Figure 9-6 and considering the A channel, the output error signal $\varDelta$ of the comparison circuit, is given by

$$\varDelta = \sin(A \pm \varDelta A)\sin(z \pm \varDelta z) - \cos\delta \sin t$$
$$= \pm\sin A \cos z \sin\varDelta z \pm \cos A \sin z \sin\varDelta A,$$

where $\varDelta A$ is the error in A; z, A, and t are computed values.

For sufficiently small $\varDelta z$ we have $\sin\varDelta z \approx \varDelta z$ and then

$$\varDelta = \pm\sin A \cos z\,\varDelta z \pm \cos A \sin z\,\varDelta A.$$

For the system given in Figure 9-7 and considering the A channel, the output error signal $\varDelta$ of the comparison circuit, is given by

$$\varDelta = [\sin(A \pm \varDelta A)](\cos\delta \sin\varphi \cos t - \sin\delta \sin\varphi)$$
$$- [\cos(A \pm \varDelta A)]\cos\delta \sin t = \pm\sin\varDelta A \sin z.$$

For $\varDelta A$ sufficiently small, $\sin\varDelta A \approx \varDelta A$ and then $\varDelta = \pm\varDelta A \sin z$.

The output signals of the comparison circuit (Figures 9-6 and 9-7) are proportional to the error in the coordinates, but also depend to a large extent on the other quantities. It is therefore more difficult to use them directly in the digital servomechanism, since the error $\varDelta$ changes in sign. The possible remedies complicate the circuits to such an extent that it is better to use variations of schemes suggested in 1, 2, 3 or 4, where the output signal depends only on the error in the coordinates and is directly proportional to it.

3 PRINCIPLES FOR THE DESIGN OF A TsSS AND SOME FUNDAMENTALS OF INFORMATION THEORY

Using the circuits discussed above (Figures 9-2, 9-3, 9-4) we can construct a digital coordinates converter servomechanism. Certain fundamentals of information theory can be used to calculate the basic parameters of the system.

Information theory can be applied to the solution of various practical problems. In particular, we can establish the frequency with which instantaneous values of a continuous signal must be read out so that the error does not exceed a fixed limit.

In Figures 9-8(a), (b), (c), A and z are plotted versus t. For the 24-hour interval, A and z may take any of an infinite number of values within the range $0° \leq A \leq 360°$, $-90° \leq z \leq +90°$.

The maximum angular error for either axis has been chosen not to exceed $1.''25$. To allow for additional errors, which may be compounded, this value has been chosen four times smaller than required. Consequently, the minimum measurable increment must not be greater than $1.''23$.

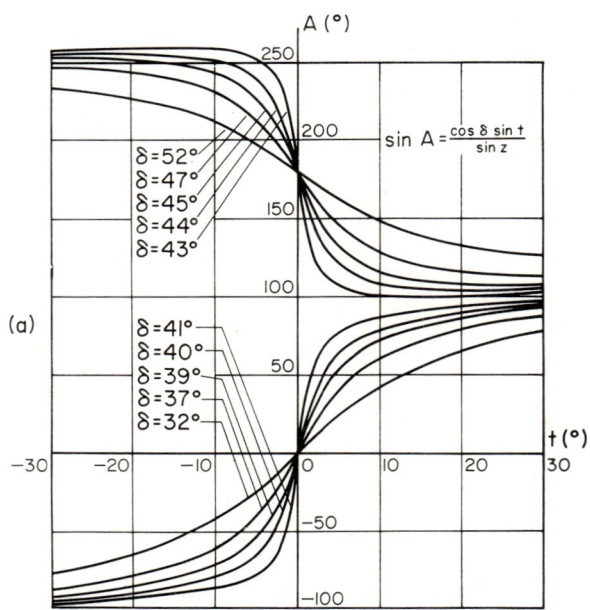

Figure 9-8(a) Azimuth A versus hour angle t for different values of declination δ.

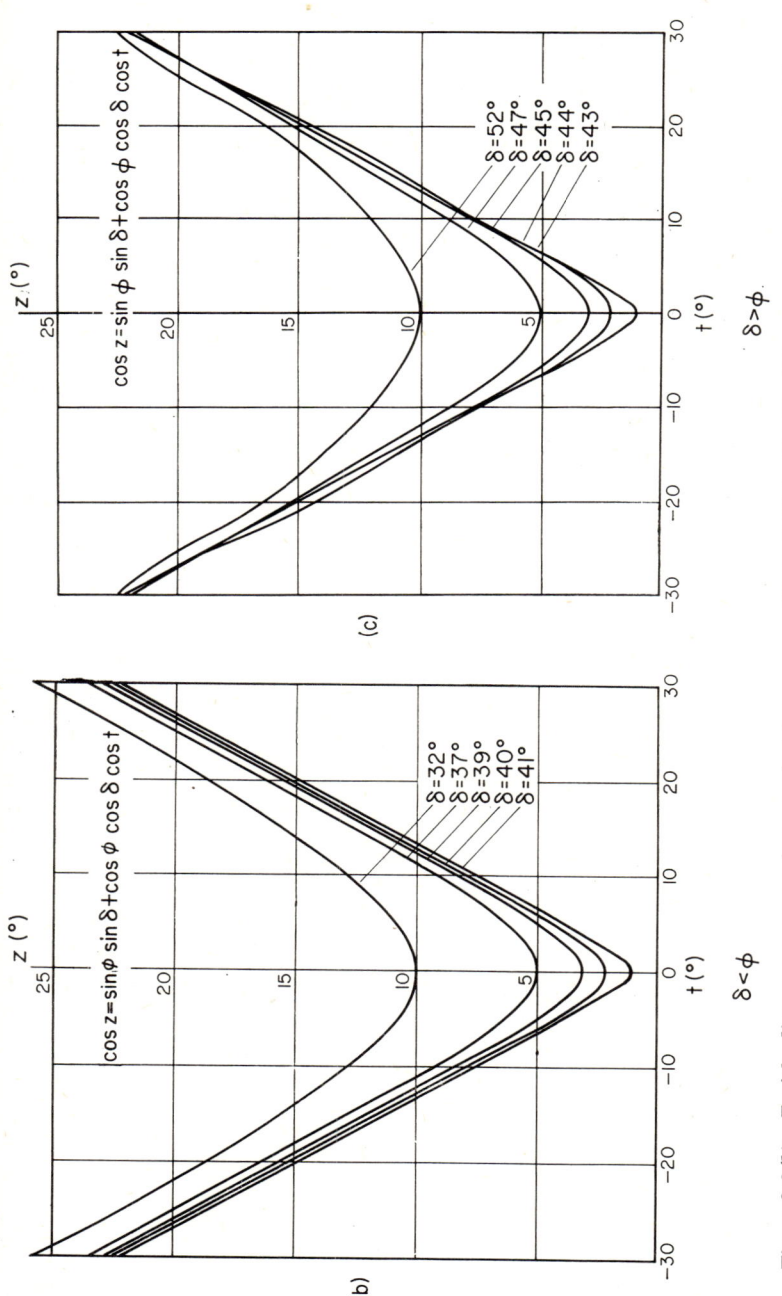

Figure 9-8(c) Zenith distance z versus hour angle t for different values of declination δ ($\delta > \varphi$).

Figure 9-8(b) Zenith distance z versus hour angle t for different values of declination δ ($\delta < \varphi$).

We shall not give calculations for the zenith distance when they are similar to those for the azimuth, since with the given range, the maximum non-uniformity of motion and, therefore, the worst conditions for setting and guiding, will occur in azimuth. Clearly the smallest interval of time required for a change in azimuth by a value equal to the resolving capability of the system is equal to the ratio of the minimum increment ΔA and the maximum rate of change in azimuth $\dot{A}_{\max}$:

$$\Delta T = \frac{\Delta A}{\dot{A}_{\max}}. \tag{9-18}$$

Figure 9-9 gives the plot of dA/dt versus zenith distance z. With the given limits for z, the guiding velocity in azimuth varies from 0 to 10 rev day^{-1}. To allow a certain margin, we calculate the system for the rate of 20 rev day^{-1} = 300 sec^{-1}, i.e. for $z \approx 2°$.

The resolving capability of the system requires a minimum measurement (or iteration) frequency of

$$f = \frac{1}{\Delta T} = \frac{\dot{A}_{\max}}{\Delta A}$$

$$= \frac{300'' \text{ sec}^{-1}}{1.''23} \approx 250 \text{ Hz}. \tag{9-19}$$

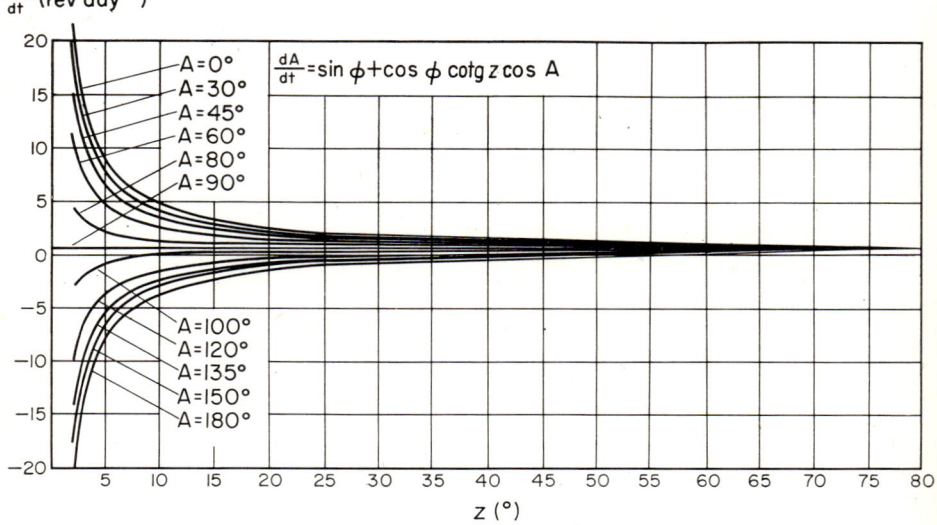

Figure 9-9 dA/dt versus z and for different values of azimuth A.

Any measurement may be considered to consist of a certain number of basic elements. If the measurement is expressed in binary code, each unit of information can be 0 or 1, and in our case the total number of possible digital azimuth values can be expressed as follows:

$$n = \frac{A_{max}}{\Delta A} + 1 = \frac{360°60'60''}{1''23} + 1 \approx 2^{20}. \tag{9-20}$$

The total information content in any measurement can be expressed in terms of the logarithm of the number of different states. Hence, the information content in a measurement is expressed by

$$H = \log n = \log\left(\frac{A_{max}}{\Delta A} + 1\right).$$

The binary code is most often used for these mathematical operations since a simple and reliable circuit can be provided, a trigger, with two stable states (0 and 1). In the by binary system the value of H is given by

$$H = \log_2 n = \log_2\left(\frac{A_{max}}{\Delta A} + 1\right) = \log_2 2^{20} = 20. \tag{9-21}$$

The number of triggers required to represent a number (i.e. the digital capacity of the storage devices, registers, etc., in the integrators) will be equal to the information content required to give the resolving capability of the control system. Hence in this case each digital integrator must have no less than 20 digits.

However, as we have already noted, under dynamic conditions the azimuth changes noticeably within the minimum time defined by Eq. (9-18). Then the rate of information flow β or the rate with which the information must be recorded in the system, is equal to $H/\Delta T$. Dividing Eq. (9-21) by (9-18), we obtain β in the form

$$\beta = \frac{H}{\Delta T} = \log_2\left(\frac{A_{max}}{\Delta A} + 1\right)\frac{\dot{A}_{max}}{\Delta A}. \tag{9-22}$$

This expression defines the rate of information flow in terms of the error of the measurement and the maximum velocity for a given measurement range.

Preliminary calculations have shown that no less than 30 integrators are required to compute A and z, the refraction and flexure corrections and certain other functions.

In sequential computation the information rate will be

$$\beta = 30 \log_2\left(\frac{A_{\max}}{\Delta A} + 1\right)\frac{A_{\max}}{\Delta A} = 30.20.250 \simeq 150.000 \text{ bits sec}^{-1}.$$

Published data shows that digital differential analyzers with magnetic drums give an operating speed not greater than 70 to 100 kHz, while those with a ferrite memory gives an operating speed of up to 1 MHz.

We could therefore use a variant of a TsDA with a ferrite memory to allow for the required information frequency. However, evaluating the information and iteration frequency given by Eq. (9-19) we notice that it is relatively high for this particular case. It is very difficult to manufacture an angular encoder to operate at the above frequency and speed of measurement. Moreover, the drive must provide a theoretically infinite range of velocities with maximum frequency of 250 Hz for comparison of actual and computed values. It is practically impossible to design a satisfactory drive system with the required dynamic performance.

Precision of practical control systems can attain a wide regulating range, 1 to 200, 1 to 500. It might seem natural to divide the entire range of speeds up to twenty revolutions per day into several subranges and to guide the instrument in the different ranges by separate drives. However, it is our opinion that it is too difficult to switch from one drive to another and to keep the motion of the telescope sufficiently smooth. In addition, the region of maximum azimuth velocities is notable for the high frequency of comparison of actual and calculated values, so that in this frequency range it is also difficult to achieve satisfactory dynamic performance of the drive.

For these reasons, the authors suggest a different method for the control in cases where high accuracy, continuity and a wide range of velocity regulation are required.

4 HIGH ACCURACY, CONTINUITY AND WIDE VELOCITY RANGE CONTROL METHOD

For digital automatic regulation systems (SARs) the iteration frequency and the frequency of comparison of actual and calculated values, as in Eq. (9-18), are determined from the formula

$$f = \frac{1}{\Delta T} = \frac{1}{\frac{\Delta y}{\dot{y}_{\max}}} = \frac{\dot{y}_{\max}}{\Delta y}, \qquad (9\text{-}23)$$

where $\Delta T = \Delta y/\dot{y}_{max}$ is the time increment corresponding to an increment Δy which is the resolving capability of the system (the limit coordinate error) and $\dot{y}_{max}$ is the maximum rate of change of the variable.

For certain applications in instruments requiring automatic regulation, the velocity, acceleration and higher derivatives vary within a wide range and attain large magnitudes. Thus if high accuracy is required, a very high iteration and comparison frequency for the SAR is necessary.

Let us see how the iteration and comparison frequency of the SAR changes if a supplementary SAR is used for the derivative of the coordinate with limit error $\Delta \dot{y}$. It is clear that the interval of time needed to give the required resolving capability in this case must be no less than $\Delta T_1 = \Delta y/\Delta \dot{y}$, since in the interval between comparisons (measurements) the rate of coordinate change is maintained with an error not greater than $\Delta \dot{y}$. The iteration and comparison frequency will then be

$$f_1 = \frac{1}{\Delta T_1} = \frac{\Delta \dot{y}}{\Delta y} = \frac{\Delta \dot{y} \, \dot{y}_{max}}{\dot{y}_{max} \Delta y} = f \frac{\Delta \dot{y}}{\dot{y}_{max}}. \qquad (9\text{-}24)$$

Similar results can be obtained for a SAR for the velocity, i.e.

$$f_2 = \frac{\Delta \ddot{y}}{\Delta \dot{y}}, \qquad (9\text{-}25)$$

where, f_2 is the iteration and comparison frequency for velocity, $\Delta \dot{y}$ is the fixed error of the SAR for velocity, and $\Delta \ddot{y}$ is the limit error of the supplementary SAR for acceleration. With a supplementary SAR for acceleration, we can obtain

$$f_3 = \frac{\Delta \dddot{y}}{\Delta \ddot{y}}, \qquad (9\text{-}26)$$

where f_3 is the iteration and comparison frequency of the SAR for acceleration with a supplementary SAR for the third derivative, and $\Delta \dddot{y}$ is the limit error for the third derivative.

Let us give an example to show how the frequency of iteration and comparison can be reduced. Suppose that we require a digital servomechanism for the function $y = 5t^2$ with an accuracy of the servomechanism of 0.001% for t between 0 and 10 seconds. The given function will then vary from 0 to 500 units, but its derivative $(dy/dt = 10t)$ will vary from 0 to 100 units per second. The iteration and comparison frequencies if one digital SAR is used for y will be

$$f = \frac{\dot{y}_{max}}{\Delta y} = \frac{100}{0.00001 \cdot 500} = 20{,}000 \text{ Hz}.$$

If a digital SAR is used for velocity, with the $\Delta \dot{y} = 0.1\%$, the frequency is

$$f_1 = \frac{\Delta \dot{y}}{\Delta y} = \frac{0.001 \cdot 100}{0.00001 \cdot 500} = 20 \text{ Hz},$$

i.e. it is reduced by a factor of a thousand.

This method can therefore be used to reduce the iteration and comparison frequency for the coordinate (or for the first, second, third, etc., derivatives) as many times as the fixed error for the first (second, third, etc.) derivative is less than the maximum value of this derivative, Eqs. (9-24), (9-25) and (9-26).

When the instrument is being guided by position, the iteration frequency and frequency of comparison of the actual and calculated values are found from Eq. (9-18) in the absence of supplementary systems. If there is a supplementary system for guiding by velocity, it can be used as an interpolator between the comparison points of the calculated and actual values of the coordinate (Figure 9-10).

Suppose that at point 1 (see Figure 9-10) the guiding system by position sets the instrument in accordance with the calculated coordinates; then until the next comparison point, the instrument will be guided by the supplementary guiding-by-velocity system with error $\pm \Delta \dot{A}/2$ determined by its accuracy.

The positioning error at point (2) will not exceed

$$\frac{\Delta A_1}{2} = \frac{\Delta \dot{A}}{2} \Delta T$$

or
$$\Delta A_1 = \Delta \dot{A} \, \Delta T = \sigma_{\max} \dot{A}_{\max} \Delta T = \sigma_{\max} \Delta A, \qquad (9\text{-}27)$$

where $\sigma_{\max} = \Delta \dot{A}/\dot{A}_{\max}$ is the total error of the guiding-by-velocity system.

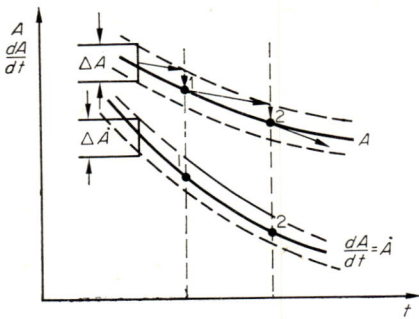

Figure 9-10 A and dA/dt versus t during automatic guidance by position and velocity.

If we assume that the error in the coordinate at point 2 is equal to the maximum error, $\Delta A_1 = \Delta A$, and the error of the guiding-by-velocity system is $\Delta \dot{A}$, the necessary interval of time ΔT_1 between comparisons, to give the resolving capability of the system, will be equal to

$$\Delta T_1 = \frac{\Delta A}{\Delta \dot{A}} = \Delta T \frac{\dot{A}_{max}}{\Delta \dot{A}} = \frac{\Delta T}{\sigma_{max}}. \tag{9-28}$$

Thus the supplementary system of guiding by velocity significantly reduces the comparison frequency, i.e. a slower digital system for the coordinate and feedback transmitters can be used in the system for the coordinate. With a lower iteration frequency (an important factor in determining angles to an accuracy of 0.0001%) or, with the same speed as before, the resolving capability of the system for the coordinate can be increased.

The following results can be given as examples. Suppose that an altazimuth instrument is to be guided by position (coordinates) with an error not exceeding 5″. Allowing for the necessary margin of error, let us calculate a system with a generating accuracy $\Delta A = \Delta z = 1.''23$. We find that the total number of discrete levels in the digital system is

$$n = \frac{360°}{1.''23} \approx 2^{20},$$

so that the number of bits

$$N = \log_2 n = 20.$$

From Eq. (9-18), the iteration period is

$$\Delta T = \frac{1.''23}{20 \text{ rev day}^{-1}} \approx 0.004 \text{ sec},$$

where 20 rev day^{-1} is the maximum velocity in azimuth. The iteration frequency is $f = 1/\Delta T = 250$ Hz.

Applying the control method we have described, we obtain the following result. We choose the actual generating error of the guiding-by-velocity system to be $\sigma_{max} = 0.001 = 0.1\%$. From Eq. (9-28) we have

$$\Delta T_1 = \frac{\Delta T}{\sigma_{max}} = \frac{0.004}{0.001} = 4 \text{ sec}.$$

Thus, the iteration frequency is reduced by a factor of 1,000.

Note that all the results were obtained for the point of maximum velocity, at which $\sigma_{max} = 0.1\%$; in all other cases the error of the SAR for velocity may be considerably greater when the resolving capability of the system is

$$\sigma = \frac{\Delta \dot{A}_{max}}{\dot{A}} 100\% . \tag{9-29}$$

If we choose 2 as a safety coefficient,

$$\Delta T_2 = \frac{\Delta T_1}{2} = 2 \text{ sec},$$

then the iteration frequency for the guiding-by-coordinate system will be

$$f = \frac{1}{\Delta T_2} = \frac{1}{2} = 0.5 \text{ Hz} .$$

This low frequency can have very little effect on the photoguiding systems, since the time constant of the positioning system is much larger than that of the photoguiding system. When a TsDA is used it is also worthwhile using it to control the instrument by velocity. Let us calculate the iteration frequency of the guiding-by-velocity system:

$$f = \frac{\ddot{A}_{max}}{\sigma_{max} \dot{A}_{max}} = \frac{450 \text{ rev day}^{-2}}{0.001 \cdot 20 \text{ rev day}^{-1}} = 0.31 \text{ Hz} .$$

A system such as this can be easily built.

Let us consider the design principles of drives for altazimuth instruments.

We mean by an altazimuth drive a unit which provides the required angular velocity for both axes of an altazimuth instrument.

Differentiating Eqs. (9-2) and (9-3) with respect to t, we obtain the relationship between the velocities in the horizontal and the equatorial coordinates.

Let us determine the accuracy to which the functions must be calculated so that the positional error of the the telescope axis will be within acceptable limits.

Modern drive mechanisms for equatorial instruments have a guiding accuracy of the order of 1×10^{-6}. Satisfactory guiding for altazimuth telescopes should have at least the same guiding accuracy, i.e. $\sigma = 0.0001\%$.

It follows that in this case the only possible system is a digital altazimuth drive mechanism. Figure 9-11 gives a block diagram of such a drive for guiding altazimuth instruments by velocity. The diagram represents a digital servomechanism that compares actual velocity values (dA/dt, dz/dt) with the respective computed values, giving as output the difference (errors) between the compared values.

The most suitable circuits for obtaining the values dA/dt and dz/dt are digital integrators.

The iteration frequency f of the guiding-by-velocity system should not be lower than

$$D = f = \frac{\ddot{A}_{max}}{\sigma \dot{A}_{max}} = \frac{450 \text{ rev day}^{-2}}{0.000001 \cdot 20 \text{ rev day}^{-1}} \approx 314 \text{ Hz}.$$

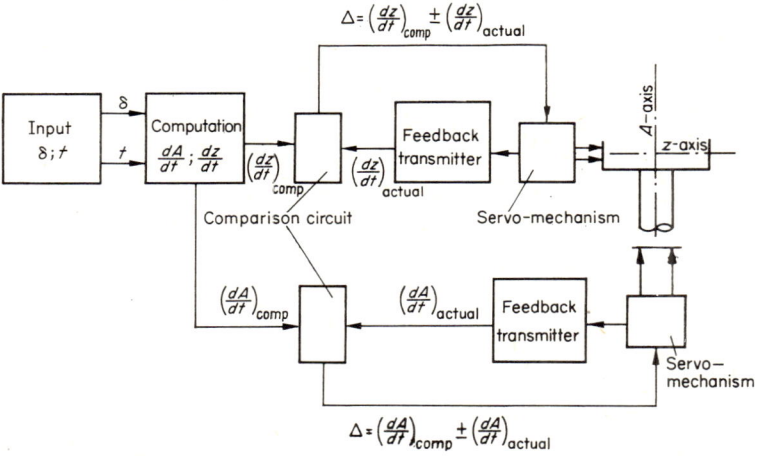

Figure 9-11 Block diagram of a guiding control (velocity) system for altazimuth instruments.

This frequency can be attained, but the drive system in this case must have strong damping as well as large amplification, i.e. it must give a very small static error and relatively small accelerations. Note also that such a guiding-by-velocity system has a lower limit to the range of steady regulation, i.e. it has a definite insensitive zone. Therefore when the instrument passes through zero velocities, the total error in position may be very large. Consequently, any guiding-by-velocity system may operate independently only in a specified range of regulation, while at zero and near-zero velocities, the accumulated error in position must be compensated by the guiding-by-position system which is also used in setting the telescope.

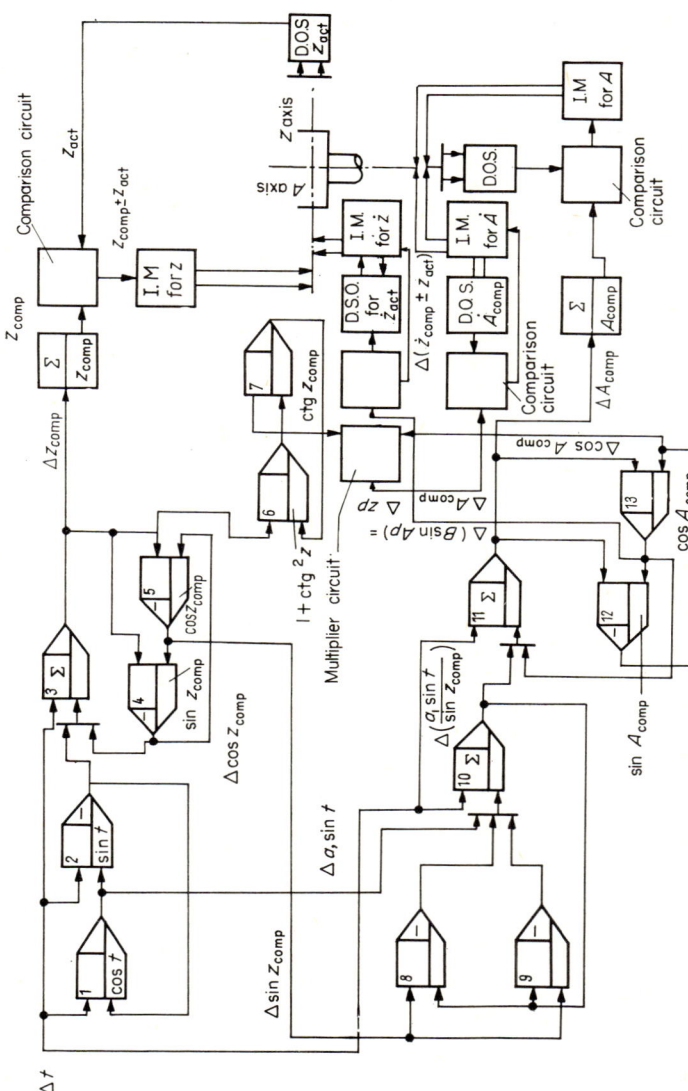

Figure 9-12 Block diagram of a guiding control (position and velocity) system for altazimuth instruments.

Using these fundamental design principles for precision digital servomechanisms, let us propose a block diagram for the control by position and velocity of an astronomical instrument (see Figure 9-12). The position is generated by the equations $z = \text{arc cos}\,(a_4 + a_5 \cos t)$, and $A = \text{arc sin}\,(a_1 \sin t/\sin z)$, and the velocity by the equations $dz/dt = b_1 \sin A$, and $dA/dt = b_2 + b_1 \cos A \cot z$. Integrators 1 and 2 give $\cos z = a_4 + a_5 \cos t$; integrators 3, 4 and 5 give $z_{\text{comp}} = \text{arc cos} \times (a_4 + a_5 \cos t)$; integrators 6 and 7 give $\cot z_{\text{comp}}$; integrators 8, 9 and 10 give $\sin A = a_1 \sin t/\sin z$; integrators 11, 12 and 13 give $A = \text{arc sin} \times (a_1 \sin t/\sin z)$; integrators 3, 10 and 11 are servointegrators; DOS is the feedback transmitter. The system operates on a real time scale, where Δt is the increment in the hour angle. The basic components of the TsSS are digital integrators. The low comparison frequency in generating the coordinates and their derivatives means that sequential operation may be used in the TsSS. Thirteen integrators in the circuit are used; eleven of them with twenty digits each, and the integrators 6 and 7 with ten. If we allow for the additional integrators required to generate a_1 to a_5, b_1 to b_2 and the functional operations on them, as well as calculation of flexure and refraction corrections, etc., the total number of integrators amounts to about 30. Using this figure, we obtain the iteration time T_{it} for a simple digital differential analyzer with magnetic drum having a track frequency $f_T = 50$ kHz is expressed by

$$T_{\text{it}} = NnT_T = 30 \cdot 20 \cdot 2 \cdot 10^{-5} = 0.012 \text{ sec},$$

where N is the number of integrators, n is the number of digits, and $T_T = 1/f_T$.

The increment in hour angle t is obtained from

$$\Delta t = \frac{86{,}164.08 \text{ sec}}{2^{20}} = 0.0821725 \text{ sec},$$

where 86,164.08 is the sidereal day expressed in seconds.

By comparing T_{it} and Δt, we see that the speed of operation of the system is seven times greater than required. The design principle for TsSSs for astronomical instruments that we have described has great merit, since circuits constructed on the basis of control by coordinates and by their derivatives give the required dynamic performance for the drive, flexibility in introducing corrections, and, most important, simplicity of basic circuits.

5 CONCLUSIONS

(1) The range of regulation in the control system for altazimuth instruments is theoretically infinite. In practice, it is possible to approach points where the velocity passes through zero. The method we have devised for reducing iteration and comparison frequencies by means of a system of control by velocity makes it possible to approach zero velocity. Simultaneously provides a sufficiently wide range of regulation, and can be used in control systems for astronomical instruments.

(2) The studies that have been made allow the selection of the optimal design for a digital servomechanism for controlling altazimuth instruments. It turns out that the setting and guiding procedure using a system of control by position can be devised by comparing the actual and calculated values of the coordinates in a TsSS, instead of functions of them.

(3) The circuits of the TsSS considered for control of altazimuth instruments with the use of a sequential digital differential analyzer is simple, and yet allows for control by position and velocity.

(4) Practical applications of this TsSS design concept can be found in astronomical altazimuth instruments, radio telescopes and similar devices requiring automatic regulation, a high level of accuracy and smoothness of operation.

REFERENCES

1. A. A. Feldbaum, *Computers in Automatic Systems* (Vychislitelnye ustroistva v avtomaticheskikh sistemakh), GIFML, Moscow (1959).
2. A. V. Shileiko, *Digital Differential Analyzers* (Tsifrovye differentialnye analizatory), IL, Moscow (1959).
3. M. Klein, G. Morgan and M. Aronson, *Digital Techniques for Computation and Control* (Tsifrovaya tekhnika dlya vychislenii i upravleniya), IL, Moscow (1960).

CHAPTER **I-10**

Automatic tube flexure compensation in stellar telescopes

G. I. GOREVA, YU. A. SABININ, P. V. NIKOLAYEV, A. V. SHUMAKHER

Institute of Electromechanics on Automation and Machine Construction Leningrad

---

The construction of systems to compensate telescope tube flexure (suggested by B. K. Ioannisiani) has been considered in recent years by the designers of large automated astronomical instruments. The flexure of the telescope tube causes the pointing of the telescope to be always somewhat lower than the setting selected at the control console. The error involved is determined solely by the zenith distance of the celestial object. The magnitude of the error in setting caused by flexure is measured by the increment in zenith distance z; this error can be obtained approximately from the expression

$$\Delta z = B \sin z, \qquad (10\text{-}1)$$

where B is a constant coefficient characterizing the rigidity of the tube and it is usually determined experimentally.

All telescopes until recently, apart from meridian circles and transit instruments, have been equatorially mounted; in these types of instruments the setting error caused by flexure is compensated by appropriate corrections in declination and hour angle. This applies not only during setting, but also during long photographic exposures since the zenith distance z changes continuously with the hour angle, thus requiring fine correction of the attitude of the telescope tube with respect to the position of the celestial object. The errors caused by flexure and by atmospheric refraction greatly complicate the very simple automatic setting procedure of equatorial telescopes. In order to compute the flexure and refraction corrections, the automatic setting system must include a coordinates converter to obtain z from t and δ, and a computer to compute Δz and work out the corresponding corrections Δt and $\Delta \delta$. Not only do large deformations of the tube have a detrimental effect in the accuracy of setting and guiding, but they also may produce poor stellar images as a result of the lack of collimation of the optical system of the telescope.

These factors show that the metallic structure of the tube must be designed to provide maximum rigidity. A special telescope tube has been designed to equalize the sag at both sides of the declination axis.* For large telescopes the weight of a rigid tube can be measured in tens of tons which increases its flexure and also overloads the support and drive mechanisms. Construction of an automatic flexure compensator would permit the use of lighter telescope tubes, subject to structural considerations only, and would simplify the control problems of the telescope.

The kind of corrections needed in t and δ to compensate for tube flexure in an equatorial telescope can be analyzed by using Figure 10-1. Let us assume that the telescope has been set by the dials to aim the point E with coordinates t and δ, but due to tube flexure the telescope points to E_1 instead.

Considering the spherical triangle E_1CE, we can easily obtain the flexure components Δt and $\Delta \delta$ from

$$\Delta \delta = -\Delta z \cos q,$$
$$\Delta t = \Delta z \sin q, \tag{10-2}$$

where q is the positional angle corresponding to the point E on the celestial sphere.

Using Eq. (10-1) and relationships from spherical trigonometry, we have

$$\sin q = \frac{\cos \varphi}{\cos \delta} \sin A,$$
$$\sin A = \frac{\cos \delta \sin t}{\sin z}, \tag{10-3}$$

where φ is the latitude of the observer and A the azimuth of the point E. object.

* See footnote p. 14. (Ed. English version.)

Automatic tube flexure compansation in stellar telescopes

We can write Eqs. (10-2) in the form

$$\Delta t = B \cos \varphi \sin t,$$
$$\Delta \delta = -B(\sin \varphi \cos \delta - \cos \varphi \sin \delta \cos t). \tag{10-4}$$

Thus, Eqs. (10-4) shows that when the error Δz is proportional to $\sin z$ as in Eq. (10-1), Δt does not depend on δ, but only on t and φ.

Since $\Delta \delta$ is measured along the arc of the great circle (see Figure 10-1), the flexure correction $\Delta \delta_y$ is given by

$$\Delta \delta_y = B(\sin \varphi \cos \delta - \cos \varphi \sin \delta \cos t). \tag{10-5}$$

The flexure correction Δt_y in hour angle is determined by the arc $a_1 a$ on the celestial equator and it is given by

$$\Delta t_y = -B \cos \varphi \sin t \sec \delta. \tag{10-6}$$

The speed corrections due to flexure for both telescope axes and for any point of the celestial sphere and for a guiding rate (in hour angle) of one revolution per sidereal day can be obtained by differentiating Eqs. (10-5) and (10-6):

$$\frac{d\Delta \delta_y}{dt} = B \cos \varphi \sin \delta \sin t,$$
$$\frac{d\Delta t_y}{dt} = -B \cos \varphi \cos t \sec \delta. \tag{10-7}$$

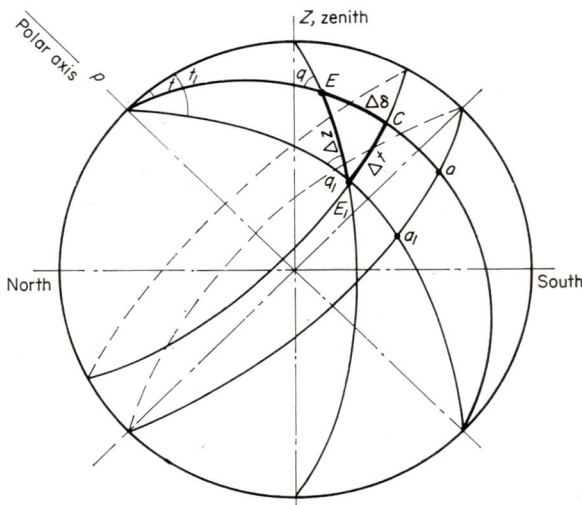

Figure 10-1 Components Δt and $\Delta \delta$ of the flexure error Δz.

Equations (10-7) show that $d\Delta\delta_y/dt$ attains its maximum value of $B\cos\varphi$ when $t = 6$ hours and $\delta = 90°$, and that $d\Delta t_y/dt$ its maximum of $B\cos\varphi\sec\delta$ at the meridian.

The minimum speed corrections for both axes can be equal to zero. This means that the speed must be regulated over a nearly infinite range. This causes problems in systems of automatic flexure compensation which continuously introduces speed corrections (depending on the coordinates of the object) into the kinematic circuits of the respective telescope axes, and which requires very high accuracy (the total error over the optical field during several hours exposure must not exceed tenths of a second).

A more promising method would be to determine the flexure by the position of a light spot; to compensate flexure the displacement of the light spot has to be compensated by adjusting one of the optical components, such as the secondary mirror of a reflector. For this purpose the telescope system must include a point light source, from which a light beam is reflected successively from a series of optical components in the telescope and brought to the crosswire of an eyepiece. The path of the beam should be chosen in such a way that when flexure occurs the light spot is displaced from the crosswire. Its position can be restored by adjustment of the secondary mirror or some other optical component when collimation of the optical system has been restored. Then the stellar images return to its previous position on the photographic plate or any other recording device. A flexure compensator such as this can be made automatic. A photoelectric servomechanism which automatically detects displacement of the light spot from the line of sight of the eyepiece could achieve the compensation by moving the secondary mirror in t and δ, in order to return the spot to its initial position. The servomechanism must not lose the light spot from its field of view during setting; after setting has been completed the light spot must fall on the crosswire with in the required accuracy. Since, in practice, the setting process can start from any initial declination δ_H and hour angle t_H, we may write

$$t = t_H + vT,$$
$$\delta = \delta_H + vT,$$
(10-8)

for the current values of t and δ, where v is the setting rate for both axes and T is the time calculated from the beginning of setting.

Then the displacement of the light spot in t and δ in the field of the photoelectric servomechanism will, from Eqs. (10-4), be

$$\Delta t_M = -n_1 B \cos \varphi \sin (t_H + vT),$$

$$\Delta \delta_M = n_2 B[\sin \varphi \cos (\delta_H + vT) - \cos \varphi \sin (\delta_H + vT) \cos (t_H + vT)],$$

(10-9)

where n_1 and n_2 are coefficients which define the relationship between the flexure of the tube in t and δ and the displacements of the light spot in the field of view of the servomechanism. The velocities $d\Delta t_M/dT$ and $d\Delta \delta_M/dT$ required for the analysis of the photoelectric servomechanism is obtained by differentiating Eqs. (10-9); the expressions are the following:

$$\left. \begin{array}{l} \dfrac{d\Delta t_M}{dT} = -n_1 vB \cos \varphi \cos (t_H + vT), \\[2mm] \dfrac{d\Delta \delta_M}{dT} = n_2 vB\{-\sin \varphi \sin (\delta_H + vT) - \cos \varphi[\cos (\delta_H + vT), \\[2mm] \cos (t_H + vT) - \sin (\delta_H + vT) \sin (t_H + vT)]\}. \end{array} \right\} \quad (10\text{-}10)$$

Figures 10-2, 10-3 and 10-4 give plots of $\Delta t_M/B$, $(d\Delta t_M/dT)/Bv$, $\Delta \delta_M/B$ and $(d\Delta \delta_M/dT)/Bv$ versus t and for $\varphi = 44°39'$, $n_1 = n_2 = 1$ and δ as parameter. We may conclude from these plots that the maximum speeds at which the servomechanism could operate for $n_1 = n_2 = 1$, $v = 60°$ min^{-1} and $B = 300''$ are

$$\frac{d\Delta t_M}{dT} = 3.7 \quad \text{and} \quad \frac{d\Delta \delta_M}{dT} = 7''.3 \text{ sec}^{-1}.$$

A photoelectric servomechanism with a half-disk mechanical chopper which satisfies the above requirements has been developed at the Institute of Electromechanics and tested on a telescope model with a tube purposely lacking rigidity. The model used was designed, manufactured and tested in cooperation with the Main Astronomical Observatory, Pulkovo (with the assistance of N. N. Mikhelson and A. V. Shumakher), and is based on the Cassegrain altazimuth telescope AP-250 (250 mm clear aperture, and 6,250 mm equivalent focal length). Added to the optical train of the telescope (Figure 10-5) are a number of components to convert it into an autocollimating system. The light from the bulb *1* is concentrated by means of condenser *2* on the focal plane of the Cassegrain system, where a small pinhole *3* is placed.

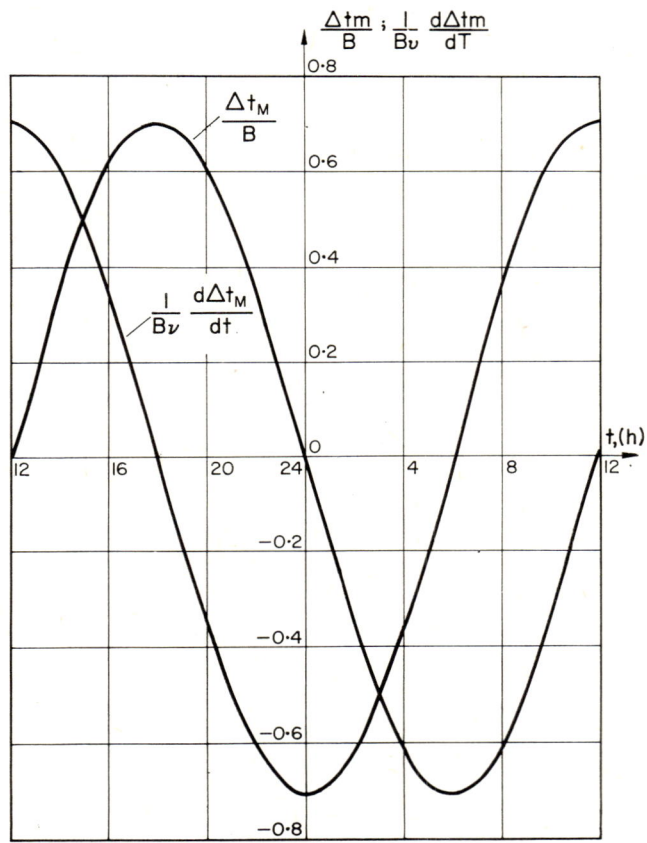

Figure 10-2 $\Delta t_M/B$ and $(d\Delta t_M/dT)/B\nu$ versus hour angle t.

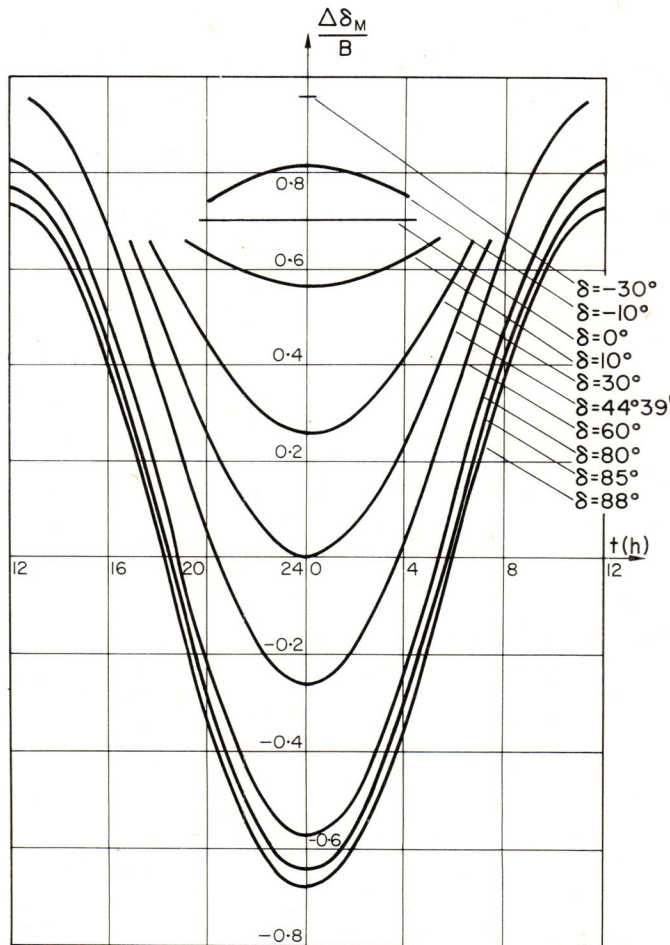

Figure 10-3 $\Delta\delta_M/B$ versus hour angle t and for different values of δ.

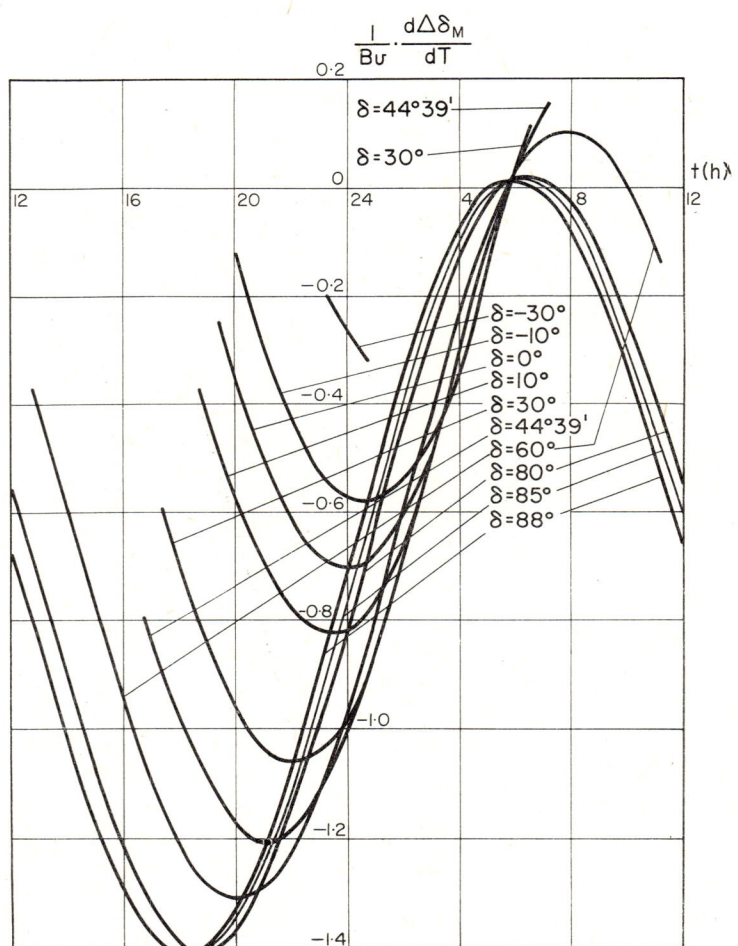

Figure 10-4 $(d\Delta_M/dT)/B\nu$ versus hour angle t and for different values of δ.

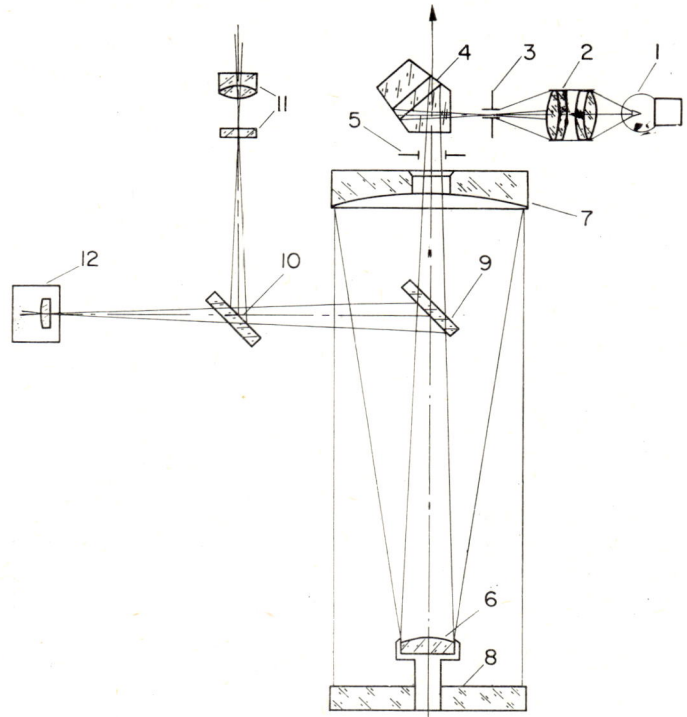

Figure 10-5 Optical system used in the AP-250 telescope tube flexure compansation experiments. The system consists of: 1, light bulb; 2, condenser; 3, pinhole; 4, pentaprism; 5, stop; 6, Cassegrain secondary mirror; 7, parabolic mirror; 8, flat mirror; 9 and 10, partially reflecting flat mirror; 11, eyepiece micrometer; 12, photoelectric head.

Behind the pinhole is the pentaprism *4*, the aperture stop *5*, the partially reflecting flat mirror *9*, and the convex hyperboloid mirror *6*. The light reflected on this mirror diverges to the concave paraboloid mirror *7*. This mirror forms a parallel beam which is reflected from the flat mirror *8*, and returns by the same path. However, part of the convergent beam reflected from mirror *6* is partially reflected by the mirror *9* through the hollow horizontal trunnion of the telescope and is focused on the plane of the light chopper of the photoelectric head *12*. Another partially reflecting flat mirror *10* is placed between the photoelectric head and mirror *9* to reflect the light to the eyepiece micrometer *11*, used to monitor visually the performance of the photoelectric servomechanism. Mirror σ may be adjusted relative to mirror *8*, and is combined with it in a single unit. When flexure of the tube takes place, the system of mirror *6* and *8* is displaced parallel to mirror *7*, which causes the light spot to depart from the axis of the chopper *12*. This produces an error signal in the servo-mechanism, which drives the servomotors returning the mirrors *6* and *8* to its initial position.

Considerable changes had to be made to the design of the telescope tube to make the flexure detectable. Figure 10-6 shows the tube made of two small thin-walled tubes *1*, the rib *2* and the rings *3* and *4*. The rib is added for rigidity and it is welded to the rings. The lower ring *3* is bolted to the middle section of the tube of the telescope. Three angle brackets *5* are attached to the upper ring *4* at 120° and the mirror cell *7* is suspended from these brackets by means of helicoidal springs *6*. The mounting with the mirrors is pulled towards the lower ring by three hinged wires *8*, whose length and tension can be regulated during adjustment of the model. The wires ensure that during flexure the upper flat mirror is displaced parallel to the lower convex one. Two identical "flexure generating mechanisms" *9* are set up on the upper ring: one operates on the plane of the figure and the other perpendicular to it. Each of these mechanisms consists of a servomotor, tachometer generator and a mechanical reducer.

To minimize the load on the motor for the main flexure generating mechanism (which generated the equivalent deformation along the zenith distance) the upper mirror is supported by a lever mechanism. Rotation of the telescope tube around the elevation axis is accomplished by the dc motor SL-260 type. Since the tube rotates only around the elevation axis, the second generating mechanism is used only to correct for deviations of the upper flat mirror in a direction parallel to the horizontal. In the optical system used, the light spot in the plane of the mechanical chopper must, in principle, move at a rate equal to the rate of motion of the system of mirrors *6* and *8* (Figure 10-5), which in this case is 0.029 mm sec^{-1} for 500 rev min^{-1} nominal speed of the servomotor. However, there was a certain misalignment of the mirrors *7* and *8* when experiments were conducted, and consequently the light spot had a nominal rate of motion which was ten times larger (0.3 mm sec^{-1}) than that of mirror *8*. Thus the operating conditions of the photoelectric servomechanism simulated on the model corresponded to a maximum rate of 0.3 mm sec^{-1} of correction for flexure in any axis. For a telescope with equivalent focal length of 6,250 mm, the above rate is equivalent to 10″ sec^{-1}. In this experiment the diameter of the light spot was equal to 0.2 mm.

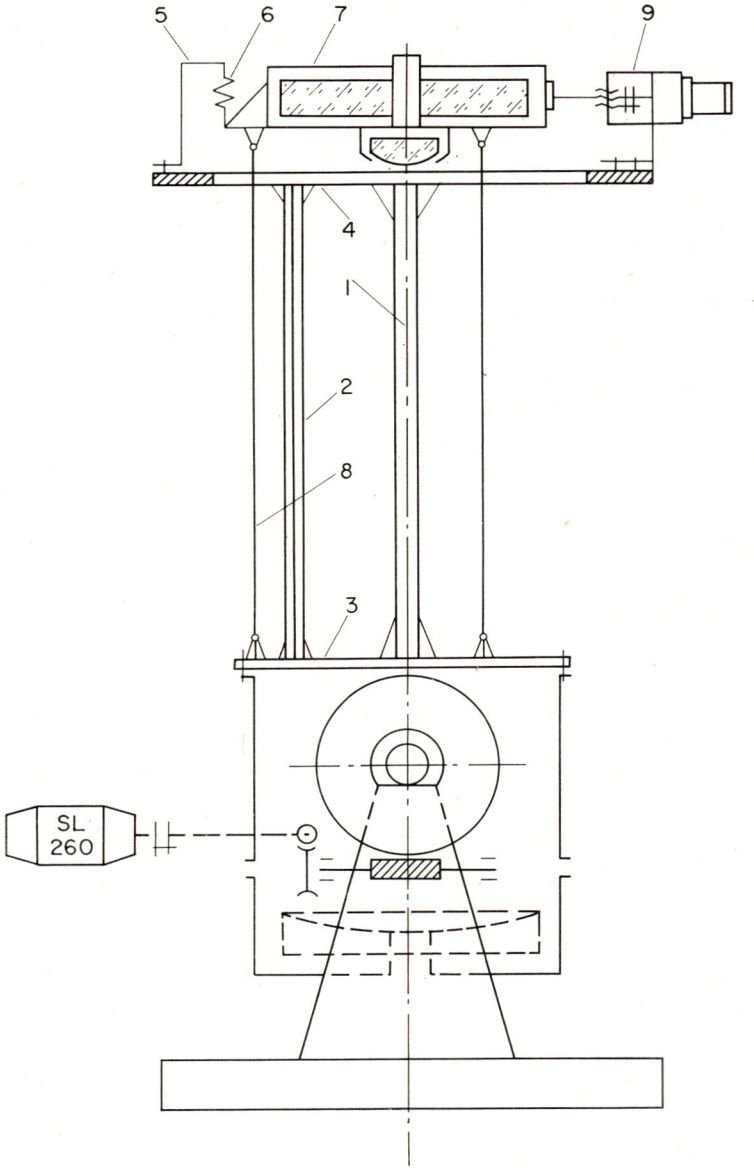

Figure 10-6 Diagram of the mechanical configuration of the telescope model. It consists of: 1, thin-walled tube; 2, rib; 3 and 4, rings; 5, angle brackets; 6, helicoidal spring; 7, mirror cell; 8, wires; 9, flexure generating mechanism.

The graph in Figure 10-7 gives the displacement Δz of the light spot in the plane of the chopper versus zenith distance z. For comparison we have also drawn the curve $\Delta z = B \sin z$ for $B = 17$ mm. The close agreement of the curves again demonstrates the validity of our assumption that the flexure in z varies very nearly as the $\sin z$ ($\Delta z = B \sin z$). The principle of operation of the photoelectric servomechanism can be understood by using the block diagram given in Figure 10-8, which corresponds to the case when the flat mirror block moves only along one axis. The light source is focused in the plane of the mechanical chopper *1*, then the light passess through the Fabry lens *2*, and an image is formed on the cathode of the photomultiplier tube *3*. When flexure occurs, the light spot is displaced from the axis of rotation of the chopper, causing the appearance of an ac error signal at the output of the photomultiplier tube. The error signal is then fed to the preamplifier *4* and main amplifier *5*, both of low frequency response; the output then is fed to the input of the phase controlled rectifier *6*, to which a reference voltage generated by the chopper is also fed. As a result of a comparison of the phases of the reference voltage and the error signal, a dc signal appears at the output of the phase controlled rectifier; the amplitude of the output is proportional to the amplitude of the error signal, an the polarity to the sign of the error.

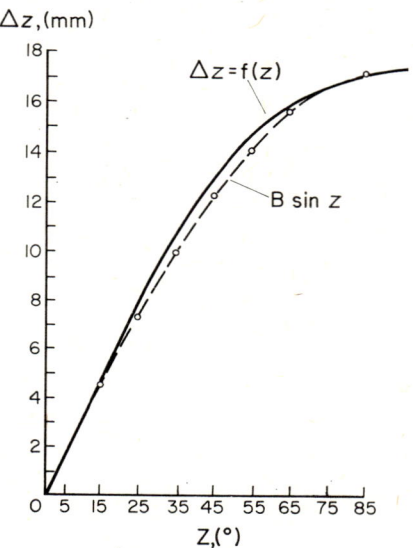

Figure 10-7 Comparison of the experimental curve $z = f(z)$ and the theoretical curve $z = B \sin z$ for $B = 17$ mm.

The dc voltage from the output of the rectifier is fed to the input of the electronic power amplifier 7. An ac power supply of 115 V and 400 Hz drives the anodes of the output vacuum tubes of the power amplifier. The output of the amplifier is the control signal for the winding of the servomotor 8 which through a reducer 9 rotates the flat mirror thus returning light spot to its original position. After the adjustment, the servomotor stops since the voltage on its control winding becomes equal to zero. If the telescope tube is in motion, the servomechanism continuously adjusts the position of the flat mirror, thus providing compensation for flexure of the tube. The ac tachometer generator 10 is attached to the shaft of the servomotor and is used to generate a negative feedback to stabilize the system.

The servomotors and tachometer generator in the model are both EM2-12 type ($P = 2$ W, $U_Y = 50$ V, $n = 5{,}000$ rev min^{-1}). As we have said, in the model, the light spot was tracked by moving the flat mirror in two orthogonal directions, and so the servomechanism was used as a two-coordinate system having two phase-controlled rectifiers, two power amplifiers and two drive mechanisms. To obtain the two components from the single ac error signal after proper amplification, two reference voltages,

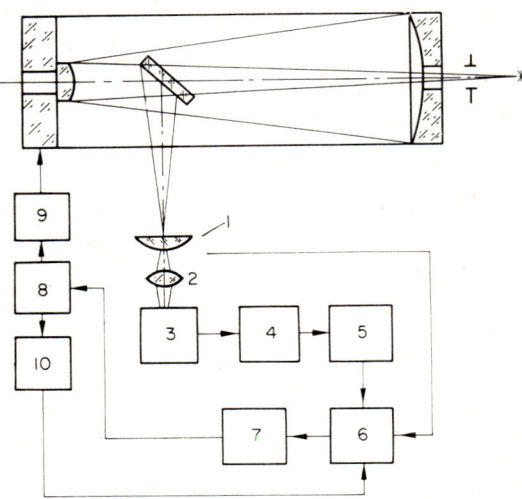

Figure 10-8 Block diagram of the experimental system for automatic tube flexure compensation (one component) in stellar telescopes: 1, mechanical chopper; 2, Fabry lens; 3, photomultiplier tube; 4, preamplifier; 5, amplifier; 6, phase-controlled rectifier; 7, power amplifier; 8, servomotor; 9, reducer; 10, ac tachometer generator.

shifted in phase by 90° with respect to one another were fed to the phase controlled rectifiers.

The light chopper in the servomechanism was made of a half-disk mounted in the hollow shaft of a special electric motor. The motor provides synchronous rotation of the half-disk with a speed of 2,250 rev min^{-1} and simultaneously generates two sinusoidal reference voltages shifted in phase by 90°. The modulation frequency of the light and the reference frequency is 37.5 Hz. An analysis of the operation of the chopper for automatic guiding units can be found in the literature [1].

The main circuit of the automatic flexure compensation servomechanism for the Z-axis (without power supply) is shown in Figure 10-9. The preamplifier uses the vacuum tube L_1 connected as cathode follower. The preamplifier with the photomultiplier tube FEU-17 are attached to the modulator flange on the telescope. The cathode follower, due to its low output impedance reduces the shunting effect of the connecting cable between the preamplifier and the main amplifier, and thus reducing cable noise problems. The vacuum tubes L_2, L_3 and L_4 are part of the main amplifier which is a narrow band RC coupled amplifier with a double T-network connected between the anode and the grid of L_2. The amplifier is tuned to a frequency of 37.5 Hz and has a transformer at its output with two secondary windings (one for each of the channels t and δ). The phase controlled rectifier uses the vacuum tube L_5 which is supplied by a square voltage generated by the triode L_9.

The reference voltage, first amplified by tube L_8, is fed to the input of L_9.

The power amplifier is a push-pull using the L_6 and L_7 vacuum tubes supplied with an ac anode voltage of 400 Hz from transformer Tr_2. The control winding of the servomotor is connected to the secondary winding of the output transformer of the power amplifier. A phase shift of 90° between the voltages of the grid and the control windings of the motor is attained by connecting it in series with the grid circuit a 0.35 μF capacitor. The phase controlled rectifier has rigid negative feedback for velocity. This signal is obtained by rectification of the ac output of an ac tachometer generator by a phase sensitive rectifier with a filter at the output to reduce ripple. To compensate for the phase shift between feedback and error signal, a phase shifter using the rotary transformer VTM-4, is introduced between the grid and the secondary windings of the tachometer generator. The laboratory tests that were made showed that when the rate of inclination of the telescope tubes is 30° min^{-1}, the light spot in the field of the

Automatic tube flexure compansation in stellar telescopes

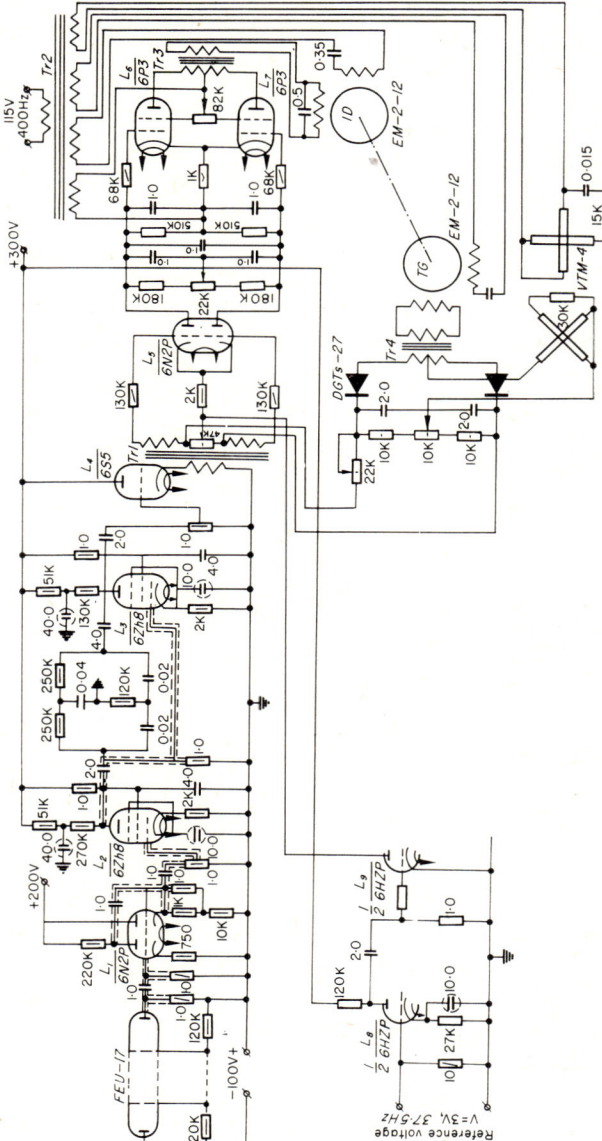

Figure 10-9 Main circuit of the automatic flexure compensation servomechanism for the z-axis.

chopper had a maximum velocity of 0.19 mm sec^{-1} which is equivalent to 6$''$.3 sec^{-1}. With this rate of inclination of the tube, the photoelectric servomechanism could reliably keep the light spot in the field of the chopper. The static error after the inclination reaches a final value (when the tube stopped) did not exceed the limit accuracy of the eyepiece micrometer, which was 0.01 mm. Figure 10-10 shows the oscillograms of the error, control and feedback signals generated by the photoelectric servomechanism. They show the voltage at the windings of the servomotor, the tachometer generator and the error signal at the output of the main amplifier. These oscillograms were obtained for compensation in z. The first oscillogram shows that the error signal during flexure compensation decreases monotonically from 1 V to 0.12 V which corresponds to an error of $0.05d$ at the chopper (d is the diameter of the light spot).

It is also clear from the oscillograms that during the flexure in z a small disturbance was caused in a perpendicular direction, due to which the motor of the A-axis contributes to the generation of the error signal. This was a result of lack of accurate adjustment in the model.

A similar photoelectric servomechanism has been used for automatic compensation of differential flexure in the ZTSAh 2.6-m telescope at the Crimean Astrophysical Observatory. In this case, when differential flexure occurs between the guider and the main telescope tube, the photoelectric servomechanism rotates a compensating plane-parallel glass plate. As a result, the light spot of the compensator returns to the axis of rotation of the chopper, and the image of the star by which guiding is carried out is displaced from the crosswire of the guider by the amount of differential flexure. The observer or the photoguider can return the image to the crosswire of the guider by resetting the telescope-guider to the proper t and δ, thus maintaining the stellar images fixed relative to the detector at the prime focus, i.e. can fully compensate for differential flexure. We would like to comment as a conclusion that the amplifier of the servomechanism could be made using semiconductors and magnetic amplifiers, and that a photoconductor or a photomultiplier tube with fewer stages of multiplication could be used if the spot is made brighter.

REFERENCES

1. Yu. A. Sabinin and P. V. Nikolayev, "A System with a semi-disk modulator of light beam for automatic guiding of telescopes", *Izv. KrAO AN SSSR*, **26**, 203 (1960).

Automatic tube flexure compensation in stellar telescopes

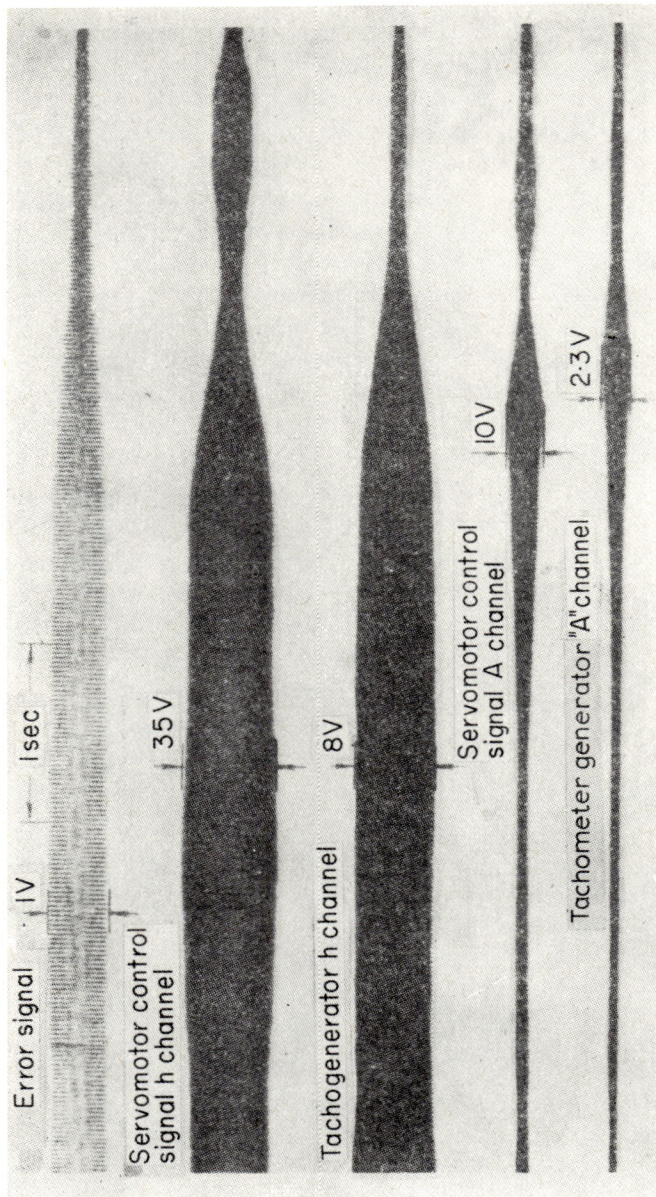

Figure 10-10 Oscillograms obtained at different stages of the automatic flexure compensation system for the "uncompensated" and "compensated" condition.

CHAPTER **I-11**

A photoelectric method to measure the "seeing" of stellar images

I. P. ROZHNOVA and YU. A. SABININ

*Institute of Electromechanics on Automation and Machine Construction
Leningrad*

In recent years the construction of giant telescopes equipped with the latest automatic control systems has received a great deal of attention in the USSR. Data about the statistical characteristics of the "seeing"* of stellar images are important to the designer of accurate guiding systems.

Research on seeing has been carried out over the years, mainly by astronomers and geophysicists; there is available today a considerable amount of experimental data that could be used, as a first approximation, for the design of automatic guiding systems. However, the approach used in these measurements depends upon the final objective: the astronomer would like to minimize the effect of seeing since, it introduces serious limitations to his observations; while for the geophysicist, it is used as a means to estimate certain parameters of the Earth's atmosphere. The available data, particularly in regard to seeing are not entirely suitable for the design of automatic guiding systems.

From the guiding systems point of view the most important data on seeing are as follows:

(1) The frequency spectrum of scintillation.

(2) The frequency spectrum of seeing.

* Throughout the specialized literature the term "seeing" does not have a uniform meaning. It has been used to describe the overall effect of star images dancing, focus drift, changes in shape, pulsation, and brightness and color scintillation. This also applies to images of extended objects. In this contribution the term seeing is used to describe the combined effect, and the degree, of the continuous changing of position, shape and size of star images and the term scintillation describes the intensity fluctuations. (Ed. English version.)

(3) The ratio of maximum and root-mean-square amplitudes of scintillation, limited to narrow bands, to the mean value of the intensity of its modulated light flux in the same frequency band.

(4) The ratio of maximum and root-mean-square amplitudes of the deviation of a star from the optical axis (due to seeing) to the mean diameter of the stellar image.

The frequency spectrum of scintillation has been studied in great detail. Its power spectrum has a maximum and constant value in the low frequency range (0–30 Hz) and falls fairly rapidly towards 100–200 Hz. This has been confirmed by numerous observations [1]. However, it is very difficult to compare the results of various observations on the ratio of the amplitudes of scintillation (maximum, mean and root-mean-square) to the mean value of the light flux. The value of this ratio is to a great extent dependent on the observing method and the characteristics of the instrumentation used [2–5].

According to L. N. Zhukova [2], the maximum amplitude of scintillation reduced to the zenith comprises 70% of the dc value of the intensity (for a clear aperture of 250 mm). According to studies made at the Institute of Electromechanics [6], the ratio of the mean amplitude of a single harmonic component ($f = 20$ Hz, $Q = f/2\Delta f = 50$) to the mean value of the total luminous flux from the star is about 5% (for a clear aperture of 200 mm). Observational results at different observatories (Dunsik, Perkins, Jungfraujoch) [2] show that the root-mean-square amplitude of scintillation at the zenith is between 5–15% of the mean value of the total brightness (for a clear aperture of 250 mm).

The different measurements of seeing [1, 7] essentially are reduced from the image "traces" on photographic plates or film. Reducing this data is a very time-consuming task, and results are always affected by subjective considerations. For this reason, attempts are being made to improve the reduction methods. Data on the frequency spectrum of seeing obtained by Mayer [7] are very interesting. According to his results, the maximum amplitudes of seeing lie in the low frequency region 0–1 Hz and fall towards 8 Hz. After the seeing recordings were reduced to the proper form, the frequency spectrum was obtained by means of an analyzer. Due to insufficient sensitivity of the film, the higher frequencies, in general, were not recorded.

The photoelectric method described below was developed at the Institute of Electromechanics. It is an attempt to develop a more reliable method for studying seeing to be used for the design of automatic guiding systems. The electronic instrumentation for data acquisition is, of course, much faster than the one using photographic methods.

The block diagram of the photoelectric system is shown in Figure 11-1. This diagram is for a pair of photomultiplier tubes (PMT); the circuit for the second pair is similar. The image of the star is formed at the vertex of a four-sided pyramid (45° sides), placed so that the t and δ axes of the telescope are parallel to two sides of the pyramid base. The pyramid is manufactured to a high degree of accuracy; the vertex has an area of not more than 5 × 5 microns. Four PMTs are used in the system, each of which is located facing the respective side of the pyramid. The PMTs on opposite sides are connected to a differential circuit. Signals from the load resistors of the PMTs are fed to the preamplifiers with an amplification gain of 10^3. Its output is then fed to the input of a balanced amplifier. If the star lies exactly on the optical axis of the telescope, identical luminous fluxes are reflected from all four sides of the pyramid and the circuit is balanced. With the slightest deviation of the star from the optical axis unbalance occurs; the signal is then proportional to the displacement from the optical axis. The seeing modulates the carrier frequency ($f = 10$ kHz), which is generated by an ac signal applied to the first stage of the PMTs.

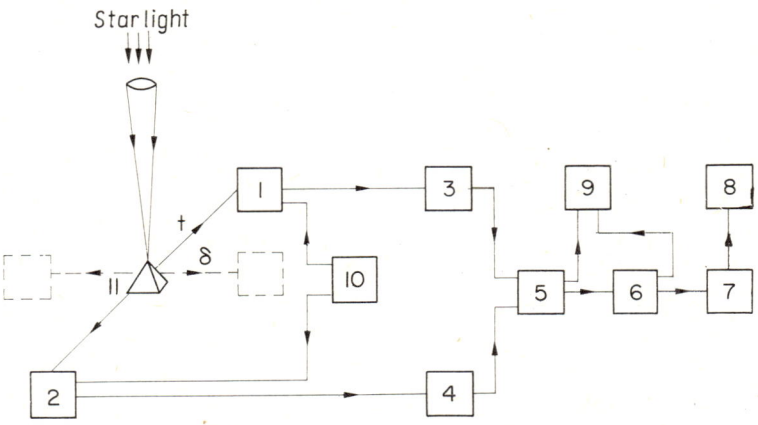

Figure 11-1 Block diagram of the photoelectric system to measure stellar "seeing": 1, 2, photomultipliers; 3, 4, preamplifiers; 5, balance amplifier; 6, linear detector; 7, frequency spectrum analyzer; 8, recorder EPP-09; 9, multichannel oscillograph; 10, frequency generator; 11, 45° four-sided pyramid.

A linear detector, rectifies the output signal which is proportional to the amplitude of the displacement of the image. The modulated signal at carrier frequency allows the use of ac amplifiers with high amplification. If the preamplifiers are placed next to the PMTs and balanced amplifiers are used, the operation of the system is stable (practically with undetectable noise).

The characteristics of the signal are measured by means of a frequency spectrum analyzer; the output is recorded by the multichannel recorder EPP-09.

In order to obtain the relationships between the amplitudes of the stellar image displacements and the diameter of the image, several recordings are made simultaneously on a multichannel oscillograph. One channel records the output signal when the circuit is in total unbalance, and the stellar image is completely on one side of the pyramid; in this case the signal is proportional to the diameter of the stellar image. The two other channels record the output signals from the balance amplifiers for the t and δ axes respectively; the outputs in this case are proportional to the displacement of the stellar image from the optical axis along t and δ. By processing the oscillograms, we can find the ratios of maximum and root-mean-square amplitudes of image displacements to the diameter. We can also use the oscillograms to obtain the phase of stellar displacement relative to an arbitrary origin (on the t or δ axis). For the design of the servomechanisms, it is important to know the number of sign reversals of the signal over a given period of time, as well as the amplitude of the displacements (i.e. how often and with what amplitude the star passes from one quadrant to the other).

Figure 11-2 is the diagram of the circuitry represented in the blocks *1, 2, 3, 4* and *5* shown in Figure 11-1. The circuitry of the other blocks are standard and do not require explanation. The four PMTs are supplied in parallel from a single high voltage dc power supply. The voltages for the different stages are obtained from a voltage divider. Experience shows that a common supply for the four PMTs does not affect the performance of the tubes. To modulate the photocurrent, an ac voltage is applied to the first stage.

Obviously it is very difficult to select four PMTs of identical characteristics, but by adjusting the gain of the amplifiers, we can equalize the overall amplification of the four channels. The gain for the channels can be chosen experimentally by sequentially placing the stellar image on all four sides of the pyramid.

The study of seeing by any method of the type discussed here involves the difficult problem of separating scintillation from seeing, since they occur simultaneously. When the star is exactly on the optical axis, identical light fluxes are reflected from each side of the pyramid; then the signals in the anode circuit of the balanced amplifier caused by scintillation is equal to zero. When the star deviates from the optical axis, a signal proportional to the magnitude of the displacement develops. This signal is also modulated by the scintillation, according to difference in the light fluxes falling on opposite sides of the pyramid. The amplitude of modulation will, in the mean, be 5–15% (difference in the light fluxes); most studies of stellar scintillation agree that on the average it is between 5 and 15% of the mean stellar brightness. We can therefore ignore in this method the error introduced by scintillation on the measurements of seeing.

This error will depend only on the magnitude of the ratio of scintillation amplitudes to the mean brightness of the stellar image and not on instrumental errors.

However, if we use this method to study seeing we can separate scintillation from seeing and thus reduce the error, as follows. We use the system to study the frequency spectrum of scintillation and to obtain the ratio of the amplitudes (maximum and root-mean-square) to the mean value of the stellar brightness.

To do this, we move the stellar image to one side of the pyramid. We have mentioned that the scintillation amplitudes are proportional to the difference in the light fluxes reflected from two opposite sides of the pyramid. Knowing this difference and the scintillation frequency spectra and the seeing from the oscillograms, we can account for the error introduced by scintillation into the characteristics of seeing.

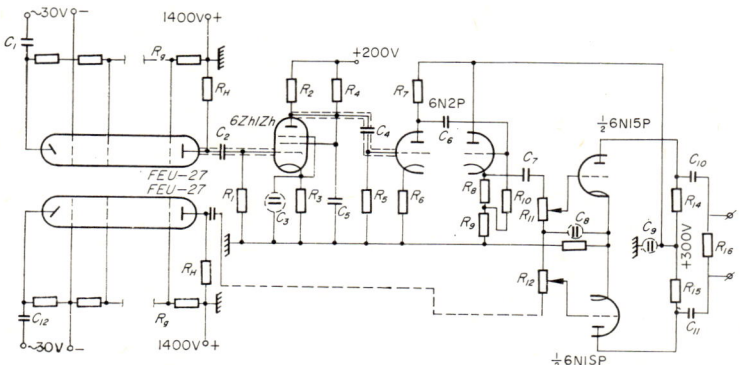

Figure 11-2 Basic circuitry of the two photoelectric channels used in the system to measure stellar "seeing".

The scintillation and seeing characteristics are recorded in succession over very short intervals of time (between one and two minutes).

Figure 11-3 shows the scintillation frequency spectra of the star α Lyrae obtained at the Institute of Electromechanics at different times and using various frequency spectrum analyzers [8]. The intervals of time analyzed range from 7 to 10 minutes (curve 1) to 20 to 30 seconds (curve 3). Statistically stellar scintillation can be described as a random stationary process.

The curves agree satisfactorily and justify the assumption that the scintillation phenomenon is relatively stationary. It is thus perfectly acceptable to examine scintillation and seeing sequentially over short intervals of time. It is also possible in principle to record the scintillation and seeing simultaneously, merely by placing a plane parallel plate (at an angle of 45°) in front of the pyramid and by splitting off part of the light beam (15-20%) to one side. If we place a PMT in the path of this beam, we can obtain the characteristics of scintillation in the usual way. It was originally intended to use this method, but at the time the experiments were being undertaken, in the summer of 1960 at the Crimean Astrophysical Observatory, the idea of splitting the light beam had to be abandoned for the following reasons: (1) the light flux falling on the pyramid was low and (2) several stellar images appeared on the focal plane due to the insertion of the plate. The amount of equipment was cut down and the number of PMTs reduced from five to four.

On the whole, the experimental work done at the Crimean Astrophysical Observatory confirmed the principles on which the method is based. The

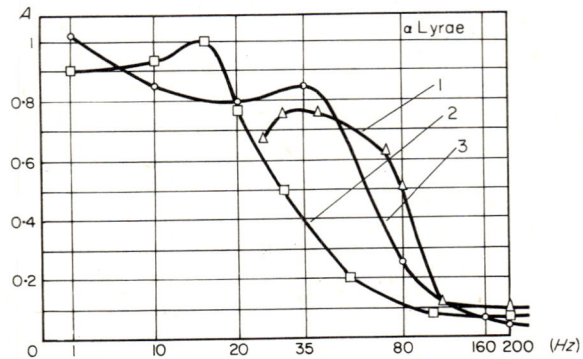

Figure 11-3 Scintillation frequency spectra of the star α Lyrae obtained under different observing conditions: curve 1, 100-mm telescope clear aperture and $z = 52°$; curve 2, 250-mm telescope clear aperture and $z = 70°$; curve 3, 250-mm telescope clear aperture and $z = 10°$.

separate units of the experimental system were tested on a laboratory model and the results show the suitability of the method to study stellar seeing for the design of automatic guiding systems.

REFERENCES

1. *Proceedings of the Conference on Stellar Scintillation* (Tr. soveshchaniya po issledovaniyam mertsaniya zvezd), Izd. AN SSSR (1959).
2. L. N. Zhukova, *A Photoelectric Method for Recording Stellar Scintillation* (Registratsiya mertsaniya zvezd fotoelektricheskim metodom), Dissertation in Physical and Mathematical Sciences, Main Astronomical Observatory, Pulkovo.
3. H. Elsässer, "Die Szintillation der Sterne", *Naturwissenschaften*, **47**, 6–10 (1960).
4. J. S. Hall, "Alternating-current techniques and sources of error in photoelectric photometry of stars", *Astronomical Photoelectric Photometry* (Ed. F. B. Wood), American Association for the Advancement of Science (1953).
5. A. H. Mikesell, "The scintillation of starlight", *Publications of the U. S. Naval Observatory*, Second Series, **17**, Part 4, Washington, D. C. (1955).
6. Study of the distribution of intensity of stellar scintillation over the frequency spectrum and establishment of the optimal frequency of modulation of the light flux in a telescope guiding system (Issledovaniye raspredeleniya intensivnosti mertaniya zvezd po spektru chastot i ustanovleniye optimalnoi chastoty modulyatsii svetovogo potoka v sisteme vedeniya teleskopa), *Report of the Institute of Electromechanics*, No. 21 (2) (1958).
7. U. Mayer, "Beobachtungen der Richtungsszintillation", *Z. Astrophys.*, **49**, 161–7 (1960).
8. R. G. Vinogradova, I. P. Rozhnova and L. N. Tikhomirova, *Harmonic Frequency Spectrum Analyzers for Non-Periodic Electrical Oscillations* (Garmonicheskiye analizatory spektra chastot neperiodicheskikh elektricheskikh kolebanii), Collection of Studies on Problems in Electromechanics, No. 4 (1960).

CHAPTER **I-12**

Programmed control of a telescope for limited observing programs

Z. H. KUBEVA and YU. A. SABININ

*Institute of Electromechanics on Automation and Machine Construction
Leningrad*

The construction of new telescopes of various types and for different purposes has brought about the need for reliable automatic control systems. It is very difficult to control large telescopes with the high accuracy required due to their small fields of view. The procedure of setting the telescope on a celestial object is particularly time consuming, since it is first necessary to calculate the actual coordinates by which the telescope must be set.

Automation of the setting and guiding processes allows more efficient use of the telescope and greatly reduces the amount of non-productive time spent by the observer. Other important advantages are gained by using systems with programmed automatic control which perform the setting and guiding tasks according to a given computer program.

The programmed control systems for a given telescope should allow the usage of universal programs which could satisfy the requirements of observing stars and other celestial bodies whose motions differ from the apparent stellar motion. At the same time the control system should introduce only small setting and guiding errors, the magnitude of which should depend on the field of view subtended by the telescope and must not exceed a few seconds of arc. Such systems are complex and in some cases can only be accomplished by using digital techniques.

A similarly pressing problem is that of designing programmed control systems for small instruments and in particular for special-purpose telescopes used for automatic observations, photometry, etc. of a limited number of bright stars.

Such telescopes can be controlled without human participation, since setting and guiding, as well as data recording, can be done automatically in accordance with the program written on punched cards, magnetic tape or any other means.

Analog systems are more suitable for special-purpose programmed telescopes working with a limited observing program and requiring comparatively low accuracy. Analog systems are much simpler, more reliable and less expensive than digital systems. For telescopes with small fields of view, digital control can be efficiently used.

The telescope programmed control system devised by the Institute of Electromechanics uses rotary transformers and other analog components and is intended for control of a telescope that measures atmospheric extinction. The observational program for atmospheric extinction uses only a limited number of bright stars in regions of the sky where there are no other bright stars at distances comparable to the field of view of the telescope. During guiding, the star brightness is measured and recorded automatically.

The required angular accuracy is within minutes of arc (between 1′ and 2′), so that the star remains in the field of view throughout the period of observation. To obtain this accuracy, positional (coordinate) control allowing for mean refraction must be used.

Thus, the programmed control system for a special-purpose equatorially mounted (parallactic) telescope must perform the following operations (Figure 12-1):

(1) Continuous computation of sidereal time s by use of the sidereal time transmitter D.Z.V.

(2) Determination of the actual value of the hour angle

$$t = s - \alpha, \qquad (12\text{-}1)$$

where α is the right ascension of the celestial object. The value of the coordinates α and δ are stored in the memory units and are input continuously into the system.

(3) Determination of the corrections Δt and $\Delta \delta$ due to mean atmospheric refraction.

(4) Computation of the apparent coordinates

$$\begin{aligned} t_T &= t + \Delta t, \\ \delta_T &= \delta + \Delta \delta, \end{aligned} \qquad (12\text{-}2)$$

and subsequent setting of the telescope according to these coordinates.

(5) Guiding of the telescope on the star; with equatorial mount, the telescope must rotate around the hour axis by the angle $t = s - \alpha + \Delta t$ and also introduce continuous corrections $\Delta \delta$ in the setting of the declination axis. The necessity to compute and feed to the telescope axes the corrections Δt and $\Delta \delta$ greatly complicate the problem. In this case only corrections for mean refraction need to be considered since instrumental errors are negligible for small telescopes.

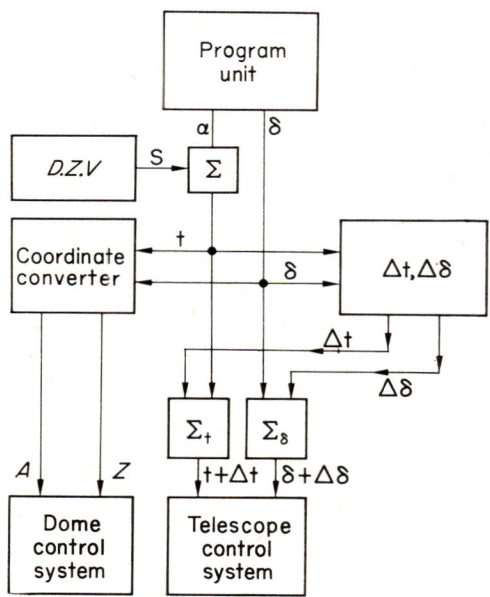

Figure 12-1 Block diagram of a programmed control system for equatorially mounted telescopes.

The formulas for corrections in t and δ to compensate for mean refraction are expressed by

$$\Delta t = -\varrho_0 \frac{\cos \varphi \sin t}{\cos \delta (\sin \varphi \sin \delta + \cos \varphi \cos \delta \cos t)},$$

$$\Delta \delta = \varrho_0 \frac{\sin \varphi \cos \delta - \cos \varphi \sin \delta \cos t}{\sin \varphi \sin \delta + \cos \varphi \cos \delta \cos t},$$

(12-3)

where φ is the local latitude and ϱ_0 is the refraction constant.

A computer has been constructed to solve the Eqs. (12-3) using rotary transformers. Its block diagram is shown in Figure 12-2. The refraction corrections are proportional to the angles of rotation of the linear rotary transformers LVT-Δt and LVT-$\Delta \delta$. The coefficients of proportionality between the angular values of the corrections and the angles of rotation of the linear rotary transformers are selected to correspond to the maximum values Δt_{max} and $\Delta \delta_{max}$:

$$M_{\Delta t} = \frac{\beta_{max}}{\Delta t_{max}} \; ; \quad M_{\Delta \delta} = \frac{\beta_{max}}{\Delta \delta_{max}} \; , \tag{12-4}$$

where $\beta_{max} = 60°$ is the maximum angle of rotation of the LVT with linear performance.

Due to the limited number of stars ($z \leq 75°$, $\delta < 75°$–$80°$) involved in the atmospheric extinction service, the refraction errors do not exceed $5'$ and in this case the coefficient of proportionality of the LVT is

$$M = \frac{[60 \times 60]'}{5'} = 720$$

which is quite sufficient.

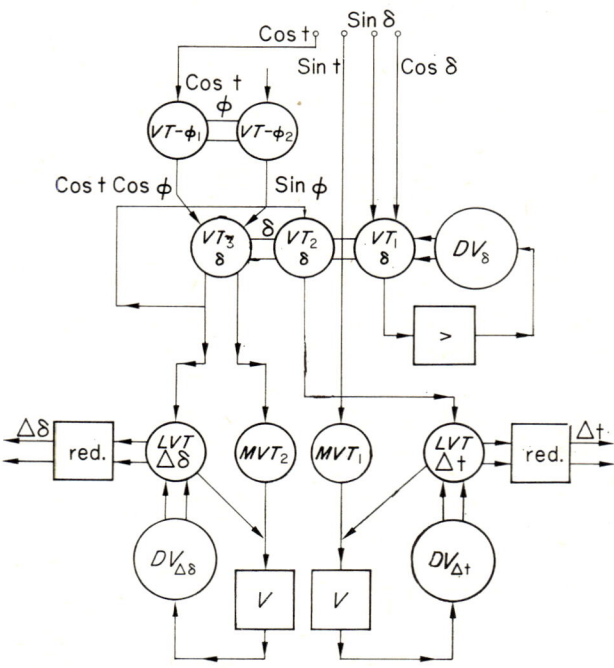

Figure 12-2 Block diagram of the analog computer for mean atmospheric refraction corrections.

The block diagram of the circuit which solves the Eqs. (12-2) is shown in Figure 12-3. The sidereal time transmitter is VT-s and its rotor is driven by the motor DV_s through a reducer at the rate of one revolution per sidereal day.

When voltages proportional to the $\sin \alpha$ and $\cos \alpha$ are fed to the windings of the stator of the VT-s the voltages at the output are proportional to $\sin t$ and $\cos t$. The angles t and δ are added to Δt and $\Delta \delta$ similarly. The angles $(t + \Delta t)$ and $(\delta + \Delta \delta)$ are generated by VT $(t + \Delta t)$ and VT $(\delta + \Delta \delta)$ transmitters which are directly coupled to the t and δ axes of the telescope.

The separate units and the system as a whole have been described in detail in [1, 2] therefore, we will only discuss accuracy, compensation of errors, and experimental results.

The accuracy with which the given coordinates are genereated allowing for refraction depends on the accuracy of the computers and the servomechanisms. Due to manufacturing limitations, the most important contribution to the total error is made by systematic errors in the VTs. These include errors in the transformation coefficients ΔK_i of the stator

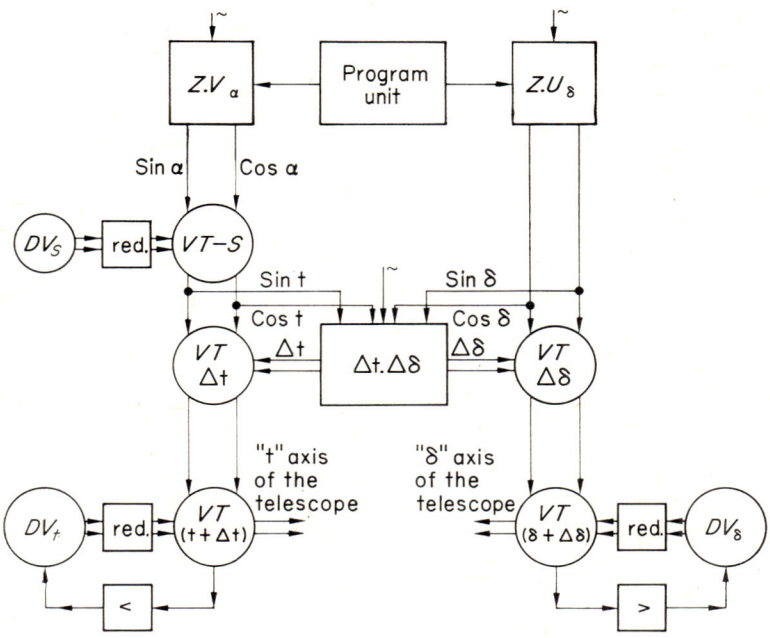

Figure 12-3 Block diagram of the system to compute the apparent coordinates t_T and δ_T.

and rotor windings and angular errors $\Delta\beta_i$ due to the fact that the windings of the stator and rotor are not perpendicular to one another.

By examining the errors in the circuit which performs the summation, we can determine the law of variation of the errors and find methods to compensate for them and thus improve the accuracy.

It can be shown by simple mathematical transformations that the systematic errors at the input of the α and δ channels can be compensated by introducing the additional angles $\Delta\alpha$ and $\Delta\delta$:

$$\alpha + \Delta_\alpha = \text{arc tan} \frac{\left(1 + \Delta K + \dfrac{\Delta K_1}{K_1}\right) \tan \alpha - \sin \Delta\beta_2}{1 + \sin \Delta\beta_1 \tan \alpha},$$

$$\delta + \Delta_\delta = \text{arc tan} \frac{\left(1 + \Delta K' + \dfrac{\Delta K'_1}{K'_1}\right) \tan \delta - \sin \Delta\beta'_2}{1 + \sin \Delta\beta'_1 \tan \delta},$$

(12-5)

where ΔK, $\Delta K'$ are the errors of the stator windings; $\Delta K_1/K_1$, $\Delta K'_1/K'_1$ the relative errors of the rotor windings of VT-α and VT-δ; $\Delta\beta_1$, $\Delta\beta_2$ are the errors due to the deviation from quadrature of the rotor windings of VT-s; and $\Delta\beta'_1$, $\Delta\beta'_2$ are due to similar deviation of the stator windings of VT-δ and VT-$\Delta\delta$.

Setting the rotors of the VTs at angles $\alpha + \Delta\alpha$ and $\delta + \Delta\delta$ respectively, is equivalent to reading the following voltages:

$$\begin{aligned} U_\alpha &= M_1 \sin \alpha; & U'_\alpha &= M_1 \cos \alpha, \\ U_\delta &= M_2 \sin \delta; & U'_\delta &= M_2 \cos \delta, \end{aligned}$$

(12-6)

at the input of the error free VT-s and VT-$\Delta\delta$.

It is clear by analysis of the circuit given in Figure 12-3 that the t and δ channels are similar, with the exception of the extra sidereal time transmitter for the hour axis. We shall therefore consider only the t channel.

The error in sidereal time and the correction Δt are small and can be ignored. Under these conditions and if Eqs. (12-6) are satisfied, the voltages in the stator windings can be written in the form

$$\begin{aligned} U_{(t+\Delta t)_1} &= C \sin (t + \Delta t) + \eta \cos t, \\ U_{(t+\Delta t)_2} &= \cos (t + \Delta t) - \eta \sin t, \end{aligned}$$

(12-7)

where

$$C = \frac{1 + \dfrac{\Delta K_2}{K_2}}{1 + \left(\dfrac{\Delta K'_4}{K_4} + \Delta K_5\right)} = \text{const}, \tag{12-8}$$

and C is a coefficient close to unity, $\Delta K'_4/K_4$ is the relative error of the transformation coefficient of the rotor windings of VT-Δt, ΔK is the error of the stator windings of VT $(t + \Delta t)$, $\eta = f(\Delta \beta_i)$ is a constant which depends on $\Delta \beta_i$, the absolute magnitude of which is taken to be the same for all the VTs in the circuit, and $\Delta \beta_i$ is the deviation from quadrature of the VT windings.

The inclination of the vector magnetic flux originated by the voltages $U_{(t+\Delta t)_1}$ and $U_{(t+\Delta t)_2}$ is expressed by

$$(t + \Delta t) - \Delta_t = \arctan \frac{C \sin(t + \Delta t) + \eta \cos t}{\cos(t + \Delta t) - \eta \sin t}. \tag{12-9}$$

Thus,

$$\Delta_t = (t + \Delta t) - \arctan \frac{C \sin(t + \Delta t) + \eta \cos t}{\cos(t + \Delta t) - \eta \sin t}. \tag{12-10}$$

This expression for Δ_t when C is nearly one, can be written as

$$\Delta_t = \left(\arctan \frac{1}{\sqrt{C}} - \arctan \sqrt{C}\right) \sin 2t + \gamma, \tag{12-11}$$

where $\gamma = \arctan \eta$ is constant. By shifting the origin of $(t + \Delta t)$ by the constant quantity γ, we obtain for the error the following expression:

$$\Delta_{t_1} = A \sin 2t, \tag{12-12}$$

where

$$A = \left(\arctan \frac{1}{\sqrt{C}} - \arctan \sqrt{C}\right). \tag{12-13}$$

Thus, the error in the solution by the computer resulting from $\Delta K \neq 0$ and $\Delta \beta \neq 0$ has a constant component and a sinusoidal component which has double frequency (see (Eq. 12-11)). In Eq. (12-12) when $C = 1$, $A = 0$. From Eq. (12-18) it is clear that by appropriate selection of the magnitudes and signs of the errors ΔK, we can make $C \approx 1$, and then the error amplitude can be greatly reduced and the accuracy of the computing circuits increased without the use of additional correction units.

The accuracy of the servomechanism which generates the angle t can be increased by use of coarse and fine rotary transformers as transmitters.

This automatic programmed system for setting and guiding a telescope in the hour angle was tested experimentally at the Crimean Astrophysical Observatory. The control accuracy achieved was of the order of $1\overset{''}{.}5$ to $2'$, and errors in determining the refraction corrections Δt did not exceed a few seconds. It can be assumed that when special correction units and coarse and fine VTs are used, at the cost of the some complication in the circuit, the errors in setting and guiding can be reduced to approximately $30''$.

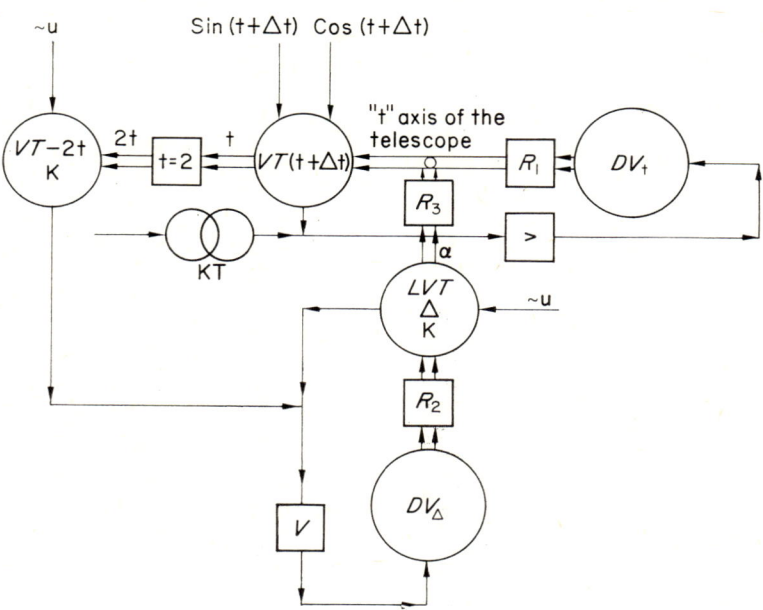

Figure 12-4 Block diagram of the unit for compensation of rotary transformer errors.

One possible variant of the unit to compensate for the error expressed by Eq. (12-12) is shown in Figure 12-4. Note that the programmed system for automatic setting on a star is suitable for telescopes other than "extinction service" telescopes. The selection of stars included in the observation program can be changed, since the rotors of the VTs can be set by a fine adjustment at any angle t or δ. On the basis of the work done, the system can be used in automatic setting on any stars with an accuracy of 1'–2' (without the use of correction systems). The basic advantage of the VT circuit for programmed control is its simplicity and reliability.

REFERENCES

1. Z. N. Mamedova, *Collection of Works in Electromechanics* (Sb. po elektromekhaniki) 5, *IEM AN SSSR* (1961).
2. N. N. Mikhelson, "A computer which corrects the position of a telescope for mean refraction", *Izv. GAO AN SSSR*, **21**, No. 162 (1958).

CHAPTER **I-13**

The electronic drive of the 48-cm reflector of the Vilnius Astronomical Observatory

V. A. YASEVICHUS

Vilnius Astronomical Observatory
Vilnius

Refurbishment of the 48-cm telescope at the Vilnius Astronomical Observatory (Lithuania) is now in the process of completion. Under this program an electronic drive was designed and built.

We had a 20-W synchronous motor at our disposal. It was calculated that a single-stage push–pull circuit using 6P3S vacuum tubes in class AB_1 would deliver to the motor a power in the order of 20 W when the grids of the tubes were excited with approximately 20 V. We selected a balance multivibrator as an oscillator, as it has advantages over the L–C and R–C coupled oscillators; the multivibrator has a simple circuit and can generate the required voltage for the grids of the 6P3S tubes. Due to the circuits in the grid of the amplifier tubes, and to the capacitor C (see Figure 13-1), the output voltage is almost sinusoidal. The generated frequency can be varied continuously between 15 and 70 Hz by a potentiometer in the grid circuit of the multivibrator. This frequency range ensures the required motor speed, and also allows guiding by using buttons K_1 and K_2 in the hand switch control box. Pressure on one of the buttons accelerates or retards the speed of the drive mechanism.

In photographic work the drive mechanism performed satisfactorily, maintaining a stable speed.

The main advantage of this drive mechanism over a gear mechanism is the simplicity of manufacture and use, and its low cost (not including labor, the cost was about 40 rubles).

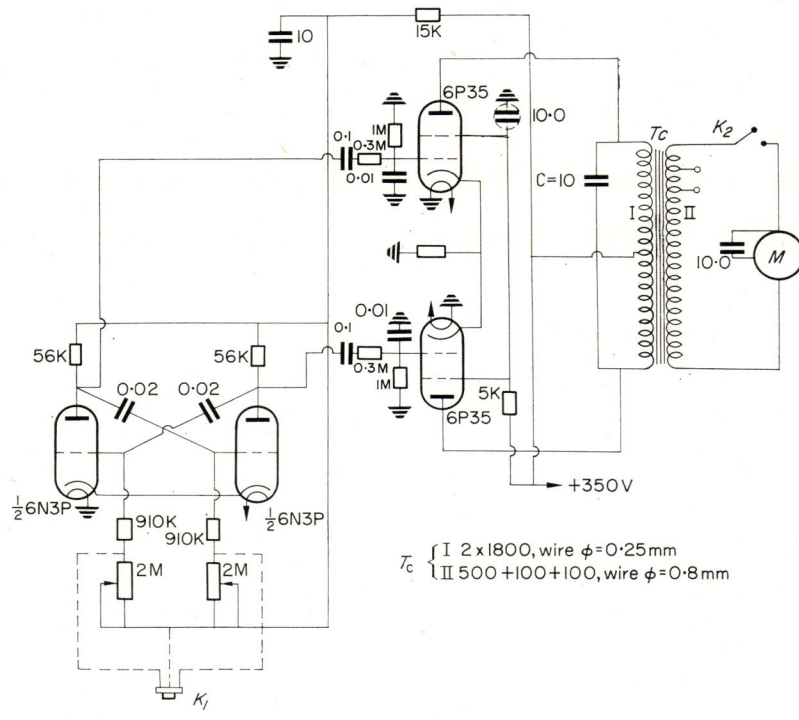

Figure 13-1 Circuit of the multivibrator and power amplifier of the electronic drive for the 48-cm reflector at the Vilnius Astronomical Observatory.

REFERENCES

1. V. I. Kalinin and G. M. Gershtein, *Introduction to Radiophysics* (Vvedeniye v radiofiziku), Gostekhizdat, Moscow (1957).
2. *Radio*, 7 (1958).

CHAPTER **I-14**

Performance of an iris photometer

V. S. AVEDISOVA

Shternberg Astronomical Institute
Moscow

The use of a photometer with an iris diaphragm was first suggested by Siedentopf in 1934, and the basic idea has remained unchanged since then, although there have been modifications and improvements to the design. The fact that almost all foreign observatories interested in stellar research have a photometer of this kind demonstrates the value of the concept.

The schematic of the basic iris photometer system developed at the Shternberg Astronomical Institute is shown in Figure 14-1. The light from

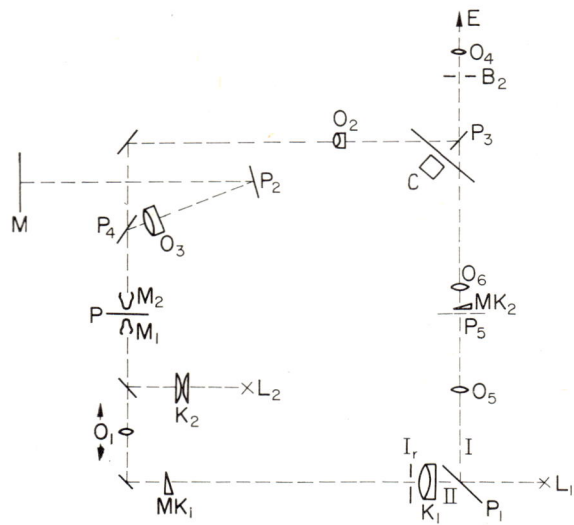

Figure 14-1 Schematic of the iris photometer developed at the Shternberg Astronomical Institute.

a filament bulb (ribbon type) L_1 is split by a partially reflecting mirror P_1 placed at an angle of 45°. The reference beam I (10%) is directed to the photomultiplier E (RCA-931 A-type), while the beam II passes through the iris diaphragm I_r (with 12 blades), and the photographic plate P. The beams (I and II) reach the photomultiplier alternately by the action of a mechanical chopper C which rotates at 33.2 rev sec^{-1}. The ac signal from the photomultiplier is amplified and fed to the "magic eye" indicator (EM-71). Measurements with the iris diaphragm are made by balancing the two beams. Due to the time constant of the human eye we cannot see the 33-Hz modulation frequency on the EM-71; instead, we see a light sector with constant aperture proportional to the amplitude of the ac signal. The accuracy in matching the two beams by this null method is 0.1%.

The reading scale is proportional to the aperture of the iris diaphragm; the relationship between the readings and the stellar magnitude is almost linear. The coordinates of the star to be measured on the photographic plate can be read from dials to an accuracy of 0.1 mm. The intensity of both beams can be adjusted: the reference beam I passes the linear wedge MK_2 and the diaphragm P_5; beam II passes the iris diaphragm and the linear wedge MK_1. The combination of the lens O_1 and the micro-objective M_1 acting as an inverted microscope projects the reduced image of the iris on the photographic plate P. Using a set of micro-objectives M_1, we can vary the area projected on the plate within a fairly wide range. The microobjective M_2 and the lens O_3 project the image of the plate and the iris on the screen M. Depending on the type of micro-objective M_2, an area of 3 or 2 mm is isolated on the plate, with magnification of 38 × or 53 × respectively.

When the whole screen has to be illuminated, an additional bulb L_2 is turned on. The measuring process consists in finding the star on the screen and centering it in the field; then the bulb L_1 is switched on and the image of the star appears on the screen encircled by the iris diaphragm; by turning the diaphragm knob we can reduce it in size until the magic eye indicates the null; the reading of the setting of the iris is the measurement.

A photometer of this kind has many advantages over those with a constant diaphragm:

(1) The null method of measurement secures against variation in the operating conditions of the photomultiplier tube and the amplifier and also provides the possibility of using ac amplification.

(2) The iris diaphragm allows a large range of stellar magnitudes to be measured (a range of ten magnitudes is easily handled) without changing setting of the instrument, whereas photometers with a constant diaphragm and under the same conditions can measure a range of only 4–5 magnitudes. An almost linear characteristic curve can be obtained over these ten magnitudes, so that the same accuracy is attained throughout. In Figure 14-2, we show the calibration curve obtained from the measurement of stars in the Pleiades cluster.

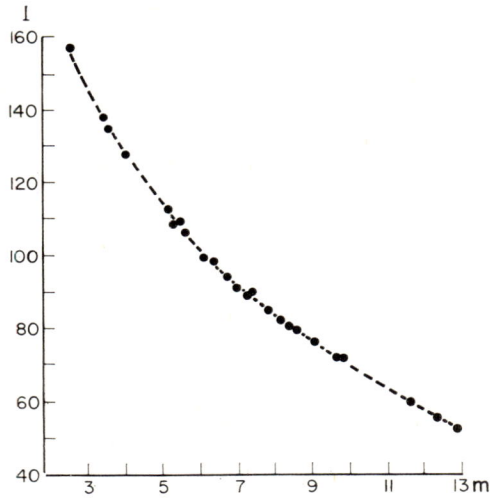

Figure 14-2 Calibration curve of the iris photometer using a plate of the Pleiades cluster taken with the 50-cm Maksutov telescope.

(3) Due to the fact that the effective diameters of the stars are measured, small variations in the background of the plate do not affect the measurements. A study was carried out to see what would happen when the background changed considerably. Artificial stars (with a range of four magnitudes) were photographed with a different background (density from 0.1 to 0.6). Measurements of these artificial star images were carried out and a comparison of the resulting magnitudes showed that measurements under those conditions are quite permissible, although the accuracy suffered to some extent in this case. The magnitudes Pv and Pg of the comparison stars N Cyg 1948 were determined from photographs taken in 1956 with the 70-cm meniscus telescope at the Abastumani Observatory. The standard region selected was the cluster M 13, which was in a different area of the plate having a different background from that of the region

under examination. The magnitudes Pv and Pg were obtained from three plates. The root-mean-square error of the measurements was equal to $\pm 0\overset{m}{.}13$, though it must be admitted that the quality of the plates left something to be desired. In particular, the focusing was unsatisfactory and experience has shown that the iris photometer is very sensitive to this.

(4) Measurements using iris photometers are of very high accuracy. The errors in the photographic plate itself limits the accuracy. The errors in the measurement consists of the centering error ε_c and the matching error ε_m; for plates of average quality the magnitude of neither one exceeds $0\overset{m}{.}02$. To determine the accuracy of the measurements, a plate of the Pleiades cluster taken with the 50-cm meniscus telescope was measured and the error in the calibration curve was found to be $\varepsilon = \pm 0\overset{m}{.}015$ and for very high quality plates it was $\varepsilon \leq 0\overset{m}{.}01$.

(5) The iris photometer can be used for plates obtained with the Maksutov and Schmidt telescopes, although the accuracy is approximately half as a result of poor image quality.

(6) The speed of measurement using the iris photometer is much faster than that using the MF-6 type photometer; between sixty and eighty stars can be measured per hour.

(7) The instrument is very convenient and simple to use.

(8) The instrument has very good measuring stability. After the instrument is switched on there is usually a slight drift in the readings, that lasts from twenty to thirty minutes.

(9) The instrument is universal. Since the continuous wedge MK_1 and the iris diaphragm are in the beam II, the instrument can also be used as a photometer with a constant diaphragm. Moreover, the iris diaphragm can be replaced by a slit for spectra measurements.

(10) Automatic recording devices and electronic counters may be used with this instrument for further processing.

The determination of the null (matching of the beams I and II) could be improved by replacing the magic eye EM-71 by the more sensitive EM-85, thus reducing the null error by 30%. Determination of the null is easily and accurately done by photoelectric means rather than visually. It would be worthwhile increasing the number of micro-objectives so that smaller magnification could be obtained on the screen, as this is sometimes necessary. In addition, it would be better if the ratio between the intensities of the beams were more flexible.

CHAPTER **I-15**

A camera with an optical compensator for observations of artificial earth satellites

KH. I. POTTER and YU. S. STRELETSKII

Main Astronomical Observatory
Pulkovo

The analysis of accurate astrometric observations of artificial earth satellites (AES) contributes to the solution of many applied problems in astronomy, geodesy and other related fields. It is generally known that observations of the orbital motion of satellites can be used in the study of the structure of the Earth's atmosphere. Unique results are given by observations of satellites to connect geodetically distant points on the surface of the earth, or to establish a uniform scale of ephemeris time, free from the effect of the non-uniform rotation of the earth. However, the accuracy of existing methods of determining the position of AES is still not sufficient to solve these problems successfully. The method we describe below improves the accuracy of observations under certain conditions, so that it almost reaches the accuracy of stellar observations.

Suppose that a rotating plane-parallel glass or the compensator for the AES motion, is placed in front of the focal plane of an astrograph with a focal length in the order of 2 to 2.5 m and corrected field of about 5° × 5° (Figure 15-1).

When the AES crosses the field, the compensator rotates at the appropriate angular velocity around an axis perpendicular to the direction of motion of the AES. In this way, we can compensate for the relative motion of the AES in the same way as the Markowitz camera [1, 2] does for observations of the moon. We obtain stellar-like images of the AES on the photographic plate, and images of reference stars; the coordinates (α, δ) of the AES can then be found by the usual methods of photographic astrometry.

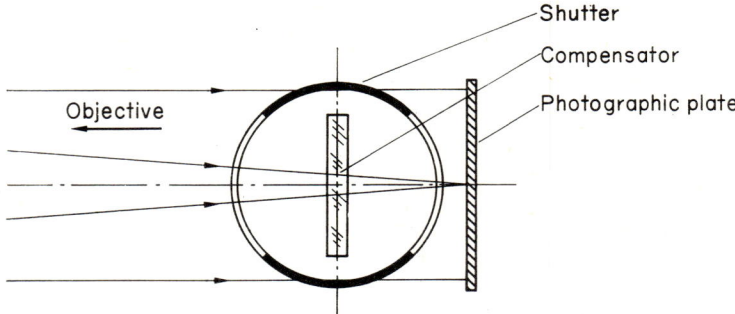

Figure 15-1 Schematic of the optical compensator for the AES motion showing the cylindrical shutter, plane-parallel glass (compensator) and photographic plate.

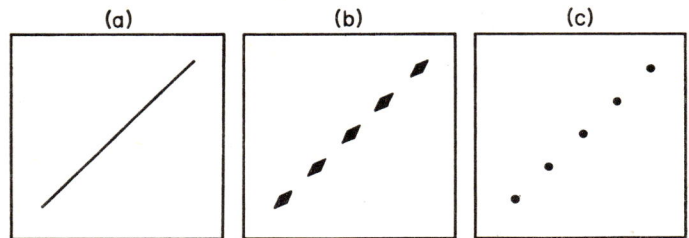

Figure 15-2 Images of and AES on the photographic plate under different observing conditions. (a) no compensation, (b) with rotating cylindrical shutter and (c) with optical compensator and cylindrical shutter rotating at the proper speed.

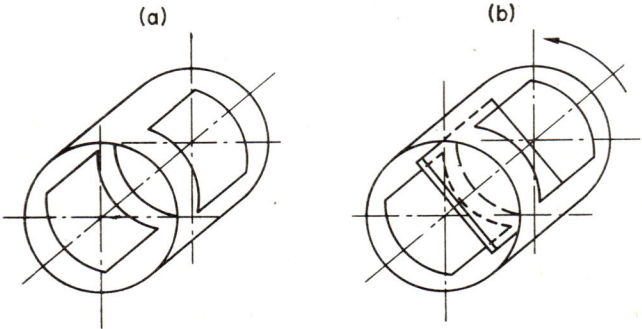

Figure 15-3 (a) Cylindrical shutter, (b) cylindrical shutter with compensator.

Let us examine the role of the compensator in more detail. If the field is not covered by the compensator, the AES leaves a linear trail on the plate (Figure 15-2(a)). If a cylindrical shutter, rotating at a suitable rate, intercepts the rays, then the satellite trail on the plate will be discontinued (Figure 15-2(b)). In this case, measurement of the resulting images would present a difficult problem, due to the unfavorable photometric structure of elongated shaped images.

Now suppose that the cylindrical shutter (Figure 15-3(a)) has a compensator of thickness d (with index of refraction n) which rotates with the shutter (Figure 15-3(b)) around an axis perpendicular to the direction of the AES motion. The compensator displaces the image in the focal plane by the amount x given by the following expression:

$$x = d \sin z \left(1 - \frac{\cos z}{\sqrt{n^2 - \sin^2 z}}\right), \tag{15-1}$$

where z is the angle of inclination of the compensator with respect to the incident rays. The rate of displacement of the image due to rotation of the compensator can be found by differentiating x with respect to time, assuming

$$z = \omega t. \tag{15-2}$$

where ω is the angular velocity of the compensator.

However, x is clearly a nonlinear function of t. In order that the velocity of displacement of the image shall be sufficiently constant, the compensator must rotate at a variable rate Eq. (15-1), or z must be restricted to small values, so that the relationship between x and z with sufficient accuracy is linear (Eq. (15-2)). The second method is preferable, especially as aberrations (astigmatism, coma) and poor focusing occur at large inclinations of the compensator. If $|z| < 20°$, then

$$x = d \frac{n-1}{n} z, \tag{15-3}$$

where x and d are expressed in millimeters, and z in radians. We may take $n = 3/2$, and then

$$x = \frac{d}{3} z = \frac{d}{3} \omega t. \tag{15-4}$$

The rate of image displacement is given by

$$\frac{dx}{dt} = \frac{d}{3} \omega \text{ (mm sec}^{-1}). \tag{15-5}$$

Let us find the angular velocity of the compensator for complete compensation of the AES relative motion. If the topocentric angular velocity of the AES is V (° sec^{-1}), and the focal length of the astrograph is F (mm) then

$$v = \frac{\pi}{180} VF. \tag{15-6}$$

For compensation of the motion of the AES we must have

$$\frac{d}{3} \omega = \frac{\pi}{180} VF, \tag{15-7}$$

from which we obtain

$$\omega = \frac{1}{120} \frac{VF}{d} \text{ (rev sec}^{-1}\text{)}. \tag{15-8}$$

If the compensator rotates at the velocity given by Eq. (15-8), a series of point images will be obtained on the plate (Figure 15-2(c)).

Let us plot the photometric profile of the images (Figure 15-4(a), (b)). Clearly the shaded areas, corresponding to the total exposure on the plate, are equal in the two graphs. Therefore, when the optical compensator is used the image will be considerably brighter, so that the limiting magnitude of the astrograph is increased. When the compensation is adequate, the images obtained are circular and easily measurable. If to this advantage we add the more accurate scale of the astrograph in comparison with standard AES cameras ($F = 250$ mm), considerably better results can be expected. The purpose of the compensator is to concentrate the light energy distributed along the elongated image (Figure 15-4(a)) into a small circular image (Figure 15-4(b)). The length of the image, S, depends on the

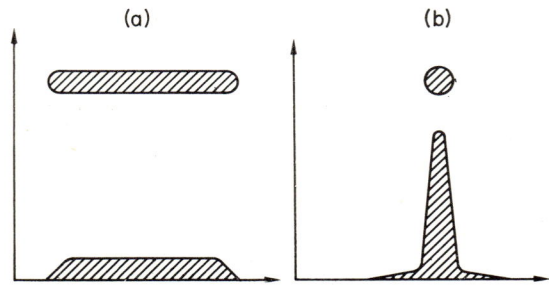

Figure 15-4 Photometric profile and shape of images: (a) elongated, (b) circular.

thickness of the compensator and on the limit angle of inclination of the compensator in the operating position. It can be shown that

$$S = \frac{\pi}{270°} z°_{max} d \text{ (mm)}. \tag{15-9}$$

Values of S satisfying this relationship and for various values of d and z_{max} are given in Table 15-1.

Table 15-1 $S = \dfrac{\pi}{270°} z°_{max} d$ (mm)

$z°_{max}$ \ d (mm)	2	4	6	8	10
10	0.2	0.5	0.7	0.9	1.2
20	0.5	0.9	1.4	1.9	2.3
30	0.7	1.4	2.1	2.8	3.5
40	0.9	1.9	2.8	3.7	4.7

How are the optimal values d and z_{max} determined? Clearly, a decisive factor is the estimate of nonlinear displacements of the images when the Eq. (15-3) replaces Eq. (15-1).

The nonlinear displacement ε is tabulated in Table 15-2, which can be used to determine the best values of d and z_{max}. It must be remembered that the diameter of stellar images on the photographic plate is usually of the order of 0.050 to 0.150 mm. Nonlinear displacements cause slight enlargement of the image (by a quantity of the order of ε given in Table 15-2); the image remains practically symmetrical relative to its photometric center in this case. The optical parameters of the compensator for the data of Table 15-2 can be taken to be $d = 8$ mm, $z_{max} = 20°$. A point to note is that the active angle of rotation of the compensator is limited by the size of the diaphragm *1* in the cylindrical section of the shutter *2* (Figure 15-5). The thickness of the compensator *3* and the cylindrical section of shutter *2* corresponds to approximately $0°5$ in the focal plane of the astrograph. The remaining part of the field *4* is not affected by the shutter and compensator. To compensate for the increase in the focal length caused by the compensator *3* a fixed plane-parallel filter *5* of the same thickness d is placed in front of the focal plane. The time when the AES and the stellar background has been photographed can be obtained by means of the contact unit *6* which transmits a signal to a printing chronograph when the rotating and fixed filters are parallel to one another. A gate *7*, which is opened by an electromagnet, establishes the correspondence between the series of images on the plate and the times recorded on the of cronograph tape. The camera plateholder *8* can be rotated by the positional angle ψ in relation to the tube of the astrograph. A synchronous motor and a worm-gear *9* in the outside of the camera rotates the compensator and the shutter. The angular velocity ω of the compensator is proportional to the topocentric angular velocity of the AES. Table 15-3 gives the values of ω calculated from Eq. (15-8) (for $F = 2,300$ mm).

Table 15-2 ε (mm)

z°_{max} \ d (mm)	2	4	6	8	10
10	0.000	0.001	0.001	0.002	0.002
20	0.004	0.008	0.012	0.019	0.021
30	0.015	0.030	0.044	0.059	0.074
40	0.037	0.074	0.110	0.147	0.184

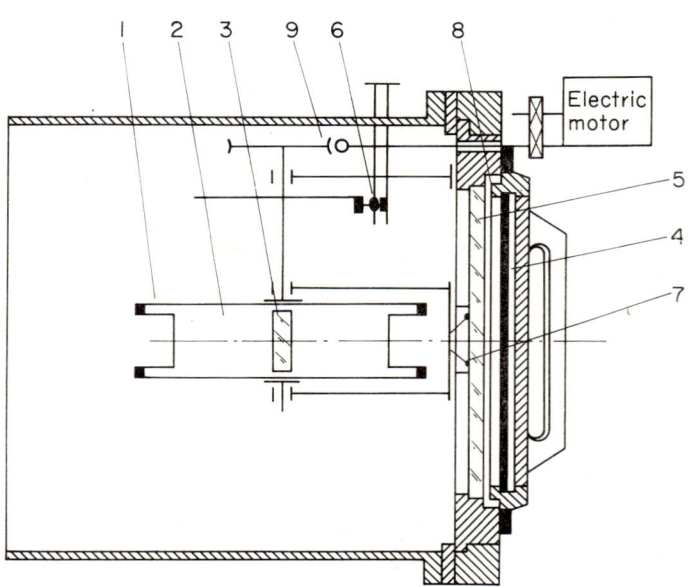

Figure 15-5 Schematic of the AES camera with an optical compensator: 1, diaphragm; 2, cylindrical shutter; 3, plane-parallel glass (compensator); 4, photographic glass plate; 5, fixed plane-parallel filter; 6, electrical contact; 7, gate; 8, plate holder; 9, worm-gear.

Table 15-3 $\omega = \dfrac{1}{120}\dfrac{VF}{d}$

V (° sec^{-1}) \ d (mm)	2	4	6	8	10
0.2	1.87	0.93	0.62	0.47	0.37
0.4	3.74	1.87	1.25	0.93	0.75
0.6	5.61	2.80	1.87	1.40	1.12
0.8	7.48	3.74	2.49	1.87	1.50
1.0	9.34	4.67	3.12	2.34	1.87

The number of revolutions of the synchronous motor is regulated continuously by a frequency generator. In order to cover the wide range of angular velocities of the AES, which requires from one revolution to 0.03 rev sec^{-1} of the motor, it is necessary to have a set of interchangeable gear reducers.

In order to carry out observations, we must know the angular velocity V and the positional angle ψ of the AES. If an image length $S = 2$ mm is reduced to a circle then an elongation of the image of 0.020 mm is caused by an error in velocity equal to $\pm 1\%$ of the true velocity and by an error in the positional angle of the order of $\pm 0°.6$. Clearly, it would not be difficult to obtain such an accuracy or even two or three times higher. At this point, we should remember that we are considering only high-altitude AESs with known oribts; only these satellites can be used to solve the problems mentioned at the beginning of this paper. Observations of short-life satellites with motions affected by non-gravitational forces (which are difficult to calculate) can be made by less accurate methods.

A series of images produced by an AES in a single pass can be recorded on a photographic plate. The distance between the images does not depend on the focal length of the objective, and is equal to $(\pi/3)\,d$, i.e. it is almost exactly equal to the thickness of the compensator. If we measure five images which are symmetric with respect to the optical center, we can expect to obtain the coordinates (x, δ) within an $0''.1$ to $0''.3$ accuracy which is much better than the one obtained by existing methods. The corresponding accuracy of the time record (when $V = 0.2°$ sec^{-1}) is 0.0003 sec. Clearly the error of a standard time service may be one order greater. This is insignificant in the solution of the geodetic problem of connecting distant points, because the observations are close in time and may be related to the transmission of the same time signals. The ephemeris time, or more exactly the difference between emphemeris time and universal time, can be determined without knowing the time more accurately than the standard time service.

We can make an approximate estimate of the expected gain Δm in stellar magnitude for observations of AES using an optical compensator. It is clear that when the image (Figure 15-4(a)) is reduced to a small circle, (Figure 15-4(b)), the value of Δm will depend on how narrow the trail image is, and consequently on how small the diameter of the circular image is. The gain Δm is defined by

$$\Delta m = 2.5 \lg \frac{S}{l}, \qquad (15\text{-}10)$$

where S is obtained from formula (15-9) and l is the width of the image trail. Taking 0.05, 0.10 and 0.15 mm as the most probable values of l, we obtain Table 15-4. For $d = 8$ mm and $z_{max} = 20°$, we obtain from Table 15-1 $S = 2$ mm. Entering Table 15-4 with $S = 2$ mm and $l = 0.05$ mm, we obtain $\Delta m = 4^m$, that is the gain between photography with compensation over photography without compensation.

Table 15-4 $\Delta m = 2.5 \lg \frac{S}{l}$

l (mm) \ S (mm)	0.2	0.5	1	2	3	4
0.05	1.5	2.5	3.2	4.0	4.5	4.8
0.10	0.8	1.7	2.5	3.2	3.7	4.0
0.15	0.3	1.3	2.1	2.8	3.2	3.6

The calculation of the limiting magnitude of the AES which can be obtained by optical compensation methods is more complicated. We can attempt to calculate it, approximately at least, for the lunar expeditionary astrograph of the Pulkovo Observatory ($D = 230$ mm, $F = 2,300$ mm) which is shortly to be equipped with a camera of this kind. The limiting stellar magnitude m_{lim} of the astrograph for an exposure of t and for a photographic emulsion of sensitivity E can be expressed by

$$m_{lim} = 2.5 \lg t \,(\sec) + 2.5 \lg E + 5 \lg D \,(\text{mm}) + 2.5 \lg P + C$$

$$(15\text{-}11)$$

where P is a quantity which characterizes the quality of the objective, the amount of light loss due to reflection and absorption, etc., and C is a constant determined by the observations. Due to the term $2.5 \lg P$, it is difficult to compare different astrographs, but for an approximate estimate of $m_{\lim}$ we may take $\lg P = 0$, i.e. we can ignore differences in the transmission factor of the different objectives.

Data about the standard Pulkovo astrograph given by N. V. Fatchikhin [3] lead to the value $C = -6.8$. To obtain the limiting magnitudes of AES moving with different angular velocities, we determine the magnitude for a total exposure t of a single AES image observed by the optical compensation method. It is easily seen that

$$t = \frac{2z_{\max} d}{3 \, VF} \text{ (sec)}. \quad (15\text{-}12)$$

The angular velocity V depends on the height of the satellite orbit; for passage across the zenith in a circular orbit

$$V = \frac{360 \sqrt{g} \, R}{2\pi H \sqrt{R + H}} \, (° \text{ sec}^{-1}), \quad (15\text{-}13)$$

where R is the radius of the earth, H is the orbital height, and g is the acceleration due to gravity. Comparing Eq. (15-12) and Eq. (15-13) for given values of $z_{\max}$, d and F we find that t is a function of H. The sensitivity of the photographic emulsions most often used in astronomy is given for different exposures by I. I. Breido [4]. In Table 15-5 we give the limiting stellar magnitude of a satellite passing across the zenith in a circular orbit at height H for different photographic emulsions (Agfa-Astro; Kodak OaO; Ilford HP 3; the film NIKFI-2 used in the photography of AES by the NAFA night aerial camera type 3 s/25 and the NIKFI-2 which turns out to be the most sensitive of all materials studied at the Main Astronomical Observatory. The values in Table 15-5 were obtained on the assumption that light losses in the objective were the same as for the standard Pulkovo astrograph. Obviously this condition will not be fulfilled; first, the expeditionary astrograph has a four element objective with air gaps, while the normal astrograph is a two-element cemented objective; second, the camera with a compensator has additional optical parts which cause light losses. Table 15-5 can be used only for preliminary estimates; more accurate data will be obtained from use of the instrument. The actual performance will probably be reduced by 1 to 2 magnitudes than those indicated in Table 15-5.

Table 15-5

Orbit height	Exp. time	Emulsion, limiting magnitude				
H	t (sec)	Agfa-Astro $m_{\lim}$	Kodak OaO $m_{\lim}$	Ilford HP 3 $m_{\lim}$	NIKFI-1 $m_{\lim}$	NIKFI-2 $m_{\lim}$
100 km	0.011	4.4	5.1	6.4	6.4	6.4
200 km	0.021	5.1	5.9	7.0	7.1	7.2
500 km	0.055	6.0	6.8	8.0	8.0	8.0
1000 km	0.11	6.7	7.5	8.4	8.7	8.7
3000 km	0.38	8.1	8.9	9.8	9.9	10.1
R	0.95	8.4	9.4	10.0	10.2	10.6
3 R	2.3	9.8	10.8	11.0	11.4	12.0
5 R	8.2	11.1	12.2	12.1	12.4	13.4
10 R	22	11.9	13.0	12.8	13.1	14.3
30 R	112	13.3	14.6	13.9	14.0	15.9
60 R	178	13.3	14.8	14.2	14.4	16.3

REFERENCES

1. W. Markowitz, "Photographic determination of the moon's position, and applications to the measure of time, rotation of the earth, and geodesy", *Astron. J.* **59** (1954).
2. Kh. I. Potter and Yu. S. Streletskii, "A camera for the observation of the moon with the normal astrograph at the Main Astronomical Observatory of the Academy of Sciences, USSR in Pulkovo", *Astron. Zh.* **36**, No. 6 (1959).
3. N. V. Fatchikhin, *Proc. of the 10th Astrometric Conference of the USSR* (1954).
4. I. I. Breido, "A catalogue of characteristics of photographic plates used in astronomy", *Izv. GAO AN SSSR*, **20**, No. 158, 156 (1958).

CHAPTER **I-16**

Observational results from the new transit instrument of the Pulkovo Observatory

N. N. PAVLOV

Main Astronomical Observatory
Pulkovo

A new transit instrument [1] was constructed at the mechanical workshops of this Observatory in 1959. This instrument greatly reduces the effect of many sources of errors which are difficult to eliminate in the reversing of transit instruments; its mechanical design is highly stable, and the support system minimizes the flexure of the horizontal axis. The trunnions and other critical parts are well protected from dust and frost, and in addition, the transit instrument is thermally insulated to reduce the effect of temperature on the observational results.

The inclination of the horizontal axis is determined by means of two highly accurate levels, which are read by eye with the aid of a terrestrial telescope to eliminate parallax. A special device is used to maintain the inclination during observations to within 0.1–0.2 of the scale value, thus reducing the errors. In addition, one and a half minutes before readings are taken, both levels can easily be raised slightly in the same direction, so that their bubbles always come to the equilibrium position from the same side and at the same speed.

The error in determining the inclination in observations of a single star proved to be of the order of ± 0.0015 sec or $0''.022$. The photoelectric transit recording system has a 0.145 sec time constant. This value is very stable in time and in practice it is independent of star brightness.

The new instrument has been mounted in a building with strong exhaust ventilation and at a considerable distance from the other observatory buildings. Observations begun in August 1959. Initially, the instrument was mounted fairly low so that the end of the dewcap of the tube, when pointing to the zenith, was 130 cm below the open observing slits. The observations obtained showed high internal accuracy, but also noticeable systematic errors of the type known as "evening signal". It has been found that these errors are caused by air vortices inside the building above the instrument, caused by the simultaneous action of the ventilation and the wind outside. The instrument was therefore raised in June 1960, and at the same time the power of the exhausters was reduced several times. The evening signal then fell by more than one and a half times.

By comparing the many corrections to the drive observed in the new transit instrument for different wind speeds, a well-defined relationship was obtained between the external random error of the drive correction and the wind velocity. For wind velocities of more than 4 m sec^{-1} the drive correction was almost 2.5 times smaller than it was in still air, which, in this case, indicates the overwhelming contribution of refraction errors as compared to instrumental errors.

On the basis of observations carried out after June 1960, it was found that deviations in the drive corrections depended on the direction of the wind, producing the "wind effect". This is similar in sign and nature to the wind effect obtained at Leningrad, Tokyo, Potsdam and Pulkovo [2–4]. In our case, this effect turned out to be also dependent on the wind speed, reaching a maximum for a winds of 2 to 4 m sec^{-1}, and falling for slight or strong winds. Since the instrument was very well protected from the direct action of the wind, and there was no possibility of abnormal refraction occurring inside the building, the most probable cause for the wind effect was abnormal refraction caused by air vortices immediately above the building. A definite relation was also discovered between the magnitude of the external random error of the drive correction and the direction of the wind. In this sense, the most accurate observations were made with west winds and the least accurate with east winds.

It is interesting to note that for the internal error, or the error in determining the drive correction for a single zenith star, the contrary occurred; the accuracy was higher for east winds. In Table 16-1 we give the external error

Table 16-1

Wind direction	No. of corrections	$E_n \cos \varphi$ (sec)	$E_1 \cos \varphi$ (sec)
easterly	53	± 0.0032	± 0.0057
westerly	44	± 0.0023	± 0.0072

values reduced to the equator ($E_n \cos \varphi$), and the internal error values for a single zenith star ($E_1 \cos \varphi$) also reduced to the equator. The E_n refers to the drive correction averaged over 24 stars, of which more than half were stars observed at the zenith.

This strange contrary effect may be a result of the nature of turbulence of the air masses for east and west winds.

The accuracy of observations on the new transit instrument is comparable with published data for the best modern zenith tubes and Danjon prism astrolabes [5, 6].

Transit instruments have several advantages over zenith tubes and prism astrolabes. Mainly, they operate over a very wide declination range simplifying the problem of relating observations made at different latitudes.

The transit instrument is most suitable for catalog observations since the connection with the fundamental system could be made very simply. Finally, by observing a wide zone of south, north and zenith stars, we can get some idea of the amount of "atmospheric inclination" causing abnormal refraction at the zenith.

REFERENCES

1. N. N. Pavlov, "Concerning a new type of resetting meridian instrument", *Izv. GAO AN SSSR*, **22**, Issue 1, No. 154, 99 (1955).
2. P. N. Dolgov, *Determination of Time by a Transit Instrument at the Meridian* (Opredeleniye vremeni passazhnym instrumentom v meridian), p. 367 (1952).
3. H. Kruger, *Vermessungstechnik*, **10**, 11 (1954); **2** (1955).
4. P. M. Afanasyeva, "The influence of wind on results of astronomical time determinations" *Proceedings of the 14th Astronomical Conference* (*1960*) (Tr. 14-i Astronomicheskoi konferentsii), p. 345.
5. W. Markowitz, *U. S. Naval Observatory, Reprint* No. 13 (1960).
6. M. Cavedor, *Bull. Ann. Soc. Suisse Chronom.*, **4**, 145–8 (1958).

CHAPTER **I-17**

A standard wedge level-tester

L. A. SUKHAREV

Main Astronomical Observatory
Pulkovo

The precision liquid levels with scales between $0''\!.5$ and $0''\!.2$ widely used in astrometric and geodetic instruments have not been satisfactorily analyzed for two reasons: (1) the testing units, as a rule, possess significant intrinsic errors which complicates and delays the process of analysis; and (2) the temperature coefficient of the testing unit is not measured and in practice is taken equal to zero without justification.

Liquid levels will probably be replaced soon by levels of other types, such as those based on Leger's vertical elastic pendulum principle or by others with an equally well-defined principle thus demanding greater accuracy on the tester. For these reasons work has been done at the Astrometric Laboratory of the Main Astronomical Observatory during the last five years to improve and standardize level-testers. Two models of a standard wedge level-tester have been constructed. They consists of two main sections: the level-tester itself with a wedge slope-generating unit and corresponding reading scale, and the optical standardization apparatus. Standardization is carried out by means of an interferometer incorporated in the tester, or by the combination of a standard reflecting angle (fused silica) and an autocollimator. We found that in our case the second optical arrangement was better than the first, therefore we shall discuss this one only (Figure 17-1).

The base of the instrument consists of a beam G one meter long with three leveling screws. On this base are mounted the adjustable stand S with the standard angle T and a stand with a ball-socket joint B for the lever A. Also mounted on the same base is the case L which holds the screw-wedge unit to generate the slope for A and the reading head N to measure the angle of inclination. Another ball-socket V attached to A rests on the wedge K. The lever A is an autocollimator with an 80 mm clear aperture and 800 mm focal length objective O, and the micrometer eyepiece P. The weight of the lever itself is over 12 kg. A fork mount support at B prevents the lever from displacing sideways. Two brackets D are rigidly connected to the top of lever A on which the test level E is mounted.

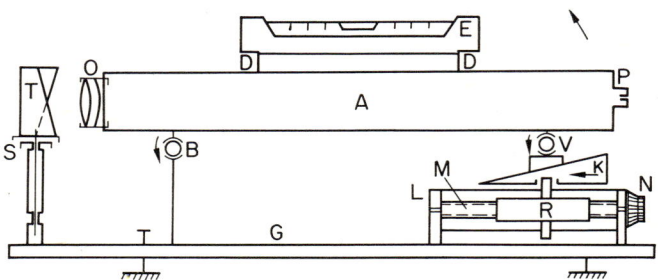

Figure 17-1 Diagram of the standard wedge level-tester developed at the Main Astronomical Observatory, Pulkovo.

Before testing begins, the scale value of the reading unit must be determined by rotating the head N of the screw M. The screw drives the nut R which has a lower and an upper lug. The lower lug constrains the rotation of the nut with the screw by sliding it in a slot in case L. The upper lug fits in a socket in wedge K; the lower wedge surface slides on the upper sliding surface of case L. The head N stops turning when the wedge K almost reaches the extreme right-hand position, i.e. when V rests on the sharp end of the wedge. By slightly turning N in the opposite direction, we can record the reading. At this point, looking through the autocollimator eyepiece, we can see two images of the autocollimator light source, one above the other at an angular distance double the angle between the reflecting surfaces of the standard T. We set the cross hair of the eyepiece P on the lower image. Continuing to look through the eyepiece, we again turn the screw M, but in the opposite direction, until the cross hair of the eyepiece P for a given setting of R coincides with the second image. We obtain a new reading from N and substract it from the first reading.

If the standard angle T was first measured by means of an interferometer, we can divide this measurement expressed in seconds of arc by the difference in the readings (difference between the first and second setting of N) to obtain the tester scale value.

For testing, the level E is mounted on the brackets D, and the wedge K is set in the middle position. The bubble of the level is brought to the center by use of the levelling screws on the base G. By turning the screw M in the direction of decreasing readings, we move K to a position in which the bubble will be displaced to an extreme position. Then, by turning the screw in the direction of increasing readings on the head N, we raise the eyepiece end of A and then, we examine the level.

Determination of the scale value of the tester and testing of the levels must be done with the instrument placed in a thermally insulated case.

Let us now briefly discuss the consideration which led to the choice of design of the level-tester, and the method of standardization.

It is not advisable to use a screw to generate very small and accurate displacements when a small portion of the screw is used. However, in order to minimize the periodic errors, which are between 1–2 μ in modern precision screws, the measurements should be made using different portions of the screw. This would at least involve two sets of measurements shifting the initial point by a half-turn of the screw. It is much easier to manufacture an accurate wedge, since plane surfaces can be produced with an error of about 0.06 μ. A well-made wedge could operate with random deviations from a flat surface of less than 0.02 μ. For the tester under consideration, the limiting error is compounded by the error in the two wedge surfaces and the flat sliding surface of case L which must not exceed 0.10–0.15 μ or $0\rlap{.}''03$. The errors in the screw which drives the wedge are negligible.

A preliminary determination of the departure from linearity of the scale was made by measuring the slope of the lever. This was done by using vernier calipers and by simultaneously using three optimeters*. One was placed above the joint B and rested on a sensitive small lever on the upper part of A, and the other two were placed approximately above the joint V and also rested on sensitive small levers on A. The distances (600 to 700 mm) between the axes of the small levers of all three optimeters were measured by vernier calipers to an accuracy of 0.1 mm.

The results of these measurements are shown in Figure 17-2. The errors of the optimeters themselves are of the same order of magnitude than that of the errors measured, and so the results provide only indirect evidence for the expected error of the tester. We could not make more

---

* An optimeter is a lever-type optico-mechanical instrument used for measurements of small thicknesses or displacements down to 0.001 mm. (Ed. English version.)

Table 17-1 Interferometric measurements of the fused silica standard angle No. 3

Measurements by		Ye. I. Bryushkova			G. M. Timoshkova			Myao-Yun-Zhui		
Set	Base l Plates	80 mm α_{80}	70 mm α_{70}	60 mm α_{60}	80 mm α_{80}	70 mm α_{70}	60 mm α_{60}	80 mm α_{80}	70 mm α_{70}	60 mm α_{60}
I	N-1	33″.625	33″.586	33″.577	33″.632	33″.579	33″.530	33″.611	33″.579	33″.577
	N-2	.625	.553	.577	.597	.554	.597	.555	.554	.512
	Mean I	33″.625	33″.570	33″.577	33″.614	33″.566	33″.564	33″.583	33″.566	33″.544
II	N-4	33″.551	33″.562	33″.512	33″.567	33″.579	33″.559	33″.562	33″.561	33″.605
	N-5	.555	.529	.482	.576	.561	.540	.562	.545	.530
	Mean II	33″.553	33″.545	33″.497	33″.572	33″.570	33″.550	33″.562	33″.553	33″.568
	Total mean $\frac{I+II}{2}$	33″.589	33″.558	33″.537	33″.593	33″.568	33″.557	33″.572	33″.560	33″.556
		±.021	±.012	±.024	±.014	±.006	±.015	±.013	±.007	±.021
	Difference I−II	0″.072	0″.024	0″.080	0″.042	−0″.004	0″.014	0″.021	0″.013	−0″.024

Mean errore using all data:
Mean difference I−II = 0″.028 ± .017.

Mean value of standard angle for different values of l.
$\begin{cases} \alpha_{80} = 33″.585 \pm .009 \\ \alpha_{70} = .562 \pm .005 \\ \alpha_{60} = .550 \pm .011 \end{cases}$

Note. The plates N-1 and N-2 of the first set were obtained by photography of interference patterns produced by the standard angle; the camera uses an aspherical lens ($D = 120$ mm, $f = 886$ mm) and the emulsions were Agfa rapid using an exposure of 20 min. in the mercury line $\lambda = 546\ m\mu$ (12/9/58–12/11/58). The plates N-4 and N-5 of set II were taken using the "Uran" objective ($D = 100$ mm, $f = 250$ mm) and the same emulsion but with an exposure of 6 min. (4/13/59–4/14/59).

accurate measurements, because the Uversky interferometers at our disposal had too small a range, and interferometers with the necessary operating range were not available.

The fused silica standard reflecting angle was chosen, since for all practical purposes the angle is independent of temperature changes. Another factor was the comparatively simple, highly accurate determination of the angle by an interferometer using photographic recordings, i.e. to give documented results. The random errors in the measurements by this method of the standard angle number 3 are given in Table 17-1; photographic recordings have also been reduced visually using a measuring engine. Measurement accuracy of the standard angle can be increased several times if photoelectric methods are used. It will not have any effect on the systematic error which is caused by imperfections on the standard angle surface and aberrations of the autocollimator objective. The systematic error can be reduced only by improving the optical quality of these components.

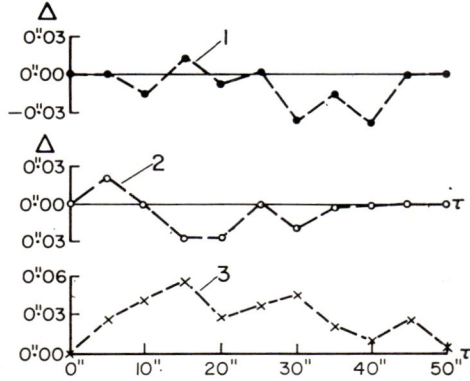

Figure 17-2 Calibration errors Δ of the wedge level tester versus level angle τ.

The basic difficulty in the standardization of the tester scale lies in the setting of the cross hair in the eyepiece P on the image of the autocollimator light source. The error of a single visual measurement using this tester with an objective 80 mm in aperture is close to $\pm 0\rlap{.}''16$ for a range equal to the standard angle.

If this angle is close to 30″, as in the usual standards, then over the entire scale of the tester (50″), the error increases to $0\rlap{.}''25$, i.e. 0.5%. In a set of measurements, the error may be reduced from two to three times, but the scale value of the tester cannot be obtained more accurately by increasing the number of measurements since, the systematic errors are the dominant ones. This can be demonstrated by comparing the scale value of the tester obtained by the method described and by using optimeters. The optimeters are mounted on two points of the lever A which are separated at an accurately known distance from one another; and the difference in height of the two points are measured. In this case, the dispersion in the measurements is as high as 0.4%, and this can be taken at present as the maximum accepted value. Application of photoelectric methods in autocollimation is not difficult and could reduce subjective as well as random errors.

The results of level studies published by the Main Astronomical Observatory after World War II [1–4] show that one can obtain a scale value with an error of 0.4 to 0.7% by a lengthy series of measurements, except for the results obtained by using method developed by Semenov [5], which are affected by errors three times smaller. The values quoted demonstrate the poor calibration methods for precision levels, particularly in regard to the length and duration of the studies that are needed. Even short experience in the use of wedge level-testers leads one to suspect that the two basic reasons for the large dispersion of errors in the calibration of screw level-testers are imperfections in the design, and insufficient thermal insulation of the level and of the tester.

Many studies of levels using wedge testers have been carried out in Pulkovo by N. R. Andreyenko and his associates. The earliest results, obtained with a unit without a thermal insulation were published in 1960 [6]. It must be noted when comparing the error in the level scale value obtained by the wedge tester and the corresponding error obtained by the screw tester, that twice as many measurements had been done with the screw level-tester than with the wedge tester.

In the study of the screw level-tester, by using different portions of the screw (180° out of phase), the effects of periodic errors were minimized. Later studies of levels using a wedge level-tester were carried out by Andreyenko at the Main Astronomical Observatory and I. F. Korbut at the Blagoveshchensky Latitude Laboratory; the unpublished results of their studies were made available to the author.

The conclusion to be drawn from these measurements can be formulated as follows: measurements with the wedge level-tester, (thermally insulated) are affected by errors which are comparable to those resulting from inaccuracies in reading the ends of the bubble and that the the time taken to examine a level is small since the tester errors are negligible.

CONCLUSIONS

(1) The standard wedge level-tester has a practically uniform reading scale, and in this respect is to be preferred to micrometer screw level-testers.

(2) The fused silica standard reflecting angle is the basis for the highly accurate scale of the tester, and for the undetectable temperature coefficient. However, photoelectric methods should be used to improve the results.

(3) The testing procedure, including calculations, takes two to four times less than with a screw level-tester, and this is a particularly important factor in the determination of the temperature coefficient of the level.

REFERENCES

1. G. K. Tsimmerman, *Tr. GAO AN SSSR*, **18** (1951).
2. P. M. Afanaseva, "A catalogue of right ascensions of 203 stars derived from observations with a new photoelectric transit instrument (1955–1956)", *Izv. GAO AN SSSR*, **21**, Issue 1, No. 160, 61 (1957).
3. A. S. Vasilev, *Tr. GAO AN SSSR*, **19** (1952).
4. A. I. Yazev, *Proc. of the 14th Astrometric Conference* (Tr. 14-i Astrometricheskoi konferentsii), p. 77 (1960).
5. L. I. Semenov, *Tr. GAO AN SSSR*, **13** (1949).
6. N. R. Andryenko, *Proc. of the 14th Astrometric Conference*, (Tr. 14-i Astrometricheskoi konferentsii), p. 268 (1960).

CHAPTER **I-18**

A reflecting astrolabe for research in fundamental astrometry

D. D. POLOZHENTSEV, KH. I. POTTER and YU. S. STRELETSKII

Main Astronomical Observatory
Pulkovo

Absolute methods for determining the coordinates of celestial objects were until recently almost exclusively confined to visual meridian circle observations. The use of photography has been limited to the compilation of differential catalogs and the solution of certain specific astrometric problems, such as the determination of parallaxes, observations of binary stars, etc. Photographic zenith tubes, photoelectric transit instruments and the Danjon prismatic astrolabe have made a definite contribution to fundamental astrometry by measuring absolute coordinates free of certain systematic errors. Thus, photographic zenith tubes give absolute coordinates independent of $\Delta\alpha_\alpha$ and $\Delta\delta_\alpha$; photoelectric transit instruments, with respect to $\Delta\alpha_\alpha$ and $\Delta\alpha_\delta$; prismatic astrolabes, with respect to $\Delta\alpha_\alpha$, $\Delta\alpha_\delta$, $\Delta\delta_\alpha$, $\Delta\delta_\delta$ (for limited α and δ). An absolute catalog requires a large number of observational data, including (in the case of meridian circle observations) determination of the latitude, the refraction constant, the absolute azimuth, as well as solar observations to obtain equinoxial corrections. The last problem is solved by comparing the results of observations using two completely different instruments, the vertical circle and the transit instrument. It is generally accepted that absolute stellar positions are not as yet accurately known.

A photographic vertical circle (the FVK) now being developed at Pulkovo under the supervision of M. S. Zverev [1] has a similar geometry to that of the visual vertical circle, but applies a modern photographic method to the recording of observations, besides introducing many design improvements. However, it has several serious drawbacks. Firstly, it cannot be used for observation of the sun or other extended sources. Secondly, it does not make use of one of the advantages of photographic methods, the integration of light and the averaging of the position of the object during exposure. The FVK records the trace of a star on the photographic plate due to its diurnal motion; subsequently, these records are measured. Shimmering of the image due to atmospheric turbulence greatly affects the accuracy of the observations; moreover, the gathering power of the telescope is considerably less than that of astrographs with objectives of the same aperture. Thirdly, the effect of flexure can be fairly large in a meniscus-type telescope.

The authors presented a report at the 15 th Astrometric Conference in Pulkovo, 13–17 December, 1960, in which they described an instrument, called a reflecting astrolabe, for photographic observation of the sun for astrometric purposes. It determines the absolute zenith distance of the sun at various azimuths and by comparing the observed zenith distance and the distance calculated from the ephemeris position of the sun, $\alpha_\odot$ and $\delta_\odot$, the errors $\Delta\alpha_\odot$ and $\Delta\delta_\odot$ can be found from the equations

$$\Delta Z_\odot = Z_{\text{obs}} - Z_{\text{calc}} = \Delta\alpha_\odot \cos\delta_\odot \sin q + \Delta\delta_\odot \cos q, \qquad (18\text{-}1)$$

where q is the parallactic angle. The least squares solution of Eq. (18-1) gives the most probable values of $\Delta\alpha_\odot$ and $\Delta\delta_\odot$. If we restrict observations to the meridian circle only, we obtain $\Delta\delta_\odot$ and in this case the instrument is equivalent to the vertical circle.

Generally speaking, the reflecting astrolabe could be used in its original form for observations of stars and other celestial objects as outlined by D. D. Polozhentsev and his colleagues. The optical scheme is shown in Figure 18-1. Starlight reflected on the mirror *1* is fed to the objective *2*; images formed at the focal plane are recorded photographically on the plate *3*. At time T_1 recorded on the chronograph an instantaneous exposure is made using the fast shutter *4*; at that time the reading Θ_I is obtained from the divided circle *5*, which has a common axis with mirror *1*. To obtain a second exposure, the mirror is turned by an angle approximately equal to the height of the object, $h = 90° - z$. The starlight then is reflected on the flat mirror *1*, the mercury level *6* and then fed to the objective *2*; at time T_2 a second star image is recorded photographically and the corresponding readings Θ_II is obtained as in the previous exposure. To obtain the true height of the object at time T_2 we measure d, the projection of the distance between images I and II (Figure 18-2) on the vertical

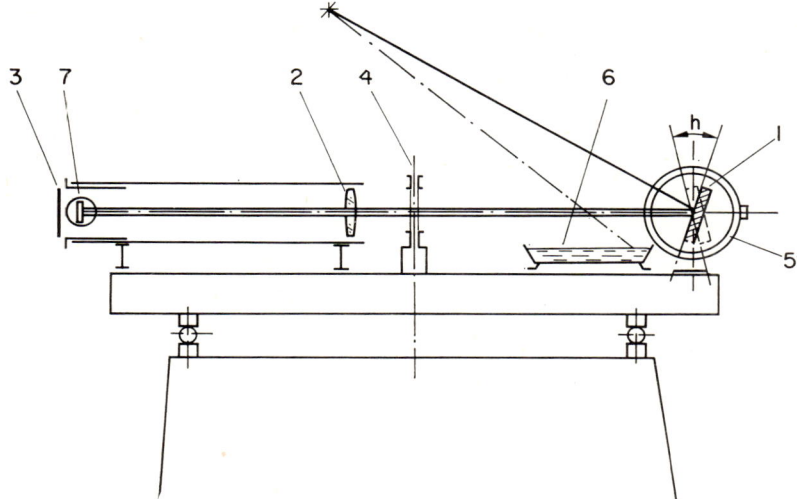

Figure 18-1 Schematic of the reflecting astrolabe for fundamental astrometry.

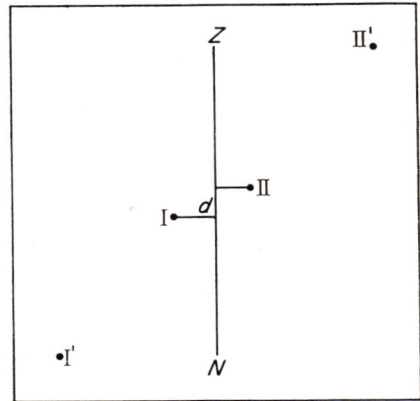

Figure 18-2 Photographic plate to measure the true height of the celestial object at a time T_2.

line ZN. The observed height will then be

$$h_{\text{obs}} = \Theta_{\text{II}} - \Theta_{\text{I}} + \tfrac{1}{2}d, \tag{18-2}$$

where d is given in to seconds of arc based on the known plate scale of the photograph. The position of the vertical line on the plate is fixed by means of two auxiliary star images, I′ is obtained before the first exposure and II′ after the second. Then, we can calculate the position of the line ZN relative to the lines I, I' or II, II' which connect the position of the star on the diurnal parallel at known times. By comparing the observed height h_{obs} and the value calculated from the formula

$$\sin h_{calc} = \sin \varphi \sin \delta + \cos \varphi \cos \delta \cos (s - \alpha), \qquad (18\text{-}3)$$

we can obtain the errors in α and δ.

In principle, therefore, we can use the reflecting astrolabe to obtain absolute zenith distances of stellar-like objects. However, observations by this method using short exposures (Figure 18-3(a)) do not give very high accuracy, due to the effect of atmospheric seeing. Neither does the variant of the method which gives the trace by the star instead of point images (Figure 18-3(b)).

We can solve the problem by using the rotating plane-parallel plate 7 (Figure 18-1) to compensate for the diurnal motion of the star over a short period of time as in the camera for observation of artificial earth satellites.

The calculation of the angular velocity, thickness of the plate and permissible angle of rotation of the plate in the usable field are similar to those for the AES camera*, and do not need to be repeated. If the compensating plate orientation is such that the axis of rotation is perpendicular to a diurnal circle and the plate is made to rotate at an appropriate speed, a series of point images of the star are obtained (Figure 18-3(c)). The exposure times can be recorded using a special contact unit, as for the AES camera.

Point images can be measured to much greater accuracy than star traces as in the method used by Zverev in the FVK. The gathering power

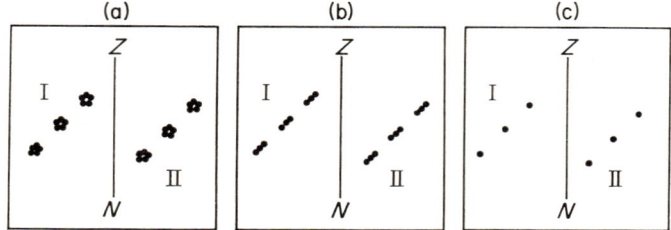

Figure 18-3 Photographs of star-like objects obtained with the reflecting astrolabe: (a) images affected by atmospheric seeing; (b) variant of method used to obtain photographs which gives traces by the star instead of point images; (c) photographs obtained using the rotating plane-parallel plate (see Figure 7-1) to compensate for diurnal motion.

* See Chapter I-15 of this book. (Ed. English version.)

of the instrument is increased by some three to four stellar magnitudes, and its scheme eliminates the problem of flexure. If observations are made exactly at the meridian, the instrument can be used as a vertical circle to determine declination but its capability is greater than this. By making observations at different azimuths, we can determine α as well as δ. We can use compensating filters of differing density to observe the moon, the larger planets and the sun, so that equinoxial corrections made from stellar and solar observations can be obtained with the same instrument, and the effect of systematic errors associated with the use of different instrument is thereby eliminated.

The reflecting astrolabe with a compensating plate therefore seems promising for research in fundamental astrometry, and we hope that the instrument will be used in the near future.

REFERENCES

1. M. S. Zverev, "A photographic vertical circle", *Izv. GAO AN SSSR*, **22**, Issue 1, No. 166, 21 (1960).

CHAPTER **I-19**

Automation of the operation of a photographic zenith tube

D. N. PONOMAREV and YE. M. LAPKIN

Main Astronomical Observatory
Pulkovo

Two photographic zenith tubes (FZT) were built by GOMZ in 1957; one for the Shternberg Astronomical Institute and the other for the Main Astronomical Observatory at Pulkovo. The Pulkovo instrument was installed in the autumn of 1957, but during use a number of serious difficulties were discovered. The design proved to be unsuitable, so that several components had to be changed, including the micrometer screw, mercury level, all the exposure and resetting mechanisms, etc. The new components were made for both FZTs, and only when the modifications were completed, in February 1961, was the photographic zenith tube handed over by GOMZ to the Shternberg Institute where observations began. Debugging, analysis and improvements of the instrument are now under way.

The FZT is basically suited for completely automatic operation: the observing cycle for a single star and the observational program for the course of an observing night requires simple procedures.

For one star, these procedures include the following sequence of operations:

(1) Start motion of the plateholder carriage.

(2) Open the shutter.

(3) Begin print of time.

(4) End print of time.

(5) Close the shutter.

(6) Stop motion of the plateholder carriage.

(7) Reset.

The above cycle is then repeated another three times.

The difference between the cycles to observe different stars lies in the length of exposure time and in the hour angles of the centers of the field. The length of a cycle varies from 78 to 208 sec; from observing one star to the next changes must be made in:

(1) Identification number of the given observing cycle, i.e. the length of exposure times and their hour angles.

(2) The time at which the cycle begins.

(3) The inital position of the plateholder carriage.

(4) The adjustment of the grids in front of the objective.

(5) The rate of motion of the carriage.

The control of the FZT has three or four possible additional commands, such as on and off ventilation, closing the shutters at the end of observations or in the event of rain, and so on.

The first control unit (manufactured by GOMZ) to command the different functions during the cycle was mechanically operated, but was changed for an electrical control unit later on. The basic component in the new control system is the step switch (ShI). Each of its terminals is connected to the corresponding terminal on a board. It is driven by a clock; the terminals are numbered sequentially to correspond to the number of seconds elapsed since ShI was triggered. To trigger at the correct time the motion of the plateholder carriage or for example the opening of the shutter, a given relay is energized by connecting it to the proper terminal.

The mechanism is only semi-automatic, since the cycle and the set of commands which change from star to star is switched on manually. A large department at the Shternberg Institute is now working on a project for complete automation of the instrument. The main problem is to trigger the cycle at the preselected time. Switching on the other commands during the cycle is a much simpler task.

For the control of photographic zenith tubes different solutions have been devised in various observatories. The one used at the Shternberg Institute is shown schematically in Figure 19-1. In this control unit there are two similar systems of counter banks: one for storage and the other for time counting. Each system consists of six counter banks corresponding to seconds, tens of seconds, minutes, tens of minutes, hours and tens of hours. The command can be given by pressing a button, or it can be read from punched cards which are fed in automatically. The time signals (seconds) are given by Time Service clocks. When the counts in the time counter matches with the counts in the storage counter bank, the observational cycle is triggered. As an example, shown in Figure 19-1 are the contacts which are closed for a command to trigger the cycle at 11 h, 30 min, 41 sec.

All the banks, in both the time and storage counters, are exactly the same and differ only in the number of digits read, which is equal to ten for the seconds, minutes and hours, to six for the tens of minutes and tens of seconds, and to three for the tens of hours. Corresponding banks are interchangeable.

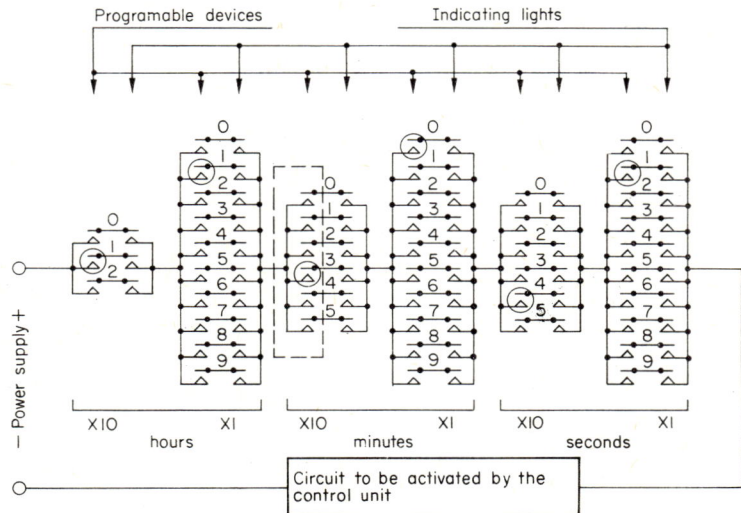

Figure 19-1 Schematic of the control unit for the FZT used at the Shternberg Astronomical Institute. The part of the schematic blocked in (broken line) corresponds to the right portion of the circuitry shown in Figure 19-2. The circles indicate the contacts that should be closed to trigger the observing cycle at 11 h 30 min 41 sec.

The circuit of an individual bank (that for tens of minutes or tens of seconds) is shown in Figure 19-2. The bank consists of the six relays R_0, R_1, R_2, R_3, R_4, and R_5 which are switched on in sequence. The seconds, minutes and hours banks have ten such relays.

At the beginning of the observing cycle one of the relays (R_3 in Figure 19-2) is triggered. An input pulse applied to the circuit causes operation of the next relay (R_4 in Figure 19-2). Each relay in the circuit is locked by its own contacts both from the command pulse and from the dc supply, so that when the pulse ends, the relay remains locked. When another pulse is applied, the next relay is triggered, interrupting the locking of the previous relay and when this pulse ends, the first relay is released. Between pulses only one relay in the bank remains excited. In our example R_5 is followed by R_0 and the whole cycle is repeated periodically sending a single pulse to the next bank when the last relay R_5 is de-energized.

The group of contacts placed at the right of the circuit given in Figure 19-2 corresponds to the contacts in Figure 19-1 blocked in by a broken line. The banks with three or ten relays are similar in their principle of operation. Additional commands can be read from punched cards and the corresponding unit can be activated through a relay.

The control system that we have described can be used whenever a pulse is required after a given set of events has occured (in the case of the FZT, after a given number of seconds).

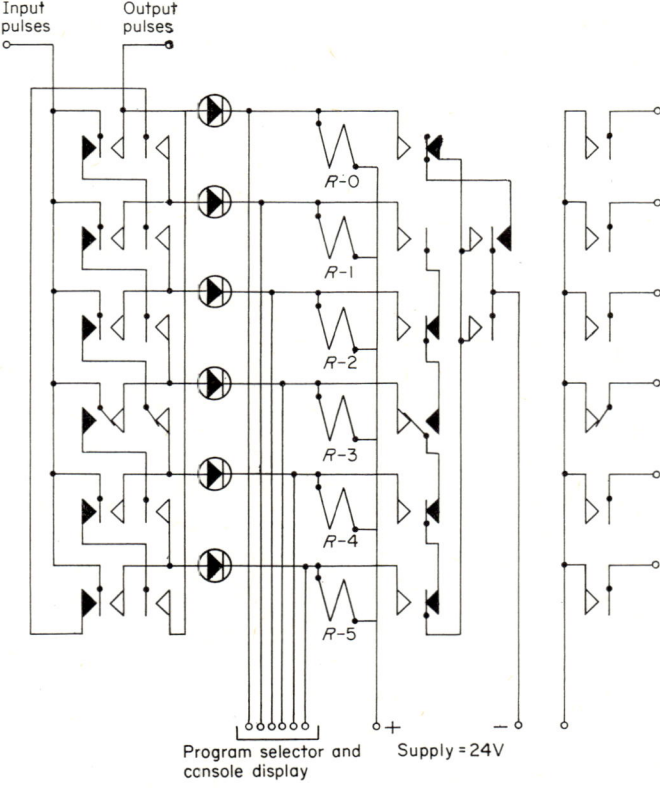

Figure 19-2 Circuitry of an electromechanical counter bank for tens of minutes (or tens of seconds) of the control unit for the FZT.

CHAPTER **I-20**

The control system of the Pulkovo photographic zenith tube

V. A. NAUMOV

Main Astronomical Observatory
Pulkovo

1 INTRODUCTION

A new instrument, the photographic zenith tube, has been widely used for astrometric determination of time and latitude. The Pulkovo photographic zenith tube (FZT) was manufactured in Leningrad in 1957. Observation with this instrument consists of relatively simple operations: switching the on and off motion of the photographic carriage, opening and closing the shutter, turning the objective with the carriage by 180°, etc.

In order to automate the process of observation of a single star and to program observations of a list of stars, two automatic mechanisms were developed during 1958–9. These mechanisms, based on step switches, are mounted in the main control console of the FZT which is placed in an observatory building 110 m from the FZT.

2 AUTOMATIC OBSERVATION CYCLE FOR ONE STAR

The automatic cycle mechanism (MATs) is a programmed unit with a 156-sec cycle which sends commands to the slave mechanisms of the FZT for observation of a single star. It consists of a step-switch (ShI) with 52 positions and a relay switch with three positions. The ShI has four decks with 52 terminals in each deck, and four corresponding insulated wipers. After the MATs have been switched on, second pulses from a crystal clock are fed to the coil of the ShI electromagnet, the wipers of the switch make one step per second, moving from one terminal to the next. The relay switch consists of a single array of terminals and a single wiper.

When the ShI passes the 52 nd terminal, a pulse is applied to the coil of the relay switch and then the relay wiper makes one step. Observation of a star takes 156 sec; during this time the ShI completes three revolutions, the relay takes three steps, and then both switches come to the initial position and stop. In the course of the first revolution of the ShI, the positive of the power supply is connected to the wiper of the first deck of the ShI across the first terminal of the relay switch. In the course of the second revolution, the positive is connected to the wiper of the second deck, and so on. For the observational method adopted at Pulkovo, the power is supplied at the proper time to the proper relay of the slave mechanisms by the corresponding contacts of the ShI. The fourth deck of terminals is not used during the operation. By combining two step-switches we obtain a programmed mechanism for 156 sec and each of the 156 terminals can be used for other operations of the FZT.

In our MATs the following operations are performed:

(1) Start motion of the plateholder carriage.

(2) Open the shutter.

(3) Begin time recording in printing chronograph, neon tube discharges and pulses from the plateholder carriage screw.

(4) End time recording.

(5) Close the shutter.

(6) Stop motion of the plateholder carriage.

(7) Turn the objective with carriage by 180°.

(8) Pull out the tape from the chronograph to separate one cycle (exposure) from the next.

These operations are for a single exposure and take 40 seconds.

For the next exposure the operations are repeated. After the fourth exposure second pulses from the crystal clock are printed by the chronograph to determine the clock corrections; the chronograph tape is then pulled out to separate one star observation from the next. Between exposures the operating modes of the printing chronograph is switched automatically from "on" to "off" to make unnecessary any corrections due to contact width.

In order to observe a single star using the MATs, the observer must set the exposure switch for a definite exposure time and the clock must switch on the MATs at the required time according to the program. After the automatic cycle has been switched on, all the operations are done by the mechanism at fixed times to an accuracy of 0.01 seconds. The operation of all the mechanisms is monitored visually using neon tubes in the main console, or by means of audible signals.

The MATs we have described can easily be adjusted for changes in the sequence of operations or for an increase in their number, as well as for changes in the length of the observation cycle of a single star.

3 THE PROGRAMMED MECHANISM

The MATs must be switched on at the proper time to within 0.5 sec, which requires great concentration from the observer. Various programmed mechanisms have been used in different observatories for observations of a list of stars. Let us briefly describe the Neuchâtel programmed mechanism which is also based on step-switches. It contains the following components:

(1) A time counter, consisting of five step-switches which count successively up to 86,400 sec; the first ShI operates on seconds pulses from the clock mechanism, the second ShI operates on every ten steps of the first, the third on every ten steps of the second, and so on.

(2) The memory, consisting of six step-switches.

(3) A mechanism to transfer the initial data for the star from punched tape to the memory, consisting of a relay matrix and distribution selector.

The relay matrix converts the five-digit code of the punched tape into the decimal code, and the distribution selector distributes the initial data for the star among the six switches of the memory. Five of these switches are set to the time when the MATs is to be switched on. The sixth is used for programming auxiliary operations. When the time counter matches the value stored in the memory, a coincidence pulse switches on the cycle, and the initial data for the next star is then sent to the memory.

A programmed mechanism can be used in various ways. For the observational method and the number of stars listed in the Pulkovo program it has to provide the initial data, the times for switching on the MATs to within 0.5 sec and the exposure time for 172 stars; it has to program the travel span of the photographic carriage for several stars and the time recording for the selected stars.

A semi-automatic programmed mechanism (PM) for a ten-minute observation cycle was developed and built at Pulkovo in 1958; this PM uses switches. From 1960, observations were carried out using the PM with a full year program.

The Pulkovo PM has the following components:

(1) A time counter consisting of two ShI.

(2) A switchboard memory.

(3) A star counter consisting of four ShIs.

All counters operate continuously.

The first ShI in the time counter which has 50 positions, is driven by second pulses from the clock. Once every 50 sec it sends a pulse to the second ShI with 48 positions. The terminals in the first deck of both switches are terminated in the switchboard. The remaining decks are used to monitor the operation of the time counter and for other tasks. The positive of the power supply is connected to the wiper of the first ShI. The wiper of the second ShI is connected to one terminal of the relay coil which includes the MATs and the other terminal of the coil to the negative of the power supply. The period of operation of the time counter is 40 minutes.

The four ShIs of the star counter are connected in series in a single 172-position switch. When observation of a given star ends, the star counter receives a pulse and makes one step. The first 50 stars of the program are counted by the first ShI, the second 50 stars, by the second ShI, and so on. The contacts of all four decks of the star counter is terminated in the switchboard. The first two are used to program the switch-on times for the MATs; the wipers of these two decks are connected in series. The other arrays are used for programming auxiliary operations.

The number of ShIs in the star counter is determined by the number of stars in the program. An increase in the number of switches does not reduce the reliability of operation of the PM since only one star counter ShI operates for one star. In fact, the Pulkovo PM consists of three ShIs, two for the time counter and one for the star counter.

The MATs is switched on at the time of culmination of the star less half the length of the observation cycle for one star. Every switch-on time is defined by the position of the time counter switches. This can be expressed by a four-digit number, the first two digits corresponding to the position of the first switch (between 0 and 49) and the last two corresponding to the position of the second switch (between 0 and 47). To obtain this four-digit number, the largest possible multiple of the time counter period is substracted from the switch-on time and the result, in seconds, is divided by 50 to within the nearest unit. The remainder and quotient, represented by two-digit numbers, give the required four-digit number.

Let us consider an example. Suppose that the switch-on time for star No. 13 is 2 h 43 min 35 sec. Substracting the largest possible multiple of 40 from this, we obtain 3 min 35 sec or 215 sec. On dividing this by 50, we obtain the remainder 15 and quotient 04; the required four-digit number is therefore 1504. To program the star No. 13, we connect on the switchboard the 15th jack of the first time counter ShI to the 13th jack of the first deck of the star counter and the 4th jack of the second time counter ShI to the 13th jack of the second deck of the star counter (see Figure 20-1). When the time counter reaches the position 1504, that means the first switch reaches the position 15 and the second the position 4, the positive of the power supply is applied across the wipers of the star counter to the slave relay. The MATs is switched on, the wipers of the third and fourth deck of the star counter apply power through their switchboard cords to the relays of the slave mechanisms. At the end of observations, the star counter makes one step, turning off the line of the 13th star, and is set in the initial position for the next star, and so on. In the course of a complete cycle of the PM, the pair of wipers makes 172 steps, successively selecting one contact per step. Therefore, although the time counter has a cycle of 40 minutes the PM could handle a 24-hour cycle.

The Pulkovo PM completely replaces the observer, who has only to switch on the control console before observations, change the photographic plate and clean the mercury level. The last two operations are done once

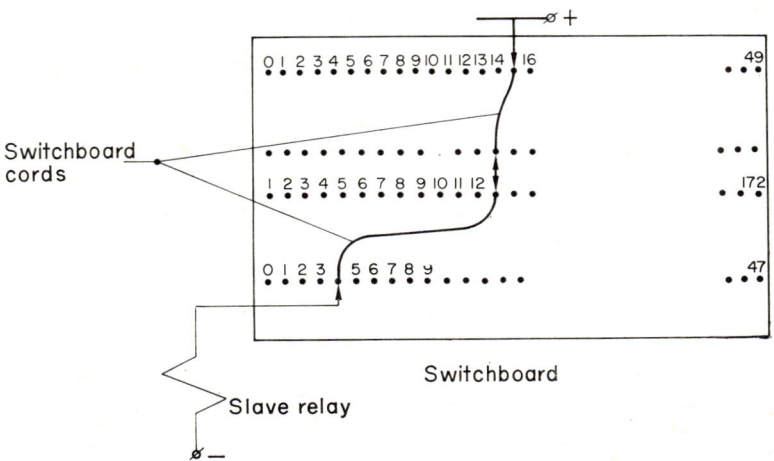

Figure 20-1 Connections of the PM to start the MATs for the observing program of the star No. 13 at 2 h 43 min 35 sec.

every six hours. The cover of the FZT building is closed and all the mechanisms are de-energized automatically by the PM.

The circuits of the Pulkovo and Neuchâtel time counters are of the same type, and due to the use of step-switches they are simple and reliable in operation. The number of switches in the Neuchâtel time counter (and correspondingly in the star counter also) can be reduced without detriment. However, the logical designs of the PMs differ basically. The Neuchâtel mechanism has a single-shot memory, containing the initial data for only one star at a time, so that the storage in the memory must be changed before another star can be observed. The fairly complex mechanism to transfer the initial data from punched tape to the memory, as well as the use of step-switches for the memory, reduce reliability. The memory of the Pulkovo PM contains the initial data for the whole list of stars, so that the circuit is simpler and the reliability greater. The use of such a memory, in turn, involves the difficult problem of retrieving the initial data for a given star. In the Pulkovo PM this retrieval process is carried out by the star counter. The duration of the operation cycle of the PM is determined only by the selected switchboard connection, and can vary within broad limits, from 5 minutes to 5 days, without the slightest change in the circuit. We can use the mechanism to program parallel observations on several instruments; for instance, at Pulkovo observations on the Mikhailov polar tube have been programmed as well as observations on the FZT.

For the programmed FZT, the basic operation is to switch on the MATs only, but in practice, particularly when a complete observing program is run, several operations are required. These can also be performed by the PM, which can be programmed to switch on several hundreds of different operations at different times over several days. The length of the operations themselves is arbitrary. In this case, the "slave relay" sends a pulse to the "star counter" (operations counter) when the operations are to begin. The switching scheme for the switch-on time and the operations themselves is unchanged. The power supply is connected to the wipers of the third and fourth decks in the operations counter at the time when the slave relay is triggered. Two different operations can be switched on at once (both the number of operations in one switching-on and the number of switchings-on can be increased by altering the operations counter circuit). The slave relay also energizes the operations counter, which makes a step when the coil of the ShI is de-energized.

The MAT and the PM have both proved to be reliable, simple and convenient in use.

CHAPTER I-21

The automatic stellar photoelectric photometer of the Latvian Astrophysical Laboratory

A. P. KUNDZIN

Astrophysical Laboratory
Riga

1 INTRODUCTION

Programs in standard and multicolor photoelectric photometry require, in general, the measurements of a large number of stars. To carry out an observation of a star, the observer must sometimes perform several operations (such as, changing the diaphragm and filters, inserting a standard light source, etc.). The observation process would take less time if these operations were performed automatically, and errors caused by fatigue and lack of concentration would be eliminated.

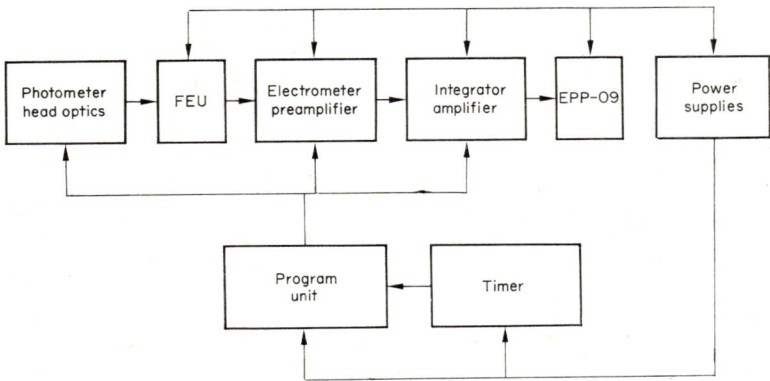

Figure 21-1 Block diagram of the automatic stellar photoelectric photometer of the Latvian Astrophysical Laboratory.

The present automatic stellar photoelectric photometer is based on a concept developed by Prof. V. B. Nikonov, and was constructed by groups at the Astrophysical Laboratory of the Latvian Academy of Sciences and at the Crimean Astrophysical Observatory, including B. P. Abrazhevsky, A. V. Bruns, A. P. Kundzin, A. I. Lobanov and V. B. Nikonov. The photometer was designed to perform automatically the observing cycle of a photoelectric or multicolor photoelectric measurement of a star.

The photometer block diagram is shown in Figure 21-1. For normal operation, the electronics and electrical units are placed in a closed cabinet which maintains the required temperature and humidity. The cabinet is in the telescope building.

2 OPTICAL SYSTEM

The photometer head is mounted on the 200 mm refractor APR-31 ($f/15$). The optical layout of the head is shown in Figure 21-2.

The eyepiece O_1 and the flat mirror 3 is used for setting the telescope on the star. The field of view of the eyepiece is about 30'.

The eyepiece O_1 is mounted on a special holder which can be moved by micrometer screws in right ascension and in declination, like those of the Ritchey plateholder. Thus, the crosswire K could be set with respect to the telescope image. The crosswire is usually set on the center of the field of view of the telescope and guided on the given star. If the star is too faint to be seen by eye through the eyepiece, the crosswire is moved away from

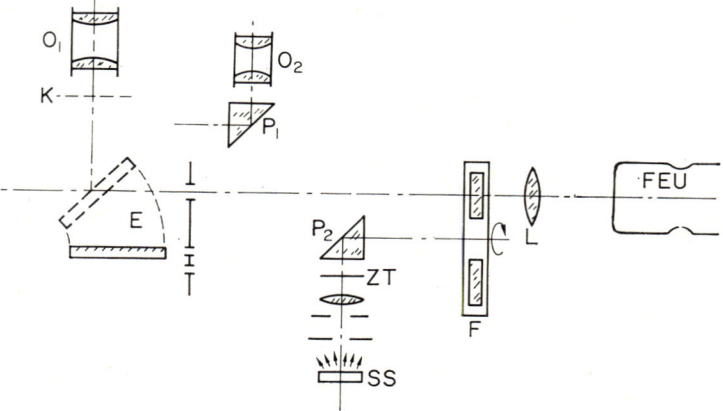

Figure 21-2 Layout of the photometer head optical system.

the center of the field and set on an adjacent visible star so that the star to be measured is in the center of the field. Knowing the coordinates of the guide star and the one to be measured and the constant of the micrometers, we can easily determine the setting of the eyepiece and the crosswire.

On the focal plane of the telescope is located a series of diaphragms D. The diaphragms are selected in pairs: one for photometry of the star and one for photometry of the background. The shapes of the diaphragms are shown in Figure 21-3(a), (b). Both diaphragms have equal areas. For photometry of the background (Figure 21-3(a)), the central disk covers the star image; for photometry of the star the diaphragm shown in Figure 21-3(b) is used. The diaphragms are changed automatically. Prior to observing, the proper pair of diaphragms are selected.

Behind the diaphragms is the monitor eyepiece O_2 and the total reflection prism P_1. With the prism inserted in the optical path and using the eyepiece, we can observe the position of the star in the diaphragm. The eyepiece is also used to check that there is no extraneous star in the background diaphragm. The filter wheel F has four positions with filters and one closed position (to measure the dark current of the photomultiplier tube). The filter wheel can be rotated automatically in either direction. The Fabry lens L projects on the photocathode of the photomultiplier FEU, the image of the telescope objective illuminated by the star under measurement.

Almost any kind of head-on photomultiplier tube may be used in the photometer. The mounting of the photomultiplier allows position adjustment parallel and perpendicular to the direction of the optical axis of the

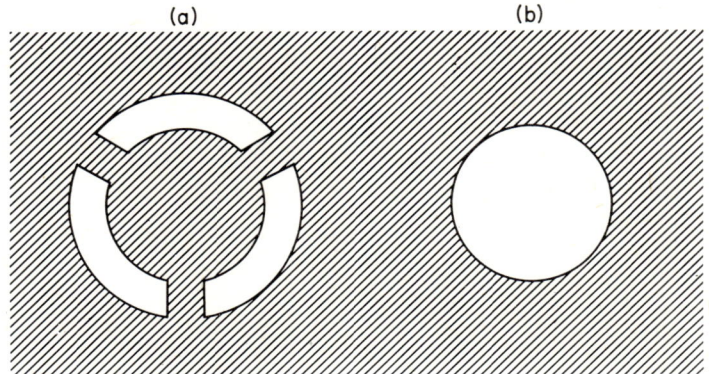

Figure 21-3 Diaphragms for the photometer head: (a) for the sky background channel; (b) for the star channel.

photometer, so that the image can be focused and centered on the photocathode. The photomultiplier is dried by silica gel and cooled by dry ice. A thermistor is placed on the glass envelope next to the photocathode to measure the temperature. The photomultipliers used are the FEU-19 and the FEU-29, as well as an experimental one (photocathode diameter 10 mm).

A luminous light source SS is used to check the absolute sensitivity of the photometer and to measure the brightness of the sky background (a phosphor excited with radioactive β emission). For photometry with the light source, the prism P_2 is inserted in the optical path. It blocks the light from the star, while the light from the source reaches the photocathode. The images of the light source and the exit pupil are projected to the same area of the photocathode surface. The prism P_2 is inserted and removed automatically.

In photometry of stars of different brightness, the sensitivity of the photometer has to be adjusted. Therefore, for photometry using the light source SS a constant sensitivity is automatically selected and the dark current recorded for this sensitivity (by automatically closing the shutter ZT in the optical system of the light source). All the automatic mechanisms may also be controlled manually.

3 THE ELECTROMETER PREAMPLIFIER AND INTEGRATOR AMPLIFIER

The photometer head together with the FEUs and an electrometer preamplifier is mounted on the telescope tube. Flexible cables supply the power and connect the head to the amplifier and program unit.

The preamplifier and integrator amplifier are similar to the one reported by Gardiner and Johnson [1] with a few changes. The preamplifier is placed in a hermetic metallic enclosure with a drier. Thus preamplifier is powered from batteries. A relay is used to switch on the fixed sensitivity for measurement of the light source SS at the same time as the prism P_2 is set into the operating position.

The integrator amplifier can operate in two modes:

(1) As dc amplifier, and

(2) as an integrator.

Any switching in the amplifier (start of integration, recording, discharge of the capacitor of the integrator, etc.) is done automatically on command from the program unit. The amplifier output drives the recorder. It is a potentiometer type EPP-09; it has a 10 mV scale, and the pen takes 2.5 sec to sweep the full scale.

Despite the large input resistances, the time constant of the amplifier is of the order of a tenth of a second, because of the strong negative feedback. Thus, the time constant of the photometer as a whole is determined in practice by the inertia of the EPP-09.

The tubes of the amplifier undergo aging. After an hour of operation, the drift of the amplifier does not exceed 0.1 mV h^{-1} on the paper chart of the EPP-09.

Unfortunately, the integrator-amplifier in its present form does not give much gain for the integrated signal on the capacitor (compared with the dc mode of amplification) because the capacitor is partially charged by stray currents. To eliminate this effect, some of the ordinary tubes will be replaced by electrometer tubes. The amplifier is supplied from a stabilized power supply.

4 OPERATION OF THE AUTOMATIC CYCLE AND THE PROGRAM UNIT

The photometer was planned to carry out a large-scale program in multicolor photometry of variable stars. Typical automatic programs include observations of the star and sky background through various filters, observation of the light source SS and recording of the dark current. Each measurement in the automatic observation cycle is made twice, symmetrically in time with respect to the mean time of the cycle. In the data reduction the mean value of the two measurements are taken.

In the automatic cycle, the duration of a star measurement through a single filter can be selected to be 16, 32, 64 or 128 seconds. The background measurement is half as long, i.e. 8, 16, 32 or 64 sec respectively, and for the light source SS it is 16 seconds. Using the integration method, the last four seconds of each measuring cycle are used for recording the integrated signal.

With the program unit we can operate with or without measurement of the SS. The length of the cycles can be changed, the number of filters used in the automatic cycle can vary from one to four, the mode of the operation of the amplifier may be altered. The circuitry used for programming consists of a timer, step-switches, program switches and output relays. The timer is driven by a synchronous motor and generates pulses to control the program unit. The timer gives single intervals of time of up to two minutes with an accuracy of not less than one hundredth of a second (a very important factor in integration). The multicontact step-switches at given settings, switch on different combinations of the operating circuits to carry out the required observations.

The very complex system of switching and locking allows changing of the measuring time for a single observation, the number of filters used, mode of operation, measuring of the SS, etc.

The output relays control the electromagnets of the slave mechanisms.

The program switches are located in the control console of the program unit, which also contains the tubes and indicators to show the actual positions of the mechanisms in the photometer head, the "start" and "stop" buttons ("stop" for interruption of the program) and other control components. After completion of the program, the complete system is automatically reset and an audio signal is sounded. For convenience, there is a portable control console which partially duplicates the main console. The readings of the EPP-09 are also duplicated on the portable console.

After experience in using the program unit a new and improved unit was designed. Its block diagram is shown in Figure 21-4. It is planned to have an independent channel to process pulses from punched cards, so that automatic programs for complete observations can be prepared using any combination of the components.

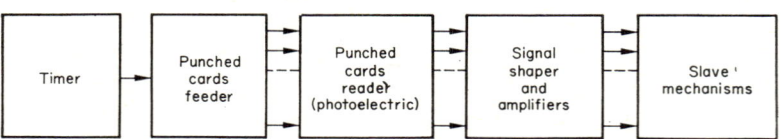

Figure 21-4 Block diagram of the new designed program unit.

5 METHOD OF OPERATING THE PHOTOMETER

The automatic photometer is operated as follows. The observing program is set up by the control toggle switches (if several stars are to be observed using the same program, the program need only be set up once). The observer then sets the telescope on the star and selects the sensitivity of the photometer to the proper value by means of a switch in the portable console; when the "start" button is pressed, operation of the automatic program and the recording on the paper chart begins.

After the program has been completed, the observer sets the telescope on the next star, selects the sensitivity and again starts the automatic program.

6 CONCLUSION

Speeding up observations with the automatic photoelectric photometer is particularly important due to Latvian weather conditions, where we have only few photometric nights. Rapid observation of variable stars and comparison stars also reduces errors due to changes in atmospheric extinction.

Moreover, the automatic photometer, together with the automatic setting systems for telescopes provides experience for the design in the near future of completely automatic observatories.

REFERENCES

1. A. J. Gardiner and H. L. Johnson, "Integrator-type dc amplifier", *Rev. Sci. Instr.*, **26**, No. 12, 1145–7 (1955).

CHAPTER I-22

An automatic photoelectric polarimeter and other instruments

L. V. KSANFOMALITI

Abastumani Astrophysical Observatory
Kanobili Mount

During recent years, electronic instruments for the visible and near infrared regions have been under development at the Radio Astronomy Laboratory of the Abastumani Astrophysical Observatory. The following very brief description of some of the new instruments indicates possible methods of increasing the efficiency of telescopes of moderate size and of automating the observing process.

1 AUTOMATIC PHOTOELECTRIC POLARIMETER

Our purpose in designing the polarimeter was to have an instrument which could perform fast measurements and, in contrast to the existing instruments [1], would give a direct indication of the amount of polarization (in percentage) and the angle of polarization (in degrees), over a wide range of object brightness. The instrument was designed to be highly sensitive and very fast (operating in a fraction of a second); it had to be able to resolve small areas on the surface of the celestial objects and in various regions of the spectrum. Also it should allow measurements with high accuracy and stability.

The instrument is used in connection with the 40-cm refractor; it was put into operation in the spring of 1960.

The specifications of the polarimeter are:

(1) Star brightness range: 5^m–11^m.

(2) Minimum time constant: 0.10 sec.

(3) Mean of three readings for a 7^m star and an 8% of polarization: 0.85%; and for 6.2%: 6.2%.

(4) Accuracy in the position angle measurement for a 7^m star: not less than $\pm 3°$.

(5) Largest field of view (diaphragm): $1'$ (2 mm).

(6) Smallest field of view (diaphragm): $1''50$ (0.05 mm).

(7) Range of time constants: 0.1 to 20 sec.

One problem was the calculation of the amount of polarization P_{mea} from the measurements of the intensity $b_{\max}$ in the plane of vibration and the intensity $b_{\min}$ at right angles obtained through the rotating analyzer. The amount of polarization P_{mea} is expressed by:

$$P_{\text{mea}} = \frac{b_{\max} - b_{\min}}{b_{\max} + b_{\min}}. \tag{22-1}$$

The other problem was to measure the position angle φ which had to be read from a certain reference plane.

If the analyzer rotates with the angular velocity ω, the polarized light received on the cathode of the photomultiplier tube FEU is modulated. Then the voltage output U_n of the FEU is expressed by:

$$U_n = U_0 \frac{B_n \cos^2 \omega t}{B_0 + B_n}, \tag{22-2}$$

where U_0 is the voltage output given by the luminous flux $B_0 + B_n$, B_0 is the unpolarized luminous flux from the celestial object, and B_n is the polarized flux.

If the amplification is automatically regulated and U_0 is kept constant regardless of the brightness of the object, and if we modulate the light signal by means of a mechanical chopper, then the polarization can be measured by a dc voltmeter.

Moreover, the angle φ can be measured as the elapsed time between the given position of the analyzer (polaroid) and the position of mean transmission. The angle φ is given as a fraction of the time T required to complete half a revolution:

$$\varphi = 2n \frac{t_c - t_\delta}{T} = \omega(t_c - t_\delta), \tag{22-3}$$

where t_c is the time at which the dU_n/dt is a maximum and t_δ is at the time that it is zero, and for the same period T.

It should be indicated that the automatic regulation system (ARU) can also be used to obtain a rough measurement of the stellar magnitude m.

Figure 22-1 shows the schematic of the optical train in the polarimeter head. The chopper M which has 144 slots and the analyzer P, modulates the luminous flux at 1,440 Hz. The chopper, the analyzer and a magnetic shielding (the shielding has two slots and it is represented in Figure 22-1 also by M) are driven by the motor DV. The light signal is thus modulated at two frequencies, one carries the brightness information of the star and the other the angle and amount of polarization. A mirror ED with a set of diaphragms is located at the focal plane; by reflexion on the mirror ED and by means of the prism and the optical system MS the observer can see through the field of the polarimeter. The sector SF holds the optical filters; the coils indicated in the schematic and the magnetic shield on the plane of the chopper M generates the reference signal for the phase controlled rectifier.

It can be demonstrated that for a given photocathode responsivity γ a minimum luminous flux F_{min}, a modulation depth r, and a photocathode current i_e generated by a single photoelectron, modulation frequencies above

$$f_{max} = \frac{F_{min}\gamma r}{i_e} \tag{22-4}$$

cannot be used, since the mean output rate of single photoelectrons becomes lower than the modulation frequency. In the polarimeter that we are discussing, the mean number of photoelectrons per second for an 11^m G 2 type star, is approximately 1,440.

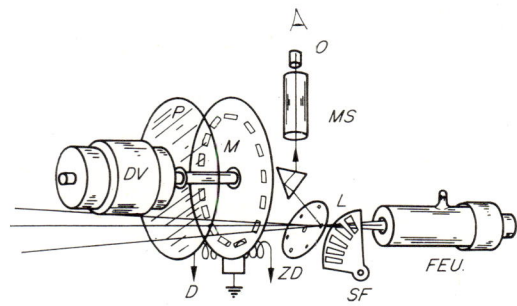

Figure 22-1 Schematic of the optical train for the automatic photoelectric polarimeter developed at the Abastumani Astrophysical Observatory: DV, electric motor; P, polaroid disk; M, light chopper (144 slots) and magnetic shield with two slots; ED, mirror with a set of diaphragms; SF, filter holder; FEU, photomultiplier tube; MS, viewing system; O, observer's eye; L, small light bulb and D, sensor for reference signal.

Since light modulation is being mentioned in this contribution, we might say that the magneto-optical modulator is the most promising. Transparent films of special ferrites give a sharp change in the transmittance for a given value of magnetic field. Mechanical modulators (rotating choppers, tuning forks, etc.) are somewhat less convenient, but very reliable. Systems using modulation by polarization are often inapplicable in principle.

Finally, modulation can be achieve also by modulating the gain of the FEU, but this method has two major disadvantages: the dark current is also modulated and when the load resistance of the FEU is large, pick-up from the modulation voltage could be much larger than the modulated signal. This method can therefore be applied only for very large light signals; in this case it is desirable to use special FEUs which are designed for this type of modulation (such as the FEU-16).

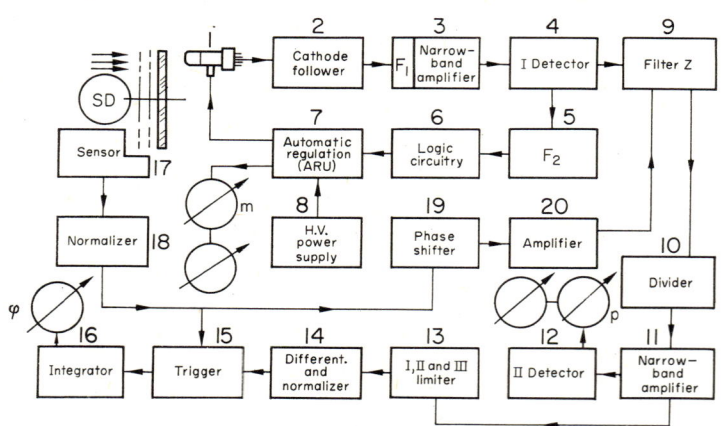

Figure 22-2 Block diagram of the automatic photoelectric polarimeter: P, amount of polarization; φ, position angle; m, stellar brightness.

The balance stage after the FEU requires compensation of the unstable dark current which has a perceptible drift; this problem can be greatly reduced by modulating the signal by means of a circuit which is analog to a mechanical chopper.

Figure 22-2 gives the block diagram of the automatic photoelectric polarimeter. The main circuit of the polarimeter is given in Figure 22-3; the FEU (L_1) is connected as previously described by the author [2], to

An automatic photoelectric polarimeter and other instruments 235

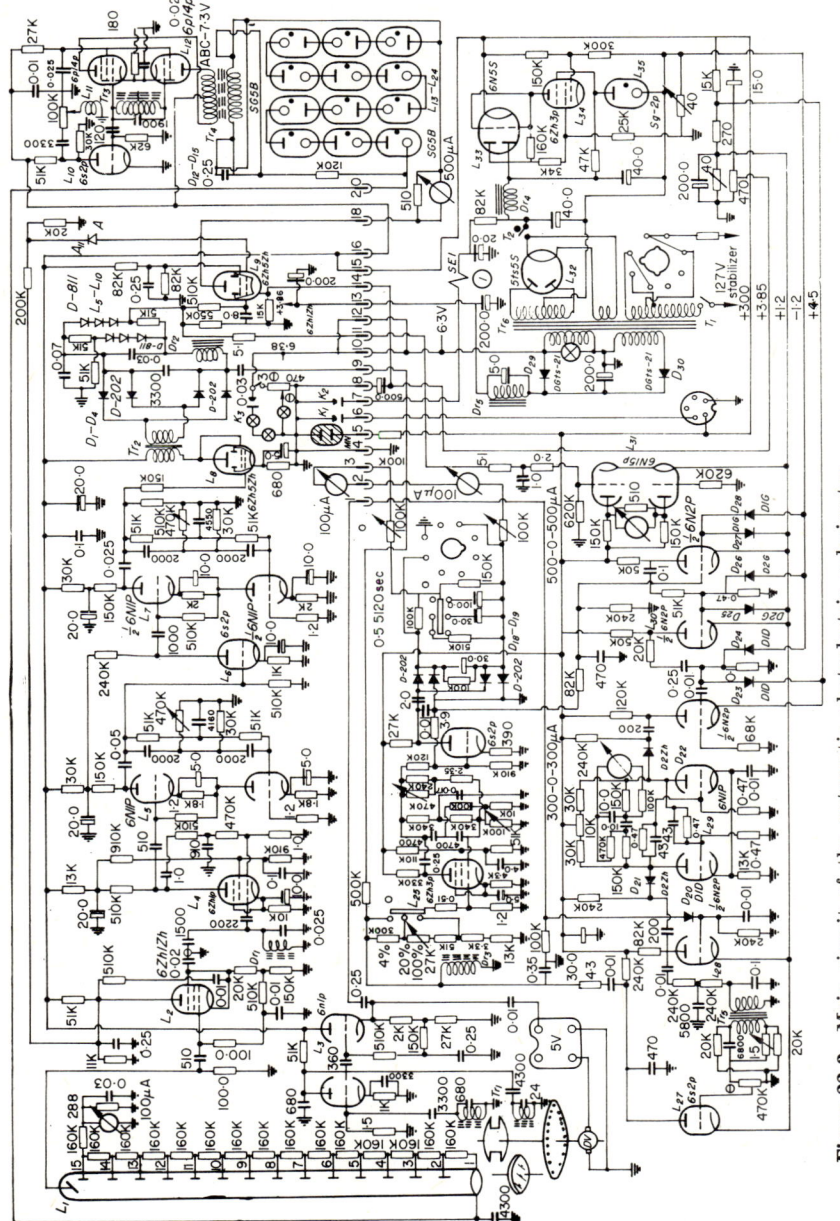

Figure 22-3 Main circuit of the automatic photoelectric polarimeter.

give the maximum signal-to-noise ratio at very high gain. To match the load, the electrometer tube L_2 is connected as a cathode follower. The narrow-band amplifier L_4–L_8 amplifies the carrier and one side band of the voltage U_n (see Eq. 22-2); the first detector uses the diodes D_1–D_4 (temperature compensated). The rectified signal is filtered and fed to a logic circuit using the Zener diodes D_5–D_{10}. The output of this stage is fed to the automatic regulating system ARU which uses the vacuum tube L_9; to the grid of this tube is connected the appropriate phasing network to prevent auto-oscillations. The output of L_9 controls the current through the voltage divider of the FEU, thus, the point of operation of the FEU depends on the input signal.

The FEU has a separate high-voltage power supply (2,100 V) which includes the vacuum tubes L_{10}–L_{12} (rf section operating at 25 kHz), a rectifier D_{12}–D_{15} and a stabilizer using glow discharge tubes L_{13}–L_{24}. The characteristics of these tubes are better than those of the SG-7s or SG-9s.

The description up to this point covers the blocks No. 1 through No. 8 of the diagram given in Figure 22-2.

The component

$$U'_n = \tfrac{1}{2} U_n, \qquad (22\text{-}5)$$

is fed through a filter to the voltage divider connected at the input of the narrow-band pass amplifier L_{25}–L_{26}; this amplifier has a strong negative feedback. The output of the amplifier is fed to the second detector D_{16}–D_{19}. Two panel meters connected at the output of this detector measures directly the amount of polarization P_{mea}.

The pulses from the three-stage limiter–amplifier L_{30}–$\tfrac{1}{2}L_{28}$, after differentiation, are fed to one of the inputs of the trigger L_{29}. The magnetic shielding driven by the motor DV, by changing the coupling between the coils Tr_1 (connected to the tube L_3) generates two pulses per one revolution of the motor. After normalization and amplification ($\tfrac{1}{2}L_{28}$) the pulses are fed to the second input of the trigger L_{29}. The panel meter in the integrating stage which is connected between the anodes of the tubes L_{31}, is graduated in degrees and gives a direct measure of the position angle [3].

From the circuit diagram (Figure 22-3), the function of the other components is self-explanatory.

The output data of the polarimeter are recorded by photographing the panel instruments with a motion picture camera in the single frame mode of operation. The camera is controlled by means of a switch and is driven by a 20-joule pulse.

2 AUTOMATIC PHOTOELECTRIC PHOTOMETER

The automatic photometer was developed specially for measuring very faint stars in connection with the 700-mm meniscus telescope. It is based on the principle of synchronous detection and integration. Since the system is more or less straight forward, we give only the block diagram (Figure 22-4).

The front optics, including diaphragms and the filters (UBV system) is similar to that of the polarimeter (see Figure 22-1). The modulator is driven by a small synchronous motor; the modulation frequency is fairly low (27 Hz) since a mechanical demodulator is used. The reference voltage is generated by an optical sensor. The output of the FEU is fed to a narrow-band amplifier; the following stage is a phase controlled rectifier using a polarized relay. Two integrating circuits are connected to the inputs of the amplifier which feeds the recorder (we plan to use the ETsPV-2 type).

A special programmer is used to control the measuring cycle: (1) record the "zero" value; (2) integrate and record; (3) record the difference between the integrated and the "zero" values; (4) reset, and (5) record again the "zero" value.

With a 5-sec time constant, 13^m stars can be measured with this automatic photometer.

The operation cycle of the photometer is triggered by pressing a button. After completing all operations of the cycle, the photometer is reset to the starting position. The gain of the instrument can be set by changing the FEU point of operation.

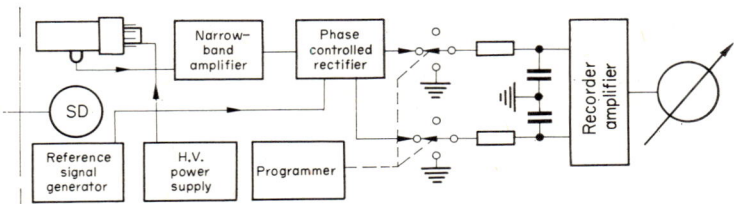

Figure 22-4 Block diagram of the automatic photoelectric photometer developed at the Abastumani Astrophysical Observatory.

3 ANALOG COMPUTER TO SOLVE THE EQUATION

$$P = \sqrt{\left(\frac{P_{\text{mea}}}{1 - km}\right)^2 - f^2(m)}.$$

In the polarimeter which we have described, it is impossible to use a phase controlled rectifier to reduce the noise level. This impossibility is due to the arbitrary position of the polarization plane which originates a do noise level at the output (Figure 22-5).

The noise level during the measurements of faint sources proves to be considerable and therefore has to be taken into account. For the reduction of our data we use the following expression:

$$P = \sqrt{\left(\frac{P_{\text{mea}}}{1-km}\right)^2 - f^2(m)} \qquad (22\text{-}6)$$

where P is the amount of polarization, P_{mea} is the measured amount of polarization, m is the stellar magnitude, $f(m)$ is the noise function of the polarimeter (see Figure 22-5) and k is a constant.

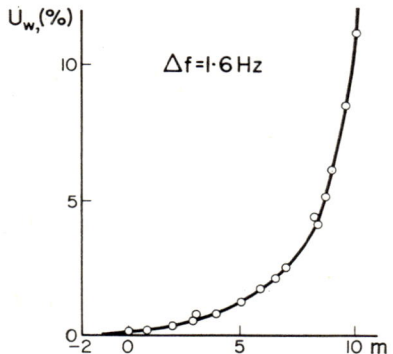

Figure 22-5 Noise function $f(m)$ given as percentage U_w of the polarimeter output signal versus stellar magnitude m and for a bandpass $\Delta f = 1.6$ Hz.

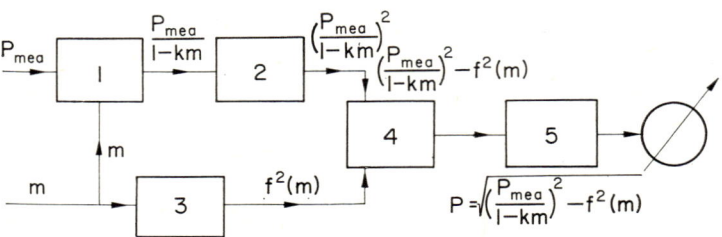

Figure 22-6 Block diagram of the analog computer to obtain directly the values of polarization P from the photoelectric polarimeter developed at the Abastumani Astrophysical Observatory.

The block diagram of the analog computer is given in Figure 22-6. The input to the computer are the values of the polarization P_{mea} and the stellar magnitude m. The circuitry represented by the block No. 1 corrects P_{mea} for the small error introduced by noise in the carrier frequency (the correction is linear because of the exponential relationship between the FEU amplification and the applied voltage). A vaccum tube 6Zh10P with two control grids, is used in this stage. The circuitry represented by the block No. 2 squares the output of block No. 1. Block No. 3 represents a functional transformer which squares the function $f(m)$ for the given value of m fed at the input. A segment-diode circuit is used in the functional transformer to approximate $f(m)$ by ten linear segments.

The block No. 4 represents an anode repeater which delivers the difference of the input signals and the block No. 5 represents a segment-diode circuit which is used to obtain the root square of the input signal. The main problem of this analog computer is the lack of stability; due to this the circuitry of the block No. 5 had to be dropped.

For this system, an automatic time constant control will be developed which will adjust the constant as a function of the stellar brightness.

4 A PARAMETRIC SPEED STABILIZER FOR A DC MOTOR

The high limiting magnitude reached by the polarimeter which we have described is achieved mainly by the narrow bandpass characteristic of the amplifier at the modulation frequency. Thus precautions have to be taken to stabilize the speed of the motor that drives the chopper. A synchronous motor fed from the line originates faulty readings due to line frequency drifts. This causes drift in the measurements of P_{mea} (between 2% to 3% of the measured value) together with large phase shifts, leading to very large drifts in the measurements of φ. The use of a local vacuum tube RC oscillator and an amplifier to drive the motor is also affected by frequency drifts. A crystal oscillator would be a solution, but it would still be desirable to be able reset the frequency for large changes in the filter resonance frequency.

All these variants were tested and a paradoxical solution was found: the dc motor D-150 with an electronic parametric frequency stabilizer. The circuitry of the stabilizer is given in Figure 22-7.

Pulses are fed to the input of the stabilizer from the pulse generator (Figure 22-3). The first stage using the tube 6Zh8 with a diode at the input circuit, amplifies the signal and makes it roughly symmetrical. The 6Zh4 tube is used as an electronic frequency meter [4]. The voltage at the detector output (D-202) is proportional to the input frequency. The voltage taken from the potentiometer 10 K is subtracted from it, and the difference is fed to the grid of the 6N15P tube which regulates the bias of the GU-50 tubes. Due to the small amplification coefficient (~ 6) of the 6N15P stage, drift has no effect on the operation of the stabilizer.

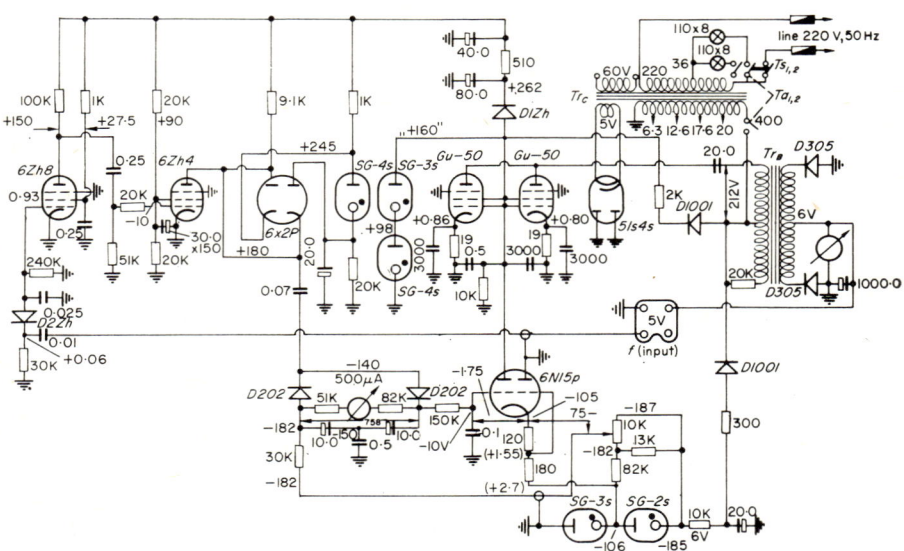

Figure 22-7 Circuitry of the parametric speed stabilizer for the modulating motor in the automatic photoelectric polarimeter.

The circuit of the output stage is unusual; the supply is ac, and the dc component is blocked by the 5Ts4S diode and the 20.0 μF capacitor. With a rectifier using D-305 diodes feeds the dc current to the motor D-150 at 5 V, 6 A.

It is clear from the analysis of the circuit that any pulse rate increase at the input of the stabilizer will correspond to a decrease in the power fed to the motor, and vice versa.

The stabilizer has excellent performance; this method of speed stabilization can be recommended in all cases where the speed of an electric motor must be maintained constant [5].

REFERENCES

1. V. N. Mikhkyura, *A Differential Recording Photometer* (Differentsialnyi samopishushchii fotometr) (Author's Abstract of his Dissertation), Leningrad (1954).
2. L. V. Ksanfomaliti, *Byull. Abastum. Observ.* (1955).
3. L. V. Ksanfomaliti, *An Automatic Electronic Polarimeter* (Avtomaticheskii elektronnyi polyarimetr), Author's Certificate No. 146070.
4. *The IC h-6 Frequency Meter: Technical Description and Instructions for Use*, MRTP (1955).
5. N. P. Barabashov, *Study of the Physical Conditions on the Moon and Planets* (Issledovaniye fizicheskikh uslovii na Lune i planetakh), Kharkov (1952).

CHAPTER I-23

A stellar photometer and a spectrometer for the 1-2.5 micron region

V. I. MOROZ

Shternberg Astronomical Institute
Moscow

1 INTRODUCTION

Astronomical observations in the infrared region of the spectrum ranging from 1 to 1,000 μ were rare until recently, despite the fact that they are of great interest. For example, interstellar dust has an absolute temperature of a few tens of degrees and its radiation peaks at about 100 μ. If we could measure the radiation of interstellar dust, our knowledge of the structure of the Galaxy would be greatly increased. Variables of long periods, stars in the M, N and R classes, radiate almost all their energy in the spectral region of wavelength longer than 1 μ. The molecules in planetary atmospheres give strong absorption and radiation in the infrared region.

Although it is difficult to carry out observations in the infrared range, there are two main reasons for the present state of infrared astronomy. Firstly, the greater part of the infrared region is blocked by the atmosphere of the Earth. The use of balloons, artificial earth satellites and rockets (particularly for the study of the nearer planets) will help to overcome this problem. Secondly, detection techniques for infrared radiation (discounting the shorter wavelength region 0.7–1.1 μ) have a higher noise equivalent power (NEP)* than that achieved in the optical and radio regions of the spectrum. It is possible that real progress will result from quantum radiophysics.

---

* The NEP is defined as the rms value of the sinusoidally modulated radiant power falling upon a detector which will give rise to an rms signal output equal to the rms noise output from the detector. (Ed. English version.)

With the present state-of-the-art of infrared technology, the most accessible region is that of the wavelength shorter than 2.5 µ, for which the detectors have relatively high detectivity (D)* and the problem of background produced by the radiation of the instrument and the sky is not as acute as in the longer wavelength region. We shall describe two instruments, a stellar photometer and an infrared spectrometer. We have used these two instruments to make some astronomical observations in this region, including the measurement of the infrared radiation from the Crab Nebula in the region 1–2.5 µ [1], measurements of the night sky spectrum [2, 3] and of the Jupiter and Saturn spectra.

2 STELLAR PHOTOMETER

The stellar photometer and the spectrometer had identical radiation detectors and amplifiers. Two types of detectors were used: a lead sulfide photoconductor [4] and a germanium cell [5]. Both were mounted in Dewars operated at the temperature of dry ice ($-73\,°C$).

The sensitive areas of the lead sulfide detectors were 0.5 mm × 0.5 mm, 2 mm × 2 mm and 3 mm × 3 mm, and of the germanium detector 2 mm × 2 mm. Cooling of the lead sulfide produces a slight shift in the spectral response towards the long-wave length cutoff and increases the responsivity † and improves the detectivity. If we consider the peak responsivity and detectivity the best samples of cooled lead sulfide detectors give a detectivity which is several times higher than that of the best uncooled samples.

Cooling the germanium detector has less effect on the spectral characteristics, but the responsivity and the detectivity are greatly improved [5], in our case by roughly 100 times. Note that, while the current noise is dominant in the lead sulfide, and it determines the detectivity, in the germanium detector there is no current noise in the absence of infrared signal.

The NEP of the lead sulfide is proportional to the length of the edge of the sensitive area (for a square detector). For the 0.5 mm × 0.5 mm models, an NEP of the order of 10^{-12} W Hz^{-1} is typical. The resistance of the detector, when cooled and in the absence of signal, varies strongly

---

* Detectivity, is the reciprocal of the NEP. (Ed. English version.)

† Responsivity is the rms signal output per unit rms radiant power incident upon the detector. (Ed. English version.)

from sample to sample. For our samples it ranged between 200 kΩ and 20 MΩ. The bias voltage used varied from 2 to 20 volts. The voltage responsivity is proportional to the supply voltage. No deviation of the signal from proportionality to light level signal was observed. For the germanium detector (area 2 mm × 2 mm) a NEP of the same order (10^{-12} W) was determined, but we are not yet able to say definitely whether improvement of the amplifier can reduce this value. In contrast, the NEP of the lead sulfide is limited definitely only by their internal noise.

The noise spectrum of the lead sulfide has the form $N(f) \sim f^{-1}$, so that it is basically not advantageous to amplify the signals from these detectors with dc amplifiers. It is better to modulate the radiation flux and amplify the ac current by means of a narrow-band amplifier tuned to the modulation frequency. This is in fact what is done in practice [4]. The higher the modulation frequency, the less noise, but the time constant of the lead sulfide limits the upper operating frequency. In addition, the responsivity of the lead sulfide is proportional to its time constant. As a result, the optimum frequency is several tens of Hz for the most sensitive lead sulfide detectors which have the best detectivity, while for ordinary ones it is much higher (several hundred Hz). In our instruments, the modulation frequency is 36 Hz. Modulation is obtained by a mechanical chopper mounted on the shaft of a synchronous motor with a speed reducer, D-104. The same modulation frequency was used for the germanium detector; in this case, in view of the absence of current noise, the choice of frequency is less important. However, since signals of the order of microvolts must be amplified and the time constant of the detector is small, it is advantageous to modulate the radiation and use ac amplifiers.

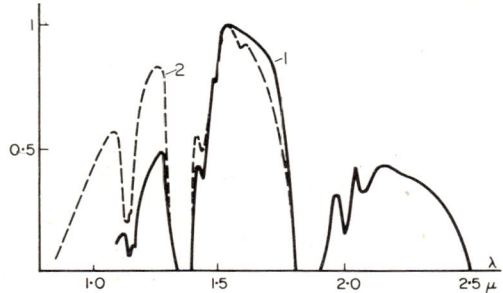

Figure 23-1 Spectral responsivity of an infrared stellar photometer taking into account the atmospheric transmittance and using lead sulfide (curve 1) or germanium (curve 2) detectors.

The spectral response of the stellar photoelectric photometer, taking into account the atmospheric transmittance, is given in Figure 23-1, (curve 1 is for lead sulfide, curve 2 for germanium). The long wavelength cutoff for the germanium detector is at 1.85 μ, and for the lead sulfide, cooled by dry ice, at 3.5 μ. In recording the radiation spectrum of the night sky we succeeded in reaching 3.4 μ [2, 3]. For photometry in integrated radiation, the region 3.0–3.5 μ provides only a small percentage of the total radiation, while the interval 2.5–3.0 μ is wholly absorbed by the atmosphere. It is thus worth filtering out the region with wavelengths longer than 2.5 μ, since this region does not contribute with signal while causing a large amount of thermal radiation noise from the sky and the telescope mirror.

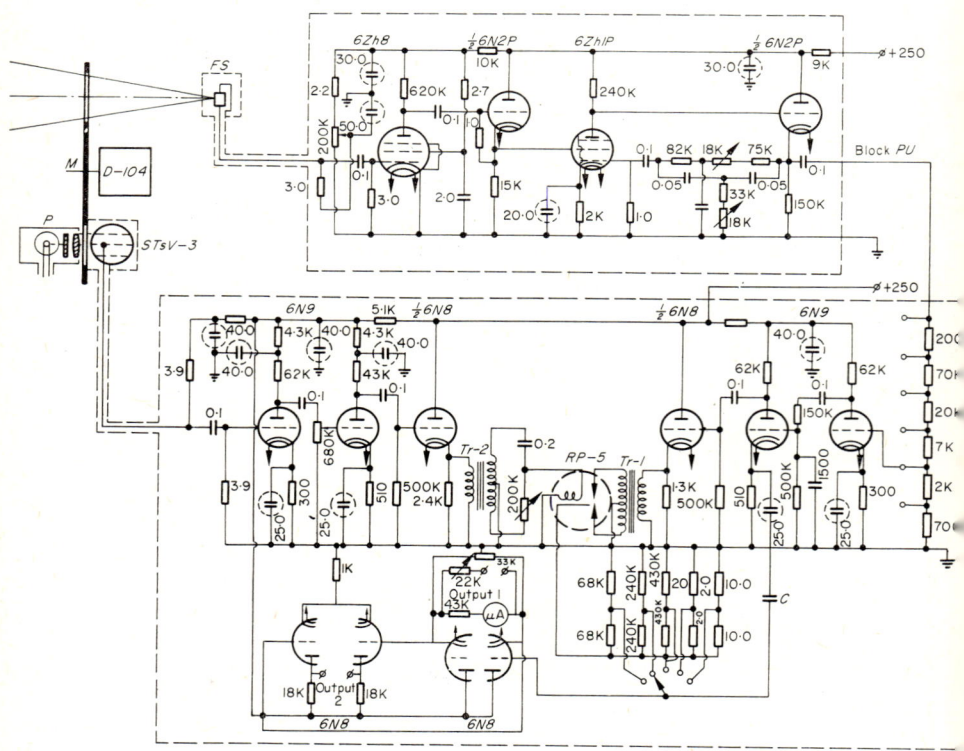

Figure 23-2 Main circuit of stellar photometer developed at the Shternberg Astronomical Institute.

The main circuit of the stellar photometer is shown in Figure 23-2. It consists of the preamplifier PU, with an amplification coefficient of 2×10^3 and 1 to 2 Hz bandpass, and an amplifier OU with amplification adjustable by steps (maximum 500). The last stage of amplification has relatively wide bandpass and has its output signal rectified by a synchronous controlled rectifier. In the narrow bandpass section of the preamplifier, a double T-filter network is connected to the negative feedback circuit, as recommended by G. V. Kirillov [6]. For synchronous rectification it uses a polarized relay RP-5. The rectified signal is filtered by an RC filter which sets the time constant. The synchronous rectifier narrows the bandpass further (to an equivalent width of $\Delta f = 1/\pi RC$). The choice of time constants ranges from 0.6 to 110 seconds. The rectified signal is fed to the differential cathode follower and the output to the potentiometer type recorder EPP-09 (or PS 1-02). A relatively coarse pen recording milliammeter (H-370) driven by an additional dc amplifier can be connected to the output. The noise level of the amplifier is determined by 3 MΩ resistance in the grid circuit of the first tube. The zero drift of the amplifier output is practically insignificant. The preamplifier and amplifier of the photometer are supplied by the universal power supply UIP-1 with the exception of the preamplifier tube filaments, which are fed from a dc source. The UIP-1 has two independent rectifiers, one is used for the preamplifier and the other for the amplifier.

The reference signal for the synchronous rectifier is generated by a photoelectric sensor consisting of a light bulb and an SbCs photocell STsV-3. The light beam from the bulb is interrupted by the same chopper used to modulate the infrared radiation.

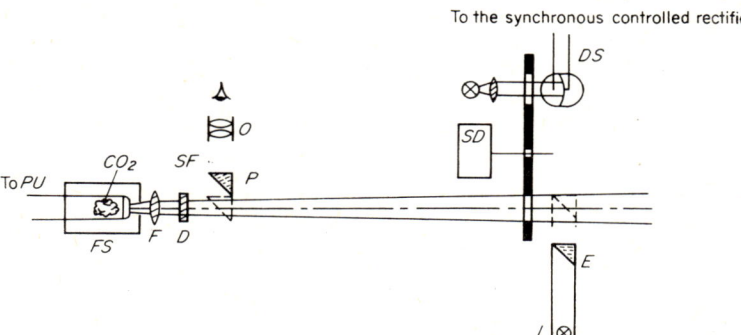

Figure 23-3 Optical layout of the infrared stellar photometer developed at the Shternberg Astronomical Institute.

If the radiation flux to be measured is sufficiently large, then synchronous rectification is unnecessary. In this case the PU can be followed by millivoltmeters LV-9, MVL-2 M, etc. To record the output signal by a recorder potentiometer type, a vacuum tube differential cathode follower or transistor emitter follower can be connected after the LV-9.

The optical layout of the stellar photoelectric photometer is shown in Figure 23-3. For setting the celestial object on the diaphragms D, there is the eyepiece O with a prism P. The size (from 2 to 1.2 mm) of the diaphragms defines the field of view of the photometer. The optical filters (SF) IKS-3 ($\lambda_{min} = 0.8\ \mu$) and coated germanium ($\lambda_{min} = 1.7\ \mu$) can be inserted in the optical path. The Fabry lens F images the mirror objective of the telescope system on the detector.

The main difficulty during observation is caused by thermal radiation from the mechanical chopper, which is always greater than the radiation of the sky and the telescope optics if the long-wavelength cutoff of the system is beyond 2.5 μ; this thermal radiation must be compensated or reduced by some method. Reduction by an optical filter with an approximate cutoff at 2.5 to 2.8 μ operates fairly effectively, reducing the background radiation from the chopper by 1.5 times. Practically any glass may be used for the optical filters (such as 5 mm K-8). However, the use of a filter does not completely solve the problem of thermal radiation from the chopper, since even with the filter the chopper gives a signal tens of times larger than the NEP for an $f/20$ optical system and a diaphragm 12 mm in diameter. For stellar photometry small diaphragms may be used, but this method is not suitable for the study of extended objects because the background and signal are reduced in proportion. To measure the flux from the Crab Nebula, we compensated the background thermal radiation from the chopper as follows. The signal from the sky background and the signal from the chopper differ in phase by 180°. Thus, by placing inside the telescope tube a light bulb L (see Figure 23-3) and by adjusting its voltage to the proper value (equivalent to increasing the sky background) both backgrounds can be mutually compensated. However, random fluctuations in the temperature of the chopper cannot be compensated and thus reduces the signal-to-noise ratio. An ideal solution would be to have a differential photometer, in which switching from the background to the background plus object is done at the modulation frequency. If combined with long-term integration, this method might improve the sensitivity threshold by one order. However, the fairly high switching frequency necessary would cause problems for the practical realization of such a system.

We used this photometer with a lead sulfide detector to measure the intensity of radiation of the central region of the Crab Nebula. The monochromatic flux reduced to the Nebula as a whole and to the wavelength 2.0 μ equals 1.3×10^{-25} W m^{-2} Hz^{-1}. Using existing data about interstellar absorption, we compared the measurements in the photographic, visual and infrared regions and were able to draw the preliminary conclusion that by the best observing method, observations both in the radio and in the optical range satisfy the spectral index $\alpha = \text{const} = 0.35$. For the value of α the optical radiation from the electrons could be disregarded and so we obtain in the Crab Nebula an estimated magnetic field of $H_1 < 10^{-4}$ G.

The total density of kinetic energy of the relativistic electrons in the Nebula is 10^{-7} erg cm^{-3}.

The radiation from the Crab Nebula in the range 1–2.5 μ was measured in 1959 [1]. The same instrument was used during 1960 for a number of other problems including an attempt to discover infrared radiation of the galactic core, measurement of the brightness of the night sky in the interval 0.8–1.7 μ, and test observations of stars in various spectral classes. In all cases we used detectors with a sensitive area of 2 mm × 2 mm. The work was first done solely with the lead sulfide, but in 1960 the germanium detector was also used. Since the long wavelength cutoff of the germanium detector is at a shorter wavelength than that for the lead sulfide detector, it is almost insensitive to the thermal radiation of the modulator. Moreover, its NEP is several times better than that of a lead sulfide of the same size.

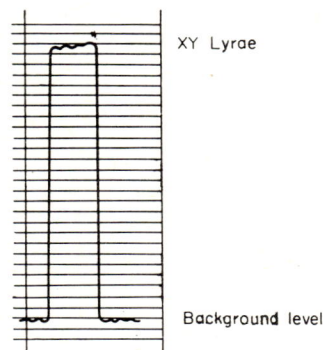

Figure 23-4 Record of the star XY Lyrae (6.5^m M8) with the infrared stellar photometer; the time constant was 2 seconds.

The limiting stellar magnitude of the photometer for M stars with the 50-inch reflector and using a germanium detector is 8–10^m, depending on the subclass. The accuracy of the measurement for stars of this magnitude is at least 0.^{m}03, for a time constant $\tau = 20$ sec. Note, however, that the limiting magnitude was calculated from measurements of brighter stars.

The recording for the star XY Lyrae (6.^{m}5, M 8) for $\tau = 2$ sec is shown in Figure 23-4. The relationship between the sensitivity and the spectral class is similar to that described in [7] (both for lead sulfide and for germanium). The limiting magnitude of our photometer is clearly higher than that of the infrared stellar photometers described by Fellgett [8] and Luncl [9], due to the use of more sensitive detectors.

3 STELLAR SPECTROMETER

For the study of the night sky spectrum [2, 3] and the spectra of planets and bright stars we have designed a spectrometer using a diffraction grating and a lead sulfide detector. The optical layout of the spectrometer is given in Figure 23-5. The diffraction grating D has 300 grooves per mm and is 150 mm × 150 mm in size; it concentrates 67% of the incident light (blaze angle 18°30′) for $\lambda = 2.05$ μ in the first order.

The autocollimating mirror Z_2 is parabolic, has a 750 mm focal length, $f/5$. From the input slit S_1 the light is reflected on the mirror Z_1 to the collimating mirror Z_2 which feeds the parallel beam to the diffraction grating. The diffracted beam is reflected by Z_2 and by the mirror Z_3 to the slit S_2, in the plane of which the spectrum is focused. The clear fused silica condenser K, placed behind the slit S_2, concentrates the beam on to the 2 mm × 2 mm size lead sulfide FS. The linear dispersion of the spectrometer is 40 Å mm^{-1}. Optical filters with $\lambda_{min} = 1.7$ μ (coated germanium, 0.5 mm thickness) and $\lambda_{min} = 0.85$ μ are use to separate orders. A chopper C is set up in front of the input slit S_1.

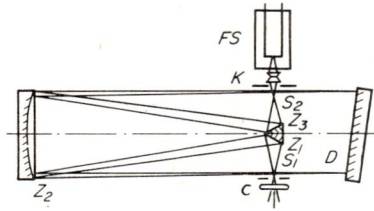

Figure 23-5 Optical layout of the infrared spectrometer using diffraction grating and developed at the Shternberg Astronomical Institut.

To scan the spectrum, the mount of the diffraction grating is rotated by a lever driven by a screw-nut mechanism. The nut cannot rotate, but uniform rectilinear motion is imparted to it by the screw. The screw is set in motion by a synchronous motor through a speed reducer. The scanning is automatically reversed by terminal switches. The range of recording is regulated within the limits 0 to 1.0 μ. For our present work we use recording rates of 0.033 μ min^{-1} or 0.0135 μ min^{-1}. The spectral width of the slits is 0.025 μ or 0.04 μ (for different regions), which correspond to a span of 6 and 10 mm on the paper chart recorder, respectively. The operating height of the slits is 15 mm. When the Jupiter spectra were recorded we used 1.5 to 2.5 mm slit height, but a satisfactory signal-to-noise ratio was obtained for Saturn with a slit of only 5 mm. For the planets we used lead sulfide detectors of 0.5 mm $\times$ 0.5 mm and a condenser lens of 10 mm diameter. The front optics of the spectrometer is the (50-in.) reflector. The autocollimation mirror of the spectrometer had to be stopped down to $f/20$ to match the telescope.

The recording of the night sky spectra was obtained using the full aperture ratio of the spectrometer and without the telescope.

The efficiency of the spectrometer versus wavelength was obtained by recording the solar spectrum in the region 1.2–2.5 μ and the spectrum of a low-temperature calibrated surface in the region 2.5–3.5 μ. Absolute standardization, when required, was obtained by means of the low-temperature calibrated source (50–60°C). The accuracy of absolute calibration was not less than 20%. The accuracy of measurements of wavelengths was ± 0.005 μ for recordings with spectral width 0.024 μ and 0.01 μ for recordings with spectral width 0.04 μ.

Portion (1.2–1.8 μ) of the recorded spectrum of Jupiter is shown in Figure 23-6. The averaged recordings for Jupiter, Saturn and the Sun are given in Figure 23-7, and they show strong absorption bands, particularly in the regions 1.6–1.8 μ and 1.9–2.4 μ.

The Jupiter spectrum agrees well with the data [10, 11], but the Saturn spectrum, by contrast, is quite different from the classical Kuiper data. This is explained by the fact that the spectrum of the rings, was superimposed on the spectrum of the disk. That means that the image of Saturn, including the whole ring system was fed in to the slit. The weak radiation in the region 2.2–2.3 μ found in the Saturn spectrum may be interpreted as reflection on the rings. The position of this maximum in the rings spectrum is a point in favor of the Kuiper hypothesis that the rings are covered by frost, if not made of ice. Comparison with laboratory studies of snow and frost shows that the spectrum of the rings is similar to that of frost. The "equivalent path" of NH_3 in the atmosphere of Jupiter for the 1.51 μ band (approximately 50 cm path under normal conditions) turns out to be an order less than for photographic spectra in $\lambda = 0.645\ \mu$. The amount of CH_4 estimated from bands in the region 1–2 μ also agrees with the estimates in the shorter-wavelength region. The difference can be explained

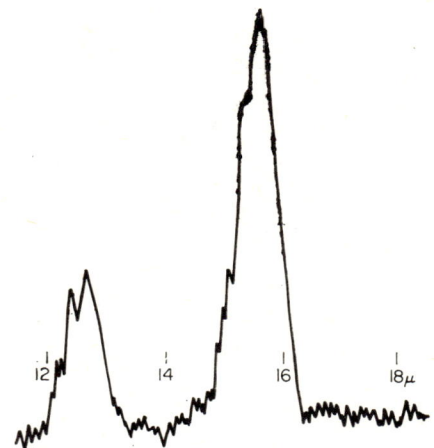

Figure 23-6 Portion (1.2–1.8 μ) of the recorded spectrum of Jupiter.

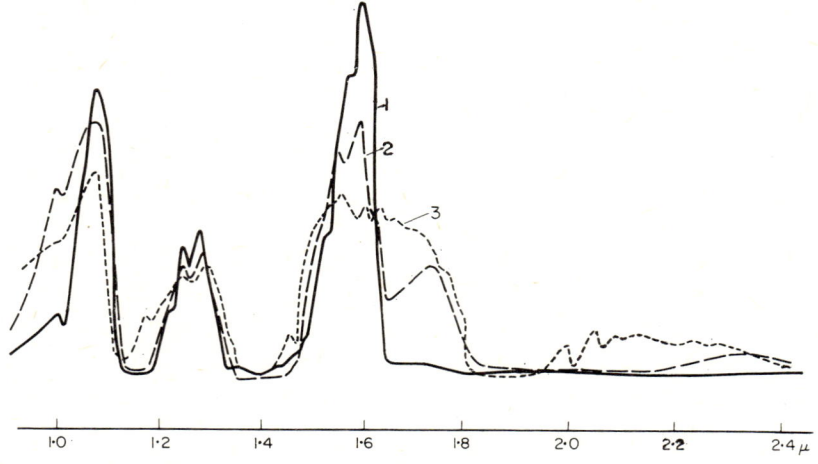

Figure 23-7 Averaged recordings of the Jupiter (curve 1), Saturn (curve 2) and solar (curve 3) spectra, showing the presence of strong planetary absorption bands.

if we assume that the NH_3 responsible for the absorption is concentrated at the height of the diffusing cloud layer, while the CH_4 is higher.

The resolution obtained with this spectrometer used in connection with the 50-inch telescope is comparable with that obtained with the prism infrared spectrometer used by Kuiper in connection with the 82-in. McDonald telescope. Note that the output may be increased three to four times in the region 1.5–1.7 μ by using the appropriate echelet.

A simplified model of this instrument is being used in connection with the solar tower of the Main Astronomical Observatory, to record high-resolution solar spectra.

4 CONCLUSION

The use of new types of infrared detectors, possibly based on the methods of quantum radiophysics, cannot be ruled out in the future, but as long as photoconductors and photodiodes are the most sensitive ones, efforts to use them more efficiently are needed. The development of instruments with automatic suppression of the background seems most promising for photometry. Without the suppression it will be impossible to make progress toward the longer-wavelength region. One possible variant of such a photometer is described in [12]. Automatic suppression of the background will permit longer integration time of the signal, which will increase the sensitivity and accuracy in spectrometric work. A more radical improvement is to be expected from methods of interferometric spectroscopy [13] which allows all the spectral information to be recorded simultaneously. The advantages of this method are undoubtedly in the infrared region, which is characterized by the high level of the irreducible detector noise [1]. This technique will be specially important for astronomical observations.

REFERENCES

1. V. I. Moroz, "The radiation flux from the Crab Nebula at $\lambda = 2\mu$ and some conclusions on the spectrum and magnetic field", *Astron. Zh.* **37**, 265 (1960).
2. V. I. Moroz, "The infrared spectrum of the night sky", *Astron. Zh.* **37**, 123 (1960).
3. V. I. Moroz, *DAN SSSR*, **126**, 989 (1959).
4. R. A. Smith, F. E. Jones and R. P. Chasmar, *The Detection and Measurement of Infrared Radiation*, Clarendon Press, Oxford (1957).
5. N. A. Vitovsky, P. N. Maleyev and S. M. Ryvkin, *Radiotekhnika i elektronika*, No. 8, 1387 (1959).
6. G. V. Kirillov, *Tr. SNTO MEI*, **4** (1950).

7. M. Neant and M. J. Bigay, "Determinations de magnitudes infra-rouge avec une cellule au sulfure de plomb", *Publ. Obs. Haute Provence*, **2**, No. 37 (1952).
8. P. B. Fellgett, "An exploration of infra-red stellar magnitudes using the photoconductivity of lead sulphide", *Monthly Notices Roy. Astron. Soc.*, **111**, 537 (1951).
9. M. Luncl, "Recherches de photometrie stellaire dans l'infrarouge au moyen d'un cellule au sulfure de plomb", *Ann. Astrophys.*, **23**, 1 (1960).
10. G. P. Kuiper, W. Wilson and R. J. Cashman, "An infrared stellar spectrometer", *Astrophys. J.*, **106**, 243 (1947).
11. G. P. Kuiper, "Infrared spectra of planets", *Astrophys. J.* **106**, 251 (1947).
12. D. J. Lovell and G. R. Miczaika, "An infrared technique for stellar photometry", in *The Present and Future of the Telescope of Moderate Size* (Ed. F. B. Wood), University of Pennsylvania Press (1958), pp. 111–28.
13. P. B. Fellget, "Investigations of image detectors", in *The Present and Future of the Telescope of Moderate Size*, pp. 51–86 (Ed. F. B. Wood), University of Pennsylvania Press (1958).
14. P. Connes and H. P. Gush, *J. Phys. Radium*, **20**, No. 11, 195 (1950).

CHAPTER **I-24**

A Fabry-Pérot etalon used with an image intensifier for the observation of faint emission objects

P. V. SHCHEGLOV

Shternberg Astronomica Institute
Moscow

Image intensifiers* with a fluorescent screen do not have high resolution and are therefore not suitable for direct stellar photography. In fact, if the resolution of the "telescope–intensifier" system is determined by the intensifier and not by atmospheric seeing, the use of such a system has a lower limiting magnitude than the "telescope–emulsion" combination. The use of this type of image intensifiers for direct photography can be justified only in the infrared region, where photographic emulsions do not have enough sensitivity. Thus, the amount of information recorded by the system is limited not by the sky background, but by the exposure time.

The main application of image intensifiers is in spectroscopy, where the recorded information is usually limited by the exposure time, rather than by the background. We have use a photocontact image converter with 0.05-mm resolution and a 10 mm useful field. The technique using this converter has proven to be very efficient for spectroscopy.

We have also used these type of image intensifiers in connection with interference monochromators. A gain of 100 in the exposure using contact photography for the H_α region led us to use these monochromators to reduce the sky background which interfered with our nebulae obser-

---

* In the English version of this contribution the terms "image intensifier" and "photocontact image converter" are used indistinctly to described the same imaging device. It consists in essence of a single stage converter in which the output of the fluorescent screen is recorded by the photographic emulsion which is brought into contact with the screen. (Ed. English version.)

vations. We used an interference filter for the H_α line with $\Delta\lambda = 40$ Å, and behind it we placed a Fabry–Pérot etalon with 0.3 mm separation between the plates and a coating with 70% reflectance. In this case, the instrumental profile was 0.8 Å wide. The equivalent bandwith $\Delta\lambda_s$ of the system can be obtained from the formula

$$\Delta\lambda_s = \Delta\lambda_f \frac{S^2}{1-r^2},$$

where $\Delta\lambda_f$ is the equivalent transmission bandwith of the filter, S and r are the transmittance and reflectance coefficients of the Fabry–Pérot etalon, respectively. In this case, $\Delta\lambda_f = 40$ Å, $S = 29\%$ and $r = 70\%$ (since absorption in multilayer dielectric mirrors is usually less than 1%), thus $\Delta\lambda_S = 7$ Å. Note that $\Delta\lambda_s$ may be reduced several times by using smaller separation between plates and higher reflectance. The transmittance of this interference monochromator (Fabry–Pérot etalon plus interference filter) is practically equal to the transmittance of the filter (about 60%) which for the given resolution is greater than in the equivalent diffraction spectrograph. The use of a comparatively narrow-band pass interference filter caused a great reduction in the equivalent bandwidth and therefore a reduction in the background, but the resolution did not reach the values obtained with slit spectrographs.

We used this system in an attempt to observe weak emission nebulae. The interference rings from the nebula NGC 7000 (the North America nebula) were recorded with an exposure of 10 minutes. For our system, the minimun detectable signal of monochromatic emission is about 10^{-5} erg cm^{-2} sec^{-1} sr^{-1} or approximately 20 times smaller than for usual methods.

With this detectivity, the detection of faint nebulae is limited by the night sky H_α emission. This emission is due to the excitation (fluorescence) of the hydrogen in the gas corona of the earth by solar L_β emission. The interference rings of this emission can be recorded by the system with one hour exposure; by using this observational technique we can detect changes in the flux of L_β solar radiation.

Observations of emission nebulae can be made only near the anti-solar point, since in this region the night sky H_α emission will be greatly reduced (practically any L_β-quantum is broken up into H_α- and L_α-quanta across several re-emissions, and there is no direct illumination of the atmosphere in this region).

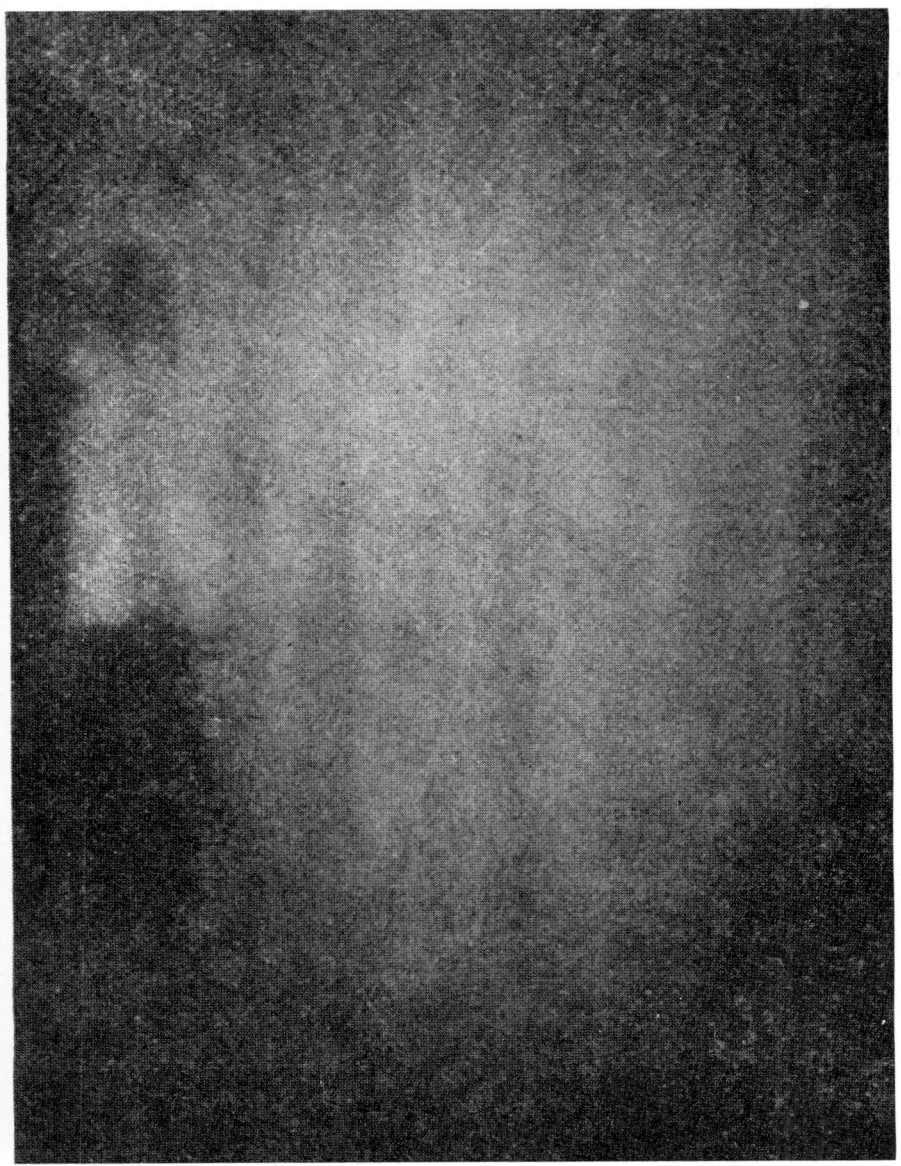

Figure 24-1 Twilight flash photograph of the 10,830 Å helium line recorded with an interference monochromator and an image intensifier coupled to it.

The mechanism for luminescence of the upper atmosphere in the 10,830 Å helium line in twilight, which N. N. Shefov described recently, is somewhat more complicated. He estimated the intensity of this line to be several kilorayleighs. Using a telescope with an 10,830 Å interference monochromator and, coupled to it, an image intensifier (a system similar to that described previously in this paper), we were able to make direct observations of the twilight flash of this line (see Figure 24-1). In subsequent photographs the interference bands were not recorded.

Thus, image intensifiers coupled to interference monochromators are a very efficient instruments for observing faint emission objects.

CHAPTER **I-25**

A recording intensity microphotometer based in the MF-4 type

E. V. KONONOVICH

Shternberg Astronomical Institute
Moscow

Photometric reduction of a large number of photographic plates takes considerable time and effort. The usual method to obtain photometric profiles point-by-point is very laborious and they seldom give the accuracy which could be obtained by photographic methods. The use of recording microphotometers of the MF-4 type does not improve the situation, because the same amount of work is required and the accuracy is reduced due to the need to allow for the transmission from a small intermediate record, such as a microphotograph. This was pointed out by G. S. Ivanov–Kholodny [1]. The various mechanical schemes [2] designed to convert the transmission record into the corresponding intensities are not often used because they save very little time and hardly improve the accuracy of the microphotometry. For the purpose of speeding up photometric measurements and to attain as high an accuracy as possible, many different designs of recording microphotometers have been suggested. These instruments automatically take into account the characteristic curve of the photographic emulsion (Figure 25-1) and perform the transformation $\log I = F(D)$.

Usually each element of the photographic image could be characterized either by its transparency T, i.e. by the fraction of incident light transmitted through the image or by the density D expressed by $D = \log 1/T$. The purpose of photographic photometry is to determine the intensities I of the light fluxes which produce the corresponding image density D on the photographic emulsion. As we know, both values are related by the expression $\lg I = F(D)$, which is obtained from the characteristic curve $D = f(\lg I)$ of the photographic emulsion from calibration. This curve for ordinary emulsions has a rectilinear ($\lg I \sim D$) portion. The relationship between the intensity and the transparency is represented by a segment of a hyperbola.

In intensity microphotometers, the required transformation ($\log I = F(D)$) can be done either by a special circuit which transforms the current proportional to the transparency into a current proportional to the intensity, or by a servomechanism which balances two light beams, one that passes through the area of the photographic plate to be measured, and the other through a unit which simulates the characteristic curve of the emulsion.

Automatic operation of the microphotometric measurements is very important and many different schemes have been proposed, sometimes ingenious, sometimes too complicated. For example, Vanderkerkhov [3] uses two ready-made instruments for this purpose: the Rosenberg microphotometer and the Zeiss recording microphotometer. The first measures density by the compensation method using a neutral wedge. The wedge is mechanically linked to a specially shaped diaphragm placed in front of the slit of the second microphotometer. With this arrangement a displacement of the wedge will be translated into a change in the height of illumination of

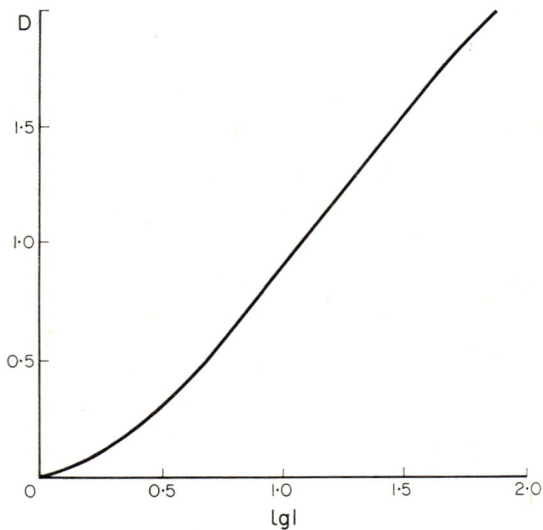

Figure 25-1 Characteristic curve of a photographic emulsion. (The characteristic curve is given, generally, as $D = \log E$, where E is the exposure expressed as the product of the intensity I and the exposure time t.) (Ed. English version.)

the Zeiss microphotometer slit; this change is proportional to I. Although, Vanderkerkhov mentions the possibility of automatic compensation it is not known whether this has been done.

The simplest scheme is described by Brown and Birthly [4], in which a steel spring is bent into the shape of the characteristic curve of the emulsion and the functional transformation is made using a contact which slides along the spring. It is very simple to design an intensity microphotometer if a ready-made functional transformer is available, as described by Oke [5]. One of the first very successful circuits is that of Minnaert and Houtgast [6], which was used for the compilation of the well-known *Utrecht Photometric Atlas of the Solar Spectrum*. This scheme has been duplicated several times, with certain modifications [7, 8]. More complicated electronic instruments were constructed later [9, 10] and these have been successfully used abroad.

Intensity microphotometers have been constructed at many soviet observatories, including Byurakan [7], Abastumani [8], Engelgardt (Kazan), the astronomical observatory of Kiev University and the Crimean Astrophysical Observatory [11]. The simplest instrument which is suitable for small laboratories is the one described by Kotlyar [12], in which, as in the Kranjc circuit [13], a recording electronic photometer is used as the functional transformer in connection with a special linear slide-wire with taps. The characteristic curve of the emulsion is given in an analog form approximated by a function divided into many sections of roughly equal length. This function is obtained by choosing potential drops at different points of the slide-wire. The same circuit was used at Göttingen [14], based on the design of Phillips [15].

The great variety of the circuits so far proposed is, to a great extent, explained by the fact that apparently none of them is fully satisfactory. Moreover, some observatories cannot construct a complicated instrument with their own resources, and are forced to restrict themselves to a simpler circuit, or merely develop the theory of the instrument [16].

Many of the circuits we have mentioned have one or more drawbacks. Some of them [9, 10] require special calibration of the emulsion, but, on the other hand, they do allow for changes in the characteristic curve caused by changes in the wavelength of the spectral region being measured. Some [11], although giving good accuracy, cannot easily be used over a wide density range. Others [5, 13], although they are simple in design, require the use of instruments which are not readily available. For instance, it is often very desirable to use an isophotometer to plot the curves for the different intensities [17–19].

An intensity microphotometer should, in our opinion, satisfy the following requirements:

(1) The instrument must be sensitive to changes of 0.1% in transparency of the photographic plate.

(2) The accuracy of the recorder should be at least 1% of the full scale. (Requirements (1) and (2) allow the instrument to be used for recording densities of up to 2.5–3 into intensities.)

(3) It should be possible to use the instrument to obtain the characteristic curve of the photographic emulsion so that the measurement errors are minimized.

(4) Resetting the instrument for a given emulsion should not take longer than 10 to 15 minutes.

(5) The instrument should permit the use of different calibration methods, such as sensitometry using a continuous wedge. Also it should have the capability to plot isophotes.

These requirements are best met by servomechanisms using two light beams, as similar as possible [9, 10, 16]. In this paper we describe our experience of constructing such an instrument. It is basically a synthesis of the Williams–Hiltner and the Minnaert–Houtgast circuits. The schematic of the instrument constructed at the Shternberg Institute is shown in

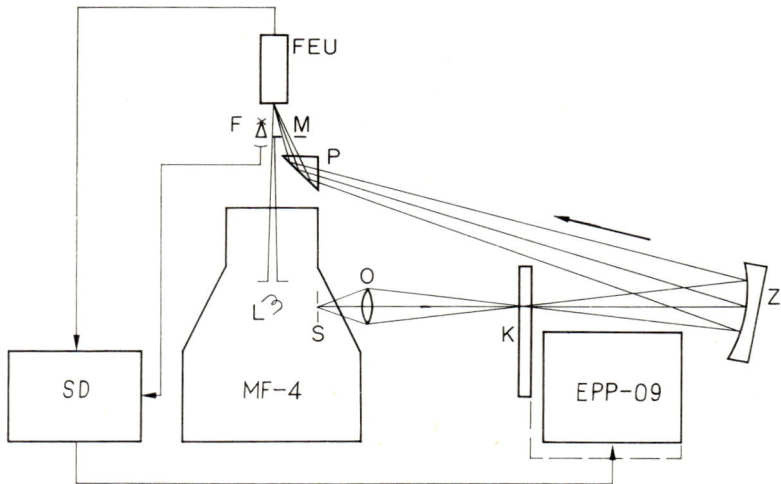

Figure 25-2 Schematic of the intensity microphotometer developed at the Shternberg Astronomical Institute and based on the MF-4 type.

Figure 25-2. It is based on a modified recording microphotometer MF-4, in which the barrier photocell has been replaced by the photomultiplier tube FEU-25, and the recording unit by the EPP-09 potentiometer type recorder. The pen takes 2.5 sec for full scale deflection. The mirror galvanometer in the MF-4 can be removed to make the installation of the modulator easier. The scheme for the transformation of the transparency into any monotonic smooth function of the transparency is a servomechanism given schematically in Figure 25-2. The servomechanism operates on the two-beam principle, one of which is the light beam which passes through the area of the photographic plate under measurement and the other passes through a special diaphragm shaped in accordance to the characteristic curve of the emulsion or some other function of the transparency. The second light beam is formed by the slit S, illuminated by the light bulb L which also illuminate the photographic plate. The objective O images the slit S on the plane of the carriage K which holds a specially shaped diaphragm similar to one of those shown in Figure 25-3. Special attention was given to the uniformity of illumination of S, which was obtained by using a two-lens condenser. The ordinates of the diaphragm (parallel to S)

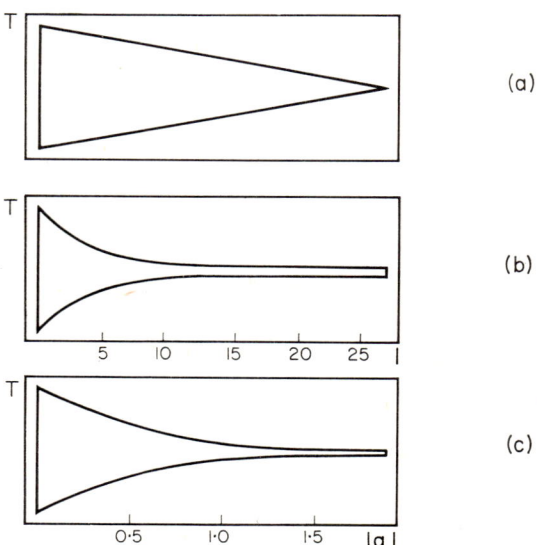

Figure 25-3 Diaphragms used in the microphotometer for the characteristic curve of a given photographic emulsion: (a) for recording transparency; (b) for transformation of transparency into intensities; (c) for transformation of transparency into the logarithm of intensities.

are proportional to the transparency, for the given emulsion, and the abscissa proportional to the transparency function (in the Minnaert–Houghast scheme [6] the transparency are the abcissas of the diaphragm). Thus a portion of the light beam which is proportional to the transparency of the image, passes through the diaphragm mounted on the carriage K. The light beam is then fed by the mirror Z and prism P to the same position as the first beam. Both light beams are modulated at 37 Hz and by the same modulator M which is a disk with rectangular teeth driven by a synchronous motor D-104 type.

We used the light modulator and synchronous detection circuit (polarizaed relay rectification) designed by V. I. Moroz, with a few modifications. After amplification and detection by the synchronous detector SD (for which a reference signal is generated by the photocell F, a light source and modulator M), a dc voltage is obtained at the output, which is proportional to the algebraic difference between the measuring and reference beams. This voltage is applied to the amplifier of the recorder EPP-09. The carriage of the recorder is rigidly connected to K by a cable through a special unit and the diaphragm, which represents the characteristic curve of the emulsion, is mounted on K.

When the photographic plate, which is mounted on the carriage of the MF-4, is displaced, the intensity of the first light beam could change due to an area of the plate with different photographic density. Thus, an error signal is generated at the output of SD; its polarity depending on whether there is an increase or decrease in the photographic density. The motor of the recorder EPP-09 will rotate in one direction or the other, depending on the polarity of the error signal, until the displacement of the carriage K with the diaphragm matches the intensities of both light beams. Thus, the pen which is rigidly attached to the carriage will record the displacement.

When the photographic plate under measurement and the paper chart recorder moves with uniform motion, the instrument records the transparency, the intensities or the logarithms of the intensities, depending on whether diaphram (a), (b) or (c) is used (see Figure 25-3). For maximum sensitivity, the instrument operates slightly under damped. In our case, the pen reaches the final position after two or three rapidly damped oscillations. However, if the slope of the characteristic curve of the photographic emulsion changes sharply, then the sensitivity in different portions of the curve has to be changed during operation. This is done in our system by voltage control on the FEU.

Preliminary studies of the instrument have shown it to be capable of repitability and highly accurate recording (2–3% of the maximum intensity). Recordings of H_α line profile in the spectrum of the solar chromosphere obtained with this intensity microphotometer are shown in Figure 25-4(a), (b) and (c). The dots represent the results of independent measurements on the microphotometer MF-2, made with the maximum attainable accuracy by the method described by Ivanov–Kholodny [1].

Setting the instrument for maximum density is done by adjusting the zero of the synchronous detector output, and for $D = 0$ (fog level of the photographic plate) by the adjustment of the slit width S of the MF-4, as well as by the introduction of neutral filters into both light beams. If after adjustment for the fog level, we introduce a neutral filter of known density D_1 into the light beam which goes through the photographic plate and the diaphragm is so shaped that the largest profile corresponds to the point D_1 on the characteristic curve, then the instrument will transform only densities $D > D_1$ into intensities, thereby expanding the scale and improving the accuracy for high densities.

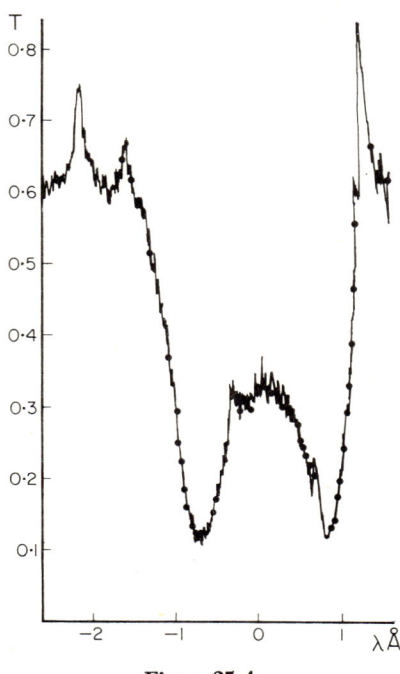

Figure 25-4a

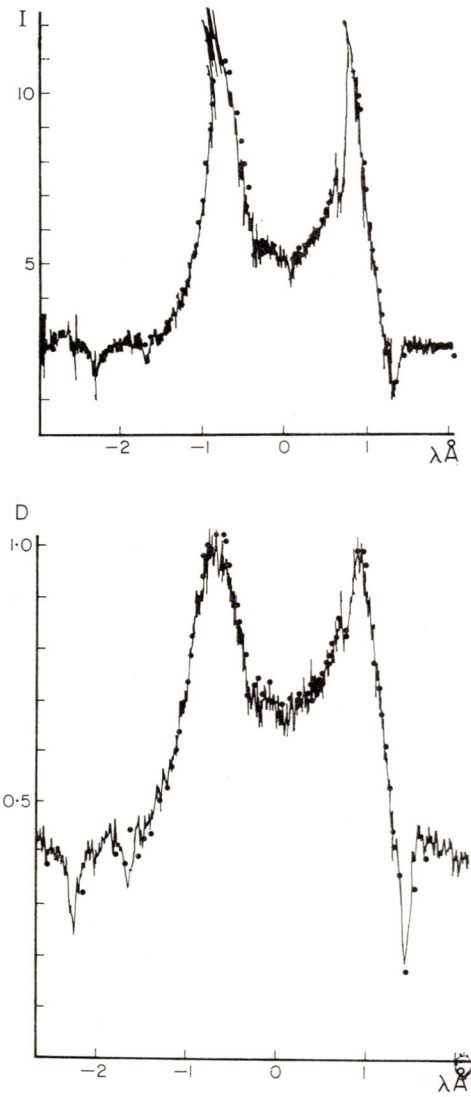

Figure 25-4 Recording of the H_α line profile of the solar chromosphere spectrum using measurements of: (a) transparency, T; (b) intensity, I; (c) density, D.

It is easy to see that if the photographic plate is mounted on the carriage K, and a narrow slit is located on the image plane of S (plane of the photographic plate) and is displaced with uniform motion along the slit S a small light spot will scan the plate and the instrument will operate as an isophotometer.

To obtain the characteristic curve of the photographic emulsion and to make the shaped diaphragm, a sensitogram obtained by a step attenuator or any other type can be measured in the instrument with a triangular diaphragm mounted on K. The displacement of the carriage is then proportional to the light flux passing through the photographic emulsion, and the instrument records the transparency. If these transparency measurements are plotted on logarithmic paper, the averaging of several measurements, the plotting and cutting out of the diaphragm take much less time, in the order of 15 to 30 minutes. The diaphragm is made of hard paper or cardboard, and is best cut with a razor blade.

A more accurate photographic method for making the diaphragms has been described by Maksutov [20]. However, if we make a sensitogram in the form of a photographic wedge, using a sensitometer similar to that described by Nabokov [21], for example, and make the wedge such that the illumination is proportional to the length, then it becomes unneccessary to construct the diaphragm and the wedge-sensitogram can be inserted directly on K.

An important advantage of this system is that since the instrument is sensitive only to the ratio of the two light beams which is detected by one photomultiplier tube only, there is no need for voltage stabilization or monitoring the operating conditions of the photomultiplier. The potentialities of the instrument are obtained at the cost of some complexity, but if the system is manufactured on a large scale by industry this will not present any problem.

REFERENCES

1. G. S. Ivanov-Kholodny, "The deviation of the profiles of emission lines of the prominences from the Doppler profiles", *Izv. KrAO AN SSSR*, **18**, 109 (1958).
2. A. Pannekoek, "Two new instruments for the reduction of stellar spectra", *Bull. Astron. Inst. Netherlands*, **9**, No. 335, 155 (1940).
3. E. Vanderkerkhov, "Procede d'enregistrement direct des densites et des intensites des spectrogrammes", *Bull. Astron. Obs. Roy. Belg. Uccle*, **4**, No. 4, 70 (1949).
4. W. N. Brown, Jr., and W. B. Birthley, "A densitometer which records directly in units of emulsion exposure", *Rev. Sci. Instr.*, **22**, No. 2, 67 (1951).

5. J. B. Oke, "A new direct-intensity recording microphotometer", *J. Roy. Astron. Soc. Can.* **51**, No. 2, 133 (1957).
6. M. Minnaert and J. Houtgast, "Direct registration of intensities in spectrophotometric work", *Z. Astrophys.*, **15**, 354 (1938).
7. G. A. Gouragadyan, "Universal microphotometer of the Burakan Observatory", *Soobshch. Byurakansk. Obs.*, **14**, 11 (1955).
8. M. V. Dolidze and L. M. Fishkova, "A supplementary device to the self-recording microphotometer 'MF-4'", *Byull. Abastum. Astrof. Obs.*, **22**, 124 (1958).
9. R. C. Williams and W. A. Hiltner, "A self-recording direct-intensity microphotometer", *Publ. Obs. Michigan*, **8**, No. 3, 45 (1943).
10. D. E. Billings, R. H. Cooper, J. W. Evans and R. H. Lee, *Electronics*, **27**, 174 (1954).
11. R. Ye. Gershberg, V. I. Pronik and S. I. Korkin, "An oscillographic attachment to the microphotometer MF-4 for recording intensities", *Izv. KrAO AN SSSR*, **22**, 166 (1930).
12. L. M. Kotlyar, "A registering microphotometer with automatic transformation of negative density to corresponding intensities", *Astron. Zh.*, **37**, 888 (1960).
13. A. Kranjc, "Un microfotometro a registrazione automatica dell'intensita", *Contrib. Osserv. Astron. Milano-Merate*, No. 121 (1957).
14. G. Bruckner, "Ein Intensitaten registrierendes Mikrophotometer und die Registrierung des ultravioletten Sonnenspektrums", *Z. Astrophys.*, **51**, No. 3, 187 (1961).
15. J. G. Phillips, "Simple self-recording intensitometer attachment for microphotometers", *J Opt. Soc. Am.*, **49**, 972 (1959).
16. M. Laffineur, *Compt. Rend.* **227**, 900 (1948).
17. R. C. Williams and W. A. Hiltner, "Some uses of the direct intensity microphotometer", *Astrophys. J.*, **98**, 43 (1943).
18. H. W. Babcock, "A contour photometer", *Publ. Astron. Soc. Pacific*, **62**, 18 (1950).
19. N. N. Mikhelson, "Recording compensated isophotometer", *Izv. GAO AN SSSR*, **19**, No. 151, 69 (1953).
20. D. D. Maksutov, *Methods for the Study of Optical Systems* (Tenevye metody issledovaniya opticheskikh sistem), Moscow (1934).
21. M. Ye. Nabokov, *Mirovedeniye*, **12**, No. 1, 44 (1923).

CHAPTER **I-26**

A spectrograph with a photocontact image converter for the observation of nebulae

V. F. YESIPOV

Shternberg Astronomical Institute
Moscow

1 INTRODUCTION

Electron telescope* systems which are now available make it possible to build instruments sensitive to the spectral region extending from the ultraviolet to 12,000 Å. The converters and intensifiers are more sensitive than the photographic emulsions commonly used in astronomical work, and so the efficiency of existing astronomical instruments can be improved. We now have image converters which fulfill the basic requirements imposed on imaging devices for astronomical instruments.

One such imaging device is the photocontact image converter, which we have used in a number of projects. The antimony-cesium (SbCs) photocathode gives a gain in sensitivity over blue sensitive emulsions by a factor of 30 to 40; multialkali photocathodes gives a gain of about 50 to 100 in the H_α region, and the cesium-oxide photocathode in the near-infrared region, has a sensitivity an order of magnitude higher than that of existing infrared photographic emulsions [1–3]. The significant increase in gain and the typical high resolution of the image converter† can be obtained only if there is optical contact between the output mica window and the photographic emulsion. Film pressure on the mica window is mecha-

---

* In the technical Russian literature the designation of "electron telescope" encompasses the systems integrated by the telescope and electronic imaging devices (e.g. image converter). The term has been introduced by analogy to the term "electron microscope". (Ed. English version.)

† In the English version of this contribution the terms "image converter" and "photocontact image converter" are used indistinctly to described the same imaging device. See footnote p. 255. (Ed. English version.)

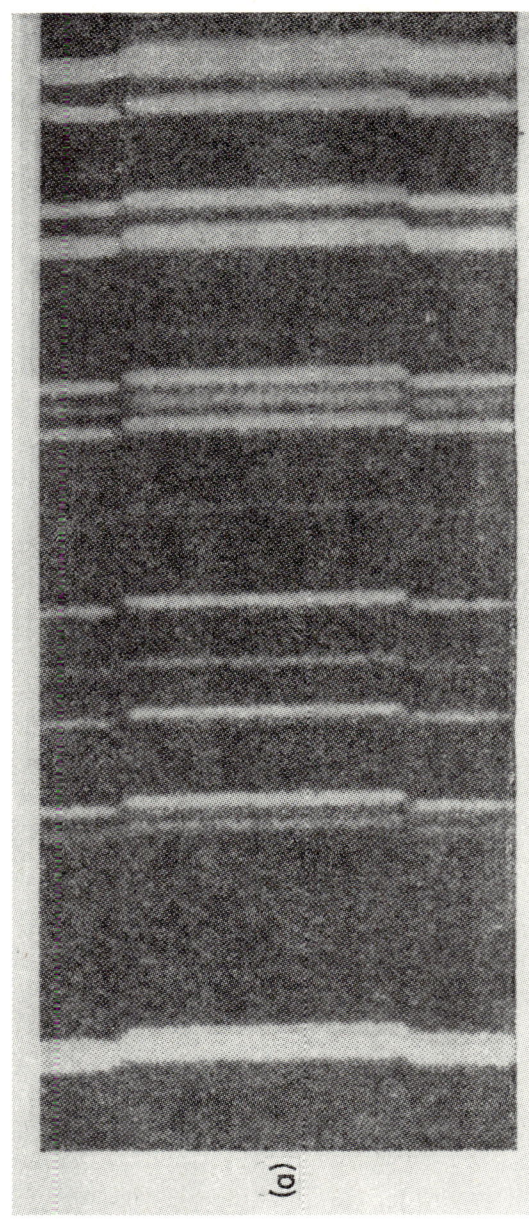

Figure 26-1a

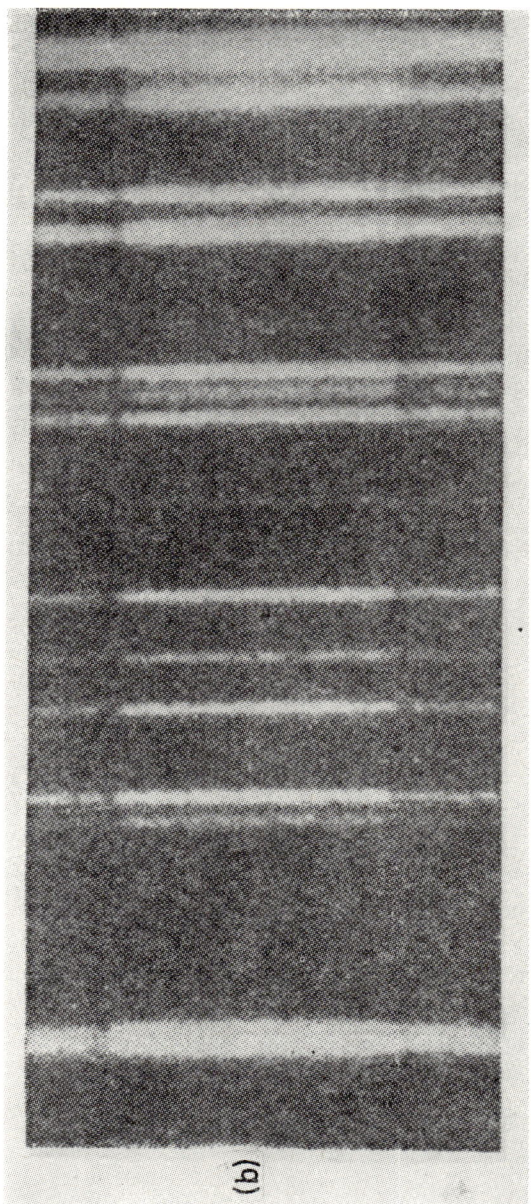

Figure 26-1 Effect of magnetic fields and shield on the image converter. Successive exposures were made after changing the attitude of the telescope-spectrograph by 3 hours in right ascension. Spectrograms obtained using the image converter: (a) without magnetic shield; (b) with magnetic shield.

nically generated and optical contact is achieved by introducing an immersion medium with its index of refraction similar to that of the mica window. In this case practically all the light output from the image on the window is transferred to the photographic emulsion; the resolution is determined by the thickness of the window. In our case the thickness is approximately $20\,\mu$ and the resolution is 0.05 mm (20 lines mm^{-1}).

High sensitivity would be impossible without the use of an emulsion with a spectral sensitivity that matches the spectral emittance of the phosphor on the mica window. The astronomical emulsion A-650 developed at the Kazan NIKFI affiliate is very effective for image converters with phosphors having maximum spectral emittance in the green. The A-650 emulsion deviates very little from the reciprocity law.

The most important aspect of any electro-optical system is the inherent background, which limits the exposure and therefore the minimum detectable signal. An image converter, with its emulsion in contact with the output window of the tube and in which special measures are taken to reduce the instrumental background, gives a noticeable background density on the emulsion only with exposures of the order of ten hours.

As the converter is simple to operate and requires no complicated accessory equipment, it may be used practically wherever ordinary photographic emulsions are used. A minor disadvantage of the converter, as well as of all units of this type, is that the magnetic fields of the earth and of the telescope mount affects it. When the electro-optical instrument moves with the telescope the electronic image is displaced due to the relative change in orientation of the image converter with respect to the magnetic field. For exposures of many hours this displacement becomes intolerably large. For example, with an exposure of three hours, shifts of 0.3 mm in the position of the spectral lines along the direction of dispersion as well as 0.1 mm in the perpendicular direction has been observed. However, these displacements can be eliminated completely by covering the converter with a magnetic shielding (Figure 26-1).

All these converters have small useful field. For photocontact converters the window is about 10 mm in diameter.

On the basis of these facts, we consider the most promising astronomical applications of these converters.

2 DIRECT PHOTOGRAPHY

As a rule, the limiting magnitude of photographic instruments depends on the sky background. In this case, image converters allow only to reduce the exposure time, but do not provide more information. An image converter with a 10 mm diameter useful field and 20 lines mm^{-1} resolution, can give about 30,000 black-and-white elements of information, whereas the ordinary 20 cm × 20 cm photographic plate can store 10^8 elements. The use of highly sensitive electro-optical instruments is therefore justified only when reasonable exposure times are not limited by the sky background. Such cases may occur when studying celestial objects in narrow spectral bands with interference filters or other types of monochromators [4].

In the study of large areas of the sky, therefore, ordinary photography has undoubted advantages over electro-optical instruments, even considering the large gain in sensitivity. The use of highly sensitive converters is recommended only for detailed study of very small isolated portions of the sky and in a given spectral regions. Since to some extent reduction in exposure times minimizes atmospheric effects, converters are recomended also for photography of planets and close binary stars.

3 SPECTROSCOPY

In many cases we are interested in isolating small regions of the spectrum or even single spectral lines. High-transmission spectrographs with small dispersions, in which the total length of the spectrum is only a few tens of millimeters, may be used for spectroscopic studies of faint objects. In these cases, it is very advantageous to use image converters as exposures can be reduced to reasonable values and, in a number of cases, new information obtained. Moreover, image converters are simply irreplaceable for near-infrared observations.

Thus, electron telescopes can be particularly effective when applied to spectroscopy, a fact which should influence the development and design of new spectroscopic instruments.

A spectrograph to be used with image intensifiers or converters must fulfil several requirements. It must be capable of operating in the entire spectral region of the image converter i.e. from the ultraviolet boundary of atmospheric transmission to 12,000 Å. The spectrograph must have a diffraction grating with the necessary dispersion in the infrared (it is more difficult to build a prism spectrograph with suitable dispersion in the infrared). The spectrographic cameras must operate over the spectral region from ultraviolet to the near infrared and must have the focal plane between 8 and 10 mm from the mounting flange. This allows the electro-optical unit to be mounted on the focal plane and also allows for possible cooling of the photocathode. The spectrographic cameras to be mounted on moving parts of telescopes must have magnetic shields. This is to eliminate the effects of changes in the relative direction of the magnetic field respect to the converter due to changes in the spectrograph attitude.

The ideal spectrograph for work with electro-optical devices would have diffraction grating, a purely reflecting collimator, removable cameras corrected for the ultraviolet, visual and near-infrared regions of the spectrum and two detachable gratings. The grating for work in the infrared, would have 600 grooves per mm and blazed in the first order; the second grating to work in the ultraviolet and in the visible would have 1,200 grooves per mm and blazed for a wide region of the spectrum.

A first approximation to such a spectrograph has been built at the Shternberg Institute. It is a high-transmission diffraction grating spectrograph to be used for observations of nebulae, and is intended to operate with photographic film in contact with the mica window of the image converter. The reflecting collimator was made at the Crimean Astrophysical Observatory by V. I. Pronik. So far, the spectrograph has operated only in the infrared. We have used a replica diffraction grating having 600 grooves per mm and blazed in the green in the second order; in the infrared it is blazed in first order.

The spectrograph has two lens-cameras, corrected for the infrared. The short focus camera ($F = 50$ mm, $f/0.85$) gives a dispersion of about 200 Å mm^{-1}. The dispersion with the second camera is 90 Å mm^{-1}. The spectrograph is mounted on the 13-inch reflector at the Shternberg Institute. A magnetic shield made of Permalloy is used to eliminate the effects of magnetic fields on the image converter (Figure 26-2).

Infrared spectra of a number of diffuse nebulae were obtained during the summer of 1960. The new spectrograph allowed us to determine the wavelengths of the forbidden infrared doublet of twice-ionized sulphur (S III) at 9,069.0 Å and 9,531.2 Å, and also the relative intensities of the lines in the spectrum of the Orion Nebula up to 11,000 Å.

It should be noted that in the absence of instrumental background, the basic limiting factor in the study of faint celestial objects in the infrared is the radiation of the upper atmosphere, which is almost 100 times more intense than in the visible (Figure 26-3).

It is proposed to use this spectrograph with a 1,200 grooves per mm grating, appropriate cameras and image converters with SbCs and multi-alkali photocathodes.

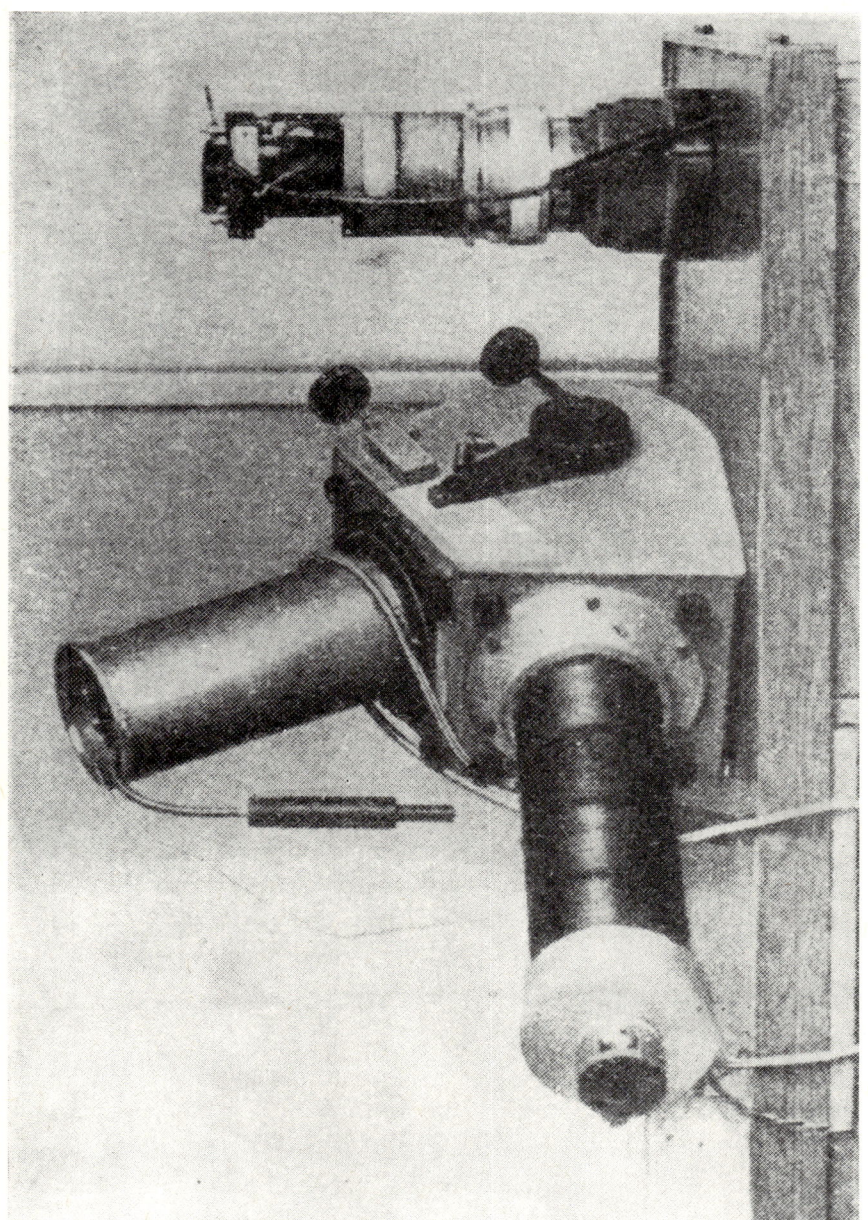

Figure 26-2 High-transmission grating spectrograph using image converter. The camera with the image converter is next to the spectrograph.

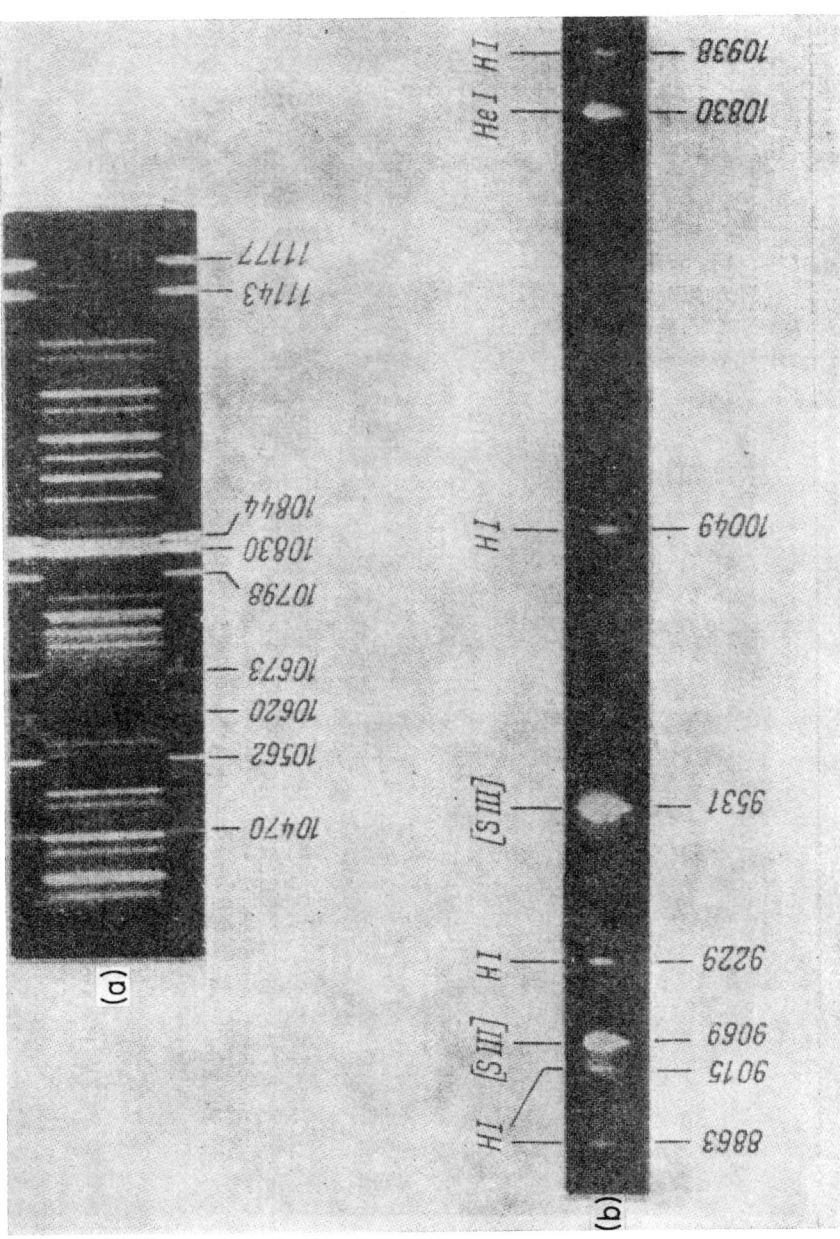

Figure 26-3 Spectra obtained using the spectrograph in connection with the image converter: (a) spectrum of the night sky obtained with a four hour exposure (the comparison spectra is also shown); (b) spectrum of the Orion Nebula in the region 9,000–11,000 Å, obtained with five minutes exposure.

REFERENCES

1. I. V. Volkov, V. F. Yesipov and P. V. Shcheglov, *DAN SSSR*, **129**, No. 2, 288 (1959).
2. V. F. Yesipov, *Astron. Zh.*, **37**, No. 3, 588 (1960).
3. I. V. Volkov, V. F. Yesipov and P. V. Shcheglov, *DAN SSSR*, **137**, No. 4, 840 (1961).
4. P. V. Shcheglov, *Astron. Zh.*, **37**, No. 3, 586 (1960).

CHAPTER **I-27**

A vacuum chamber for testing precision mirrors for astronomical instruments

V. A. SAVIN

Main Astronomical Observatory
Pulkovo

In testing precision optics, including mirrors for astronomical instruments, considerable difficulties are caused by inhomogeneities of the air in the testing room. One can observe "boiling" of the image of an artificial star similar to the "seeing" of the image of stars observed in large telescopes. In observing the shadow pattern one sees distortions which mask the true nature of the pattern given by the mirror under test.

These phenomena are caused by air currents generated by hot or cold objects. It is simple enough to locate and eliminate these currents; heating devices and cold water pipes may be removed, cold external walls, windows, or floors should be avoided, and even the space where the light rays propagate may be thermally shielded by means of a tube.

Less obvious and more dangerous are stratified layers of air, even in rooms which contain no hot or cold sources. This stratification does not allow the floor, ceiling, walls, or even the air itself to reach thermal equilibrium.

The change Δn in the index of refraction n of air versus height due to temperature changes and density changes due to reduction in atmospheric pressure also as a function of height can be expressed by the series

$$\Delta n = a(h - h_0) + b(h - h_0)^2 + c(h - h_0)^3 + \cdots, \qquad (27\text{-}1)$$

where $(h - h_0)$ is the distance between parallel rays (difference in height for horizontal beams) and a, b and c are constants.

The difference δ in the optical path for a distance L relative to a parallel path spaced by $(h - h_0)$ is given by the general expression

$$\delta = \int_L \Delta n \, dl \qquad (27\text{-}2)$$

or

$$\delta_{11} = \Delta n \, L. \qquad (27\text{-}3)$$

The first term of Eq. (27-1) which gives the linear change in Δn, is not important as it merely displaces the image. But the quadratic term affects the image with astigmatism, the cubic term with coma, and so on, i.e. the difficulties are presented by the terms of the series which give a deviation from linearity.

As an example, let us consider a case where $L = 2l = 10$ m. The temperature difference Δt of the air for a given difference in height (that we assume equal to the aperture of the mirror) gives a corresponding difference Δn in the index of refraction of air expressed by

$$\Delta n = (n_0 - 1) \frac{\Delta t}{273}. \qquad (27\text{-}4)$$

By introducing the above values into the Eq. (27-3) and taking as a difference in the optical path $\delta \leq \lambda/4$, we obtained

$$\delta_{11} = \frac{0.00029}{273} \Delta t \times 10^7 \, \mu \leq \frac{\lambda}{4},$$

which gives $\Delta t \leq 0.01\,°C$.

The change Δn due to variations in the air density versus height, to sufficient accuracy is linear, and need not worry us, even when the mirror is large.

The effect of the nonlinearity of Δn versus height is clearly noticeable in the testing of long-focus mirrors. For comparatively small optical paths (a few meters) the effect of nonlinearity in Δn is comparatively small, and to simplify the testing process we do not take it into account. But for especially accurate mirrors, such as the flat mirrors of coelostates, for which the Common test-method is used, this simplification cannot be made.

Various designs for optical testing may be devised to reduce the harmful inhomogeneities of the air in the optical path, but the best method is undoubtedly the use of a vacuum chamber.

A vacuum chamber for testing flat mirrors with aperture up to 500 mm by the Common method was constructed at the initiative of Academician

V. P. Linnik and has been used during the past few years whenever the temperature conditions make it impossible to carry out tests.

The general view of the vacuum chamber to test flat mirrors, designed by V. I. Karkalev, is shown in Figure 27-1.

The chamber consists of an iron tube of about four meters in length and walls 10 mm thick. One end of the tube is close by a bolted convex cover sealed with a rubber gasket.

On the inside of the cover (see Figure 27-2) is mounted a 400 mm aperture, 4,500 mm radius spherical mirror which is used to test flat mirrors by the Common method. The head of three screws for adjusting the spherical mirror can be seen on the cover (see Figure 27-1).

The flat mirror to be tested is mounted in the chamber at 45° angle respect to the optical axis of the spherical mirror (see Figure 27-2). At the opposite end of the chamber is mounted an observing window through which the tests are conducted. This window is made of 3 mm thick plane-parallel glass plate and is 50 mm in diameter. This window used with our spherical mirror introduces a very small amount of spherical aberration.

Figure 27-3 shows an experimenter testing a mirror in the vacuum chamber. In front of the observing window there is mounted an "artificial star" source, a Foucault knife edge tester or a microscope for examining the image of the artificial star given by the spherical-flat mirror system.

In Figure 27-3 we can see the rubber hose from the chamber to the fore pump which evacuates the chamber to a pressure between 5 to 10 mm of mercury; this pressure is reached in 15 to 20 minutes. The use of rubber gaskets and vacuum grease in the joints of the chamber maintains this vacuum for several hours.

When the flat mirror is placed in the chamber, the image of the artificial star may be outside the field of the observing window. In this case a small flashlight bulb is dropped by a wire through the air intake of the chamber (this opening, which is closed by a bolt during operation is located on top of the tube chamber not far from the mirror unit). The image of the bulb due to the reflection on the spherical mirror and the reflections on the flat mirror can be set by the adjustment screws of the spherical mirror. This adjustment can be completed when the light source, the image of the source and the artificial star are aligned approximately with the center of the observing window.

This setting by eye ensures that the image of the artificial star appears in the window of the chamber.

282 New Techniques in Astronomy

Figure 27-1 General view of the vacuum chamber for testing flat mirrors at the Main Astronomical Observatory, Pulkovo.

A vacuum chamber for testing precision mirrors

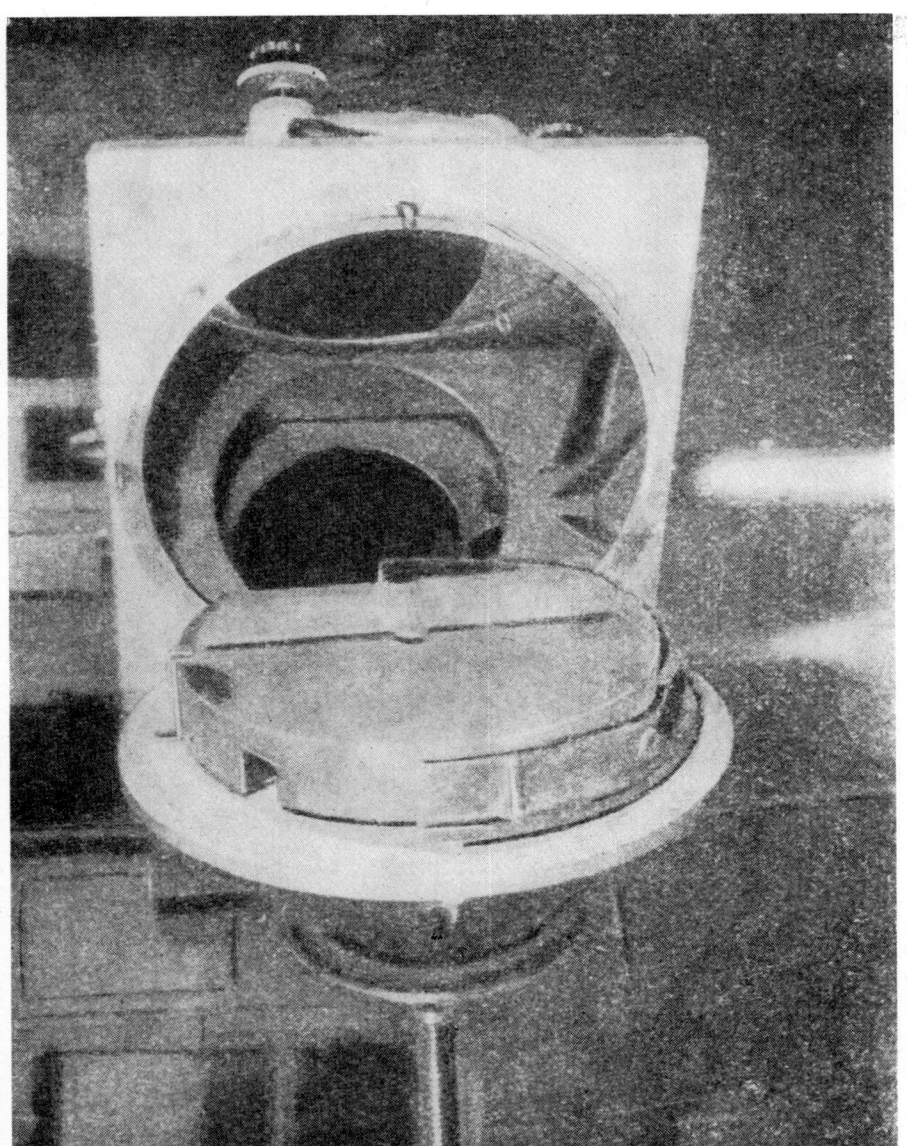

Figure 27-2 View of the chamber with the cover open, showing the spherical mirror mounted on inside at an angle of 45°.

284 New Techniques in Astronomy

Figure 27-3 Experimenter carrying out a test of a flat mirror using the chamber at the Main Astronomical Observatory, Pulkovo.

Our experience with testing mirrors in the vacuum chamber has shown it to be extremely reliable and unequivocal, not only because of the absence of air currents, but also due to the reduction in the heat transfer from the environment to the mirror by convection and radiation.

It is possible to simultaneously use two Foucault testers with vertical and horizontal edges, so that local errors in the horizontal and vertical sections of the mirror and its astigmatic errors may be determined from the shadow pattern.

CHAPTER **I-28**

An interference-polarization filter for astrophysical studies of the sun in the K-line of ionized calcium

S. B. IOFFE and N. M. DRICHKO

Main Astronomical Observatory
Pulkovo

Observations of the chromosphere of the sun in the K-line of Ca II ($\lambda = 3{,}933.7$ Å) using interference-polarization filters (IPF) are of considerable interest, but until recently a filter with the characteristics required for this wavelength did not exist.

In 1958 an experimental model of an IPF for the K-line was produced with a transmission-band half-width of about 0.5 Å (the IT-53). This filter gives high-contrast, richly-detailed pictures of chromospheric formations in the K-line in different parts of the solar disk and at the limb, and has been used for two years in routine observations at Pulkovo.

The IT-53 consists of ten interference-polarization IP stages. The first eight are made of quartz in the manner described by Lyot, while the last two contain calcite elements with additional compensatory quartz plates. The modification was made because of the impossibility of making the calcite plates in the Lyot system to the required accuracy. The peak transmission of the IPF coincides with the wavelength 3,933.7 Å at a temperature of 37.2°C; the accuracy of thermostatic control is ± 0.1°C. The angular field of the filter is about 1°.5, and the peak transmittance is about 1%. The exposure time using RF-3 film is from 0.1 to 1 second.

To carry out normal observations in the K-line together with observations in the H_α-line a set of IPFs must be provided for chromospheric telescopes of solar stations. Due to the high cost and scarcity of the crystal materials used in IPFs, this can only be done when the quartz and calcite can be replaced by cheaper, easily available artificial materials.

An experimental IPF for the K-line (the IT-58) was developed in 1960, in which the Icelandic spar was completely replaced, and the quartz partially (75%) by the artificial crystal ADP (dihydrophosphate of ammonia). The basic characteristics of the IT-58 are the same as those of the IT-53. Since ADP has a larger double refraction coefficient than quartz and calcite, the accuracy to which the operating temperature is maintained in the IT-58 is double that for the IT-53 and is $\pm 0.05°C$.

To construct an IPF using the new crystal material, we had to alter the optical design and develop new methods for assembling and tuning the IP stages for a given wavelength. With the new IP stages the tolerance in the thickness of the crystal plate is sharply reduced, and the precision finishing of the plates is substituted by their accurate orientation.

The tuning of the IP stages to a given wavelength, the matching and the orientation of the separate elements are done on special jigs.

Trials of the IT-58 at the Main Astronomical Observatory show that it gives an image of the chromospheric formations as good as the one given by the IT-53, but that in the central part of the aperture a sharp deterioration in the image quality is observed, due to defects in the cementing of the polarizers. Since this is easily remedied, and is not connected with the use of ADP, it may be concluded that the artificial crystal ADP may be used in narrow-band IPFs instead of quartz and calcite.

We would like to take the opportunity to discuss the use of ready-made "coronal" and "chromospheric" IPFs. Over the course of several years a large amount of experimental material has been obtained using these filters. At the present time, several of them are in need of repair. In particular, renovation of the polarizing filters and immersion is required. It would be advisable that the improvements in the mounting of the optical components and the thermostatic control incorporated in new IPF models be introduced to the older models when they are overhauled. We suggest that a special production team be formed for this purpose in one of the major observatories.

CHAPTER **I-29**

A slitless stellar spectrograph with guiding and spectral reference lines

V. P. LINNIK

Main Astronomical Observatory
Pulkovo

In 1959 we proposed a system using interference reference fringes for slitless stellar spectrographs. Basically, it uses interference fringes superimposed to the spectrum, like Talbot's* bands. If great accuracy is to be attained, these bands must be sufficiently narrow but, the narrower they are, the more bands there are, thus cluttering the spectrum of the star.

The system described below is free of this defect and should be useful for measurement of Doppler shifts. Figure 29-1 shows the schematic of the basic slitless spectrograph with an optical scheme to superimpose reference lines and where P represents the objective prism or, in general, the dispersion element of the spectrograph, L is a cylindrical lens, the axis of which is parallel to the direction of the dispersion and its focal length is 5 to 10 times greater than that of the objective O. A is a good quality plane-parallel plate which, not being cemented, can be attached to or removed from the lens L. The main focus of the objective O is at F, and the focus of the system lens L and objective O is at F'. At F we obtain the expanded spectrum. The image properties at F will depend upon the properties of the corresponding area of the entrance pupil; for example, the image at a will depend on the area A of the entrance pupil.

When the plate A, shown in Figure 29-1, has been positioned and its thickness has been appropriately chosen, we obtain an extended spectrum

---

* In 1837 F. Talbot discovered that if a thin glass plate is place across one half of the beam of a prism spectroscope (applies also to diffraction gratings) a series of dark bands appear across the spectrum. For the phenomenological description and explanation of the Talbot's bands see Ditchburn [1] or Longhurst [2]. (Ed. English version)

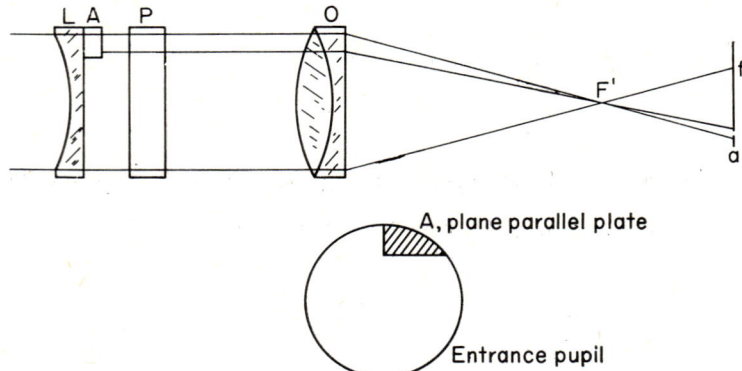

Figure 29-1 Schematic to explain the principle of operation of a stellar slitless spectrograph (for Doppler shift measurements) with an optical scheme to superimpose reference lines: L, cylindrical lens with a focal length 5 to 10 times larger than the focal length of the objective O; A, plane-parallel plate; P, objective prism or dispersion element; O, telescope objective; F, focal plane of the objective O and F', focus of the system consisting of cylindrical lens L and objective O.

of the star in the focal plane F, adjoining the spectrum intercepted by a number of Talbot interference bands. The position of these bands in the spectrum depends only on the thickness of A and the wavelength of the corresponding portion of the spectrum. Thus, these lines can be used as reference lines for determining the position of the spectral lines of the star, similar to the comparison spectra in slit spectrographs.

The Talbot's bands are not the only ones to give reference lines. Instead of the plate A, we could take a plane-parallel plate with a partially reflecting surface twice as long, but in this case the accuracy of manufacture would have to be several times higher and the light loss would be considerably greater.

If the spectrum is to be widened by moving the telescope respect to the star perpendicularly to the direction of dispersion, then we may remove the cylindrical lens (with certain restrictions). In this case the Talbot plates have to be placed on a wedge-shaped plate with the deviation angle equal to the width of the spectrum. The side of the wedge must be parallel to the direction of dispersion with sufficient accuracy. When the star moves during the exposure, by an angle equal to the deviation angle of the wedge, we obtain the spectrum of the star and adjacent its blurred spectrum with the reference lines superimpose. Displacement of the wedge relative to the spectrum will greatly affect the accuracy of measurement. However, if spectrograms taken with the same wedge setting are used for relative measurements, a rotational misadjustment of $0°.5$ will not affect the accuracy.

It can be shown that changes in the width of the comparison interference bands can cause problems in identification of the reference lines. However, the Doppler shift of spectral lines is, in general, considerably less than the distance between two reference lines, and the non-uniform distribution of the reference lines over the spectrum (the distance between bands is smaller in the blue region and larger in the red) allows them to be easily identified by the method to be described below.

A stereocomparator is useful for the reduction of spectrograms with interference reference lines. Two spectrograms are taken: one of the given star and the other of any star in the same spectral class, but with an accurately known radial velocity. Both spectrograms are put on the stereocomparator, and the shift of spectral lines is determined by matching the positions of the reference line closest to some spectral line, and then on the line. All the measurements are reduced to two readings. Generally, the shift of spectral lines stands out immediately with a stereocomparator if the radial velocity is sufficiently large.

This method is strictly differential, since there is a dense set of reference lines around each spectral line. We need, therefore, to obtain only one spectrogram of a star with known radial velocity for each spectral class and then use this set for all future measurements. Small changes in the scale of the spectrograms will not affect the accuracy of measurement since the positions of the reference lines will not change in relation to the adjacent spectral lines. The use of a stereocomparator to reduce spectrograms obtained with an objective prism will be particularly useful in view of the ease with which the spectral lines are located. This applies specifically to the search for stars with considerable radial velocity.

By using a small part of the pupil (about one-tenth of its area), we can guide the spectrum by tracking the image of a star in the focal plane of the camera.

Figure 29-2 gives the schematic of the slitless spectrograph developed at the Main Astronomical Observatory. It consists of a collimator O_1 and O_2, a reflecting grating C and a camera lens O_3, as well as the additional components which form part of the reference lines and guiding systems.

In the model we have constructed, the focal length of the collimator is 250 mm and has a relative aperture $1:5$. The diffraction grating has

600 grooves per millimeter and blazed in the first order at about 4,000 Å. The camera of the spectrograph with 300 mm focal length gives a dispersion of 50 Å per millimeter at 4,000 Å.

The collimator consists of two positive achromatic lenses: the lenses O_2 and O_1 placed near the star image S. The lens O_1 can be moved along the collimator axis for focusing, and perpendicular to this axis for guiding.

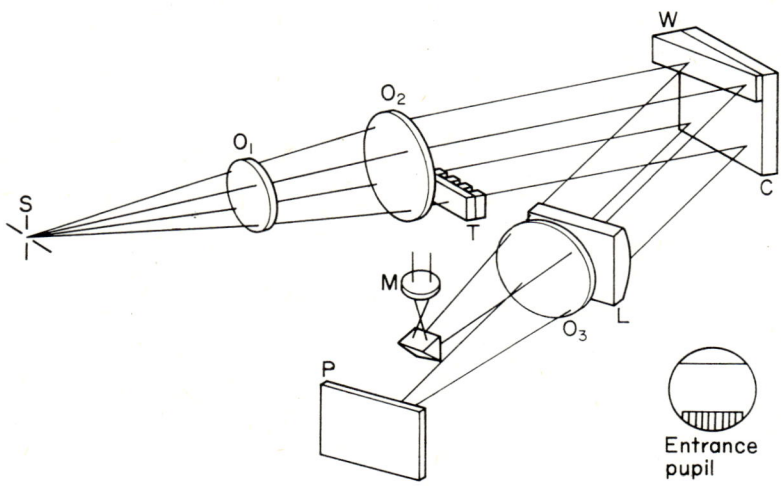

Figure 29-2 Schematic of the slitless stellar spectrograph developed at the Main Astronomical Observatory: S, star image; O_1 and O_2, collimator; T, series of plates of uniform thickness mounted on a plane-parallel plate; C, diffraction grating; W, reflecting wedge; L, cylindrical lens; O_3, camera objective; M, right-angle prism and collimating lens; P, photographic plate.

The lens is mounted in a steel holder and can be adjusted by means of two electromagnets along two axes perpendicular to the optical axis. The driving force of the electromagnets is controlled by adjusting the current by means of a potentiometer. Behind O_2 is the system of plates T which gives the reference lines (Talbot's bands). Instead of a single plate occupying half the pupil as in the usual Talbot system, we have a system made up of a series of plates of 1.5 to 2 mm width and about 1 mm uniform thickness. The plates are 10 mm long, and they are placed across the pupil parallel to one another at intervals equal to their width. This is done by fitting them into contact with a plane-parallel plate. The Talbot bands produced by this system give greater contrast. Moreover, the stellar spectrum which is superimposed on the system of interference bands is generated by a system with equivalent narrow aperture (1.5 to 2 mm) and thus is significantly blurred due to diffraction. In this way we obtain a clear interference pattern on a very blurred background. Since the brightness of the interference maxima is twice as intense as the ordinary spectrum background, we can use twice as small a pupil area to obtain a photographic density equal to that of the basic spectrum.

In our instrument the reference lines system occupies one-fifth the height of the pupil, or about one-eighth of its area, so that only a very small part of the energy falling on the telescope objective is used for the reference lines. Furthermore, since the image of the reference lines is formed by at least twice as small an aperture as that of the telescope, the effect of atmospheric turbulence on the reference lines will be correspondingly smaller.

Finer reference lines may be obtained at the cost of making the system more complex, for example, by using plates whose thickness varies in arithmetic progression.

Placing the system which produces the reference lines in front of the dispersion element makes the location of the reference lines on the spectrum dependant only on the wavelength and the optical thickness of the plates.

The diffraction grating C and the camera are so placed that the first-order spectrum for 4,000 Å will be at the center of the photographic plate P. The reflecting wedge W is firmly fixed to the edge of the grating which corresponds to the part of the pupil opposite the reference lines system. The angle of the wedge is so chosen that the reflected beam gives an image of the star in the focal plane of the camera above the central portion of the spectrum. Not far from the focal plane of the camera, a beam reflected from the wedge is in turn reflected by a prism to the microscope used for guiding. The cylindrical lens L in front of the camera objective is used, as already stated, to extend the spectrum and increase the distance between the images of the reference lines and the stellar spectrum. The lens L clears the beam reflected on the wedge W which is use for guiding. The spectrograph, as usual, is set so that the direction of the spectral lines coincides with that of the diurnal motion of the star on the plate. In this case, very small changes in the position of the star can conveniently be corrected by fine guiding the position of the lens O_1 controlled by electromagnets and damped by rubber washers of special shape. Instead of the microscope, we could use a fairly simple photoelectric guiding system in which the stellar image is projected on either side of an aluminized prism (depending on the shift of the image), and which reflects the light to one of two correspondingly positioned photocells. The current from the photocells, after processing, is used to make the corrections by properly setting the lens O_1.

Since the light beams in the spectrograph and guiding section of the instrument follow the same path, both beams will be equally affected by temperature changes. Therefore, to a great extent, the performance of the instrument will be independent of temperature changes when fine guiding is used as described.

REFERENCES

1. R. W. Ditchburn, *Light*, 2nd edn, Blackie (1963), pp. 329–31.
2. R. S. Longhurst, *Geometrical and Physical Optics*, Longmans (1962), pp. 109–11.

QUESTIONS, ANSWERS AND DISCUSSIONS

1.a FOR CHAPTERS I-1 THROUGH I-4

N. N. Mikhelson Have any attempts been made to reduce the sky background?

V. B. Nikonov Not so far, but it is possible to do this and it should be done in the near future.

V. A. Savin Were photographs taken with the telescope in various attitudes?

V. B. Nikonov Yes, the telescope operates well in any attitude.

P. V. Shcheglov In what photometric system were the stellar magnitudes given?

V. B. Nikonov In the mean B, close to the photoelectric, the limiting magnitude is 19th, with an exposure time of less than 2 seconds.

Yu. N. Semenov Were the photographs taken with the guiding system?

V. B. Nikonov No, because the exposures were of the order of 1 second.

V. A. Savin What was the flexure of the telescope tube for different attitudes of the telescope?

B. K. Ioannisiani The uncompensated flexure of the tube was 0.3 mm, but there is sufficient rigidity in the system for the residual flexure to be compensated; offset guiding is planned to be used with the plateholder.

N. N. Mikhelson What is your opinion of the control systems?

V. B. Nikonov They are still in the debugging stage, and testing has not yet been completed.

V. A. Savin Were the conditions for testing the mirrors at the instrument plant sufficiently accurate?

V. V. Oshurko The operating conditions were not good enough and must be improved, but nevertheless there is no doubt that the mirror is well made.

V. B. Nikonov During the tests at the factory there were very strong vibrations making it very difficult to observe a point source; during tests the telescope gave images 10μ in diameter, i.e. $0.''2$ for a point source.

B. K. Ioannisiani The question of assessment of the quality of the mirror is premature since astronomical tests are not yet complete, but the first impressions are good.

V. F. Sitnik Have the advantages of vacuum spectrographs been definitely established?

P. V. Dobychin No, but some experimenters prefer to work with them.

A. A. Kalpnyak For which region is the spectral long-wavelength apparatus designed?

P. V. Dobychin For the near infrared region.

D. S. Usanov What is the reason for limiting the use of stainless steel in astronomical instruments?

P. V. Dobychin There is no special limitation; it is used whenever appropriate.

P. P. Dobroravin Do the 450-mm telescopes change their settings for latitude?

V. V. Aleksandrov Yes, within small limits (using a special wedge).

G. F. Sitnik What type of telescope will the amateur telescope be?

V. V. Aleksandrov A Newtonian telescope with 100-mm metallic primary mirror. It will cost about 120 rubles.

B. K. Ioannisiani What type of mirror will it have?

V. V. Aleksandrov Spherical.

B. K. Ioannisiani Should amateur telescopes be of the meniscus type?

V. V. Aleksandrov Yes, an amateur meniscus telescope is also being manufactured.

N. N. Mikhelson What kind of mount is used for the amateur telescopes?

V. V. Aleksandrov Altazimuth-type, but an equatorial version will also be made.

V. V. Oshurko Why does the amateur telescope have a metallic mirror?

V. V. Aleksandrov It is intended for inexperienced observers who have no experience with optics.

D. S. Usanov What metal will be used?

V. V. Aleksandrov Chromium steel.

M. S. Zverev Will the telescope retain its properties over a long period?

D. D. Maksutov Yes, its basic properties will be retained.

P. V. Dobychin   What type of glass is used for the correcting plate of the Piazzi–Smyth?

D. D. Maksutov   It is made of K-8 glass and is non-achromatic.

B. K. Ioannisiani   What effect has the lack of centering of the system on the image quality?

D. D. Maksutov   The centering conditions are not very strict: the centering requirement is the same as in the 700/980-mm telescope.

P. V. Shcheglov   What is the future for telescopes of this type?

Yu. S. Streletskii   The answer to that question can only be given after it has been manufactured, debugged and tested.

V. B. Nikonov   When will the PM-700 telescope be installed?

Yu. S. Streletskii   Very soon, the construction of the building is somewhat delayed.

N. N. Pavlov   I would like to congratulate the writers of the reports on the completion of the Shain telescope. During debugging the problem of thermal and atmospheric conditions must be solved. The Shain telescope will give us sufficient experience for other large telescopes. The astrometrists must also work on this problem, which is a difficult one but which will necessarily be successfully solved.

D. S. Usanov   Stainless steel should be more extensively used in astronomical instruments, particularly for the trunnions, mirrors and other components.

D. Ya. Martynov   It is impossible for amateur astronomers to obtain glass blanks for mirrors and abrasives for processing. It would be desirable if the Astrosoviet could assist the All-Union Astronomical and Geodetic Association in improving the situation.

QUESTIONS, ANSWERS AND DISCUSSIONS
1.b   FOR CHAPTERS I-5 THROUGH I-14

N. S. Zhurkin   What is the accuracy of the system under dynamic conditions?

P. V. Nikolayev   The accuracy of setting is not less than 7'; the accuracy of guiding has not yet been determined as the telescope is not yet installed.

Ye. M. Neplokhov Is the guiding system of the telescope overloaded?

P. V. Nikolayev Two systems are used for tracking, both of which can be disconnected by a turn of the VT stator.

A. Ye. Solomonovich When did you begin work on the system?

P. V. Nikolayev Four years ago.

Ye. M. Neplokhov How do the two systems operate?

P. V. Nikolayev In sequence, first the setting system and then the photoguider.

N. S. Zhurkin How near can the telescope approach the zenith?

P. V. Nikolayev To within $5°$.

P. P. Dobronravin Can the photoguider interact with the setting system?

V. P. Yegorov No, the system is similar in this respect to that for an equatorial mount.

B. K. Ioannisiani What advantages has your system over the ZTSh system?

V. P. Yegorov It is simpler, cheaper and readily available; the operation of the two systems is similar and their accuracy is comparable.

B. P. Obrazhevsky What is the difference between the system for guiding the dome and that for the telescope?

V. P. Yegorov The first is simpler and less accurate; a feature of it is that it allows the installation of the telescope off the dome axis of symmetry.

Ye. M. Neplokhov The angle p is a function of what?

V. P. Yegorov It is a function of the hour angle t and the declination δ.

G. L. Bruk What is the range of regulation?

V. P. Yegorov It is up to ten revolutions per day.

A. Ye. Solomonovich How do you input the initial data?

V. A. Myasnikov It can be done in several ways. If the program of observations is known, the inital data can be stored on a magnetic drum or other storage device.

L. V. Ksanfomaliti What type of generator is used for the reference voltages?

P. V. Nikolayev It is a light source and a photodetector.

B. K. Ioannisiani The difference between this system and that in the ZTSh is that in the latter flexure is allowed, and it is not compensated for; the system under discussion corrects for flexure and centering of the optical components is maintained.

P. P. Dobronravin What is the shimmering frequency spectrum?

I. P. Rozhnova So far, observations have been made photographically, so that only shimmering with frequencies of up to 2–3 Hz have been studied; the method would not be suitable for frequencies of more than 100 Hz.

O. B. Vasilev How is the star kept at the vertex of the pyramid?

I. P. Rozhnova It is not.

L. V. Ksanfomaliti Why is electronic modulation used?

I. P. Rozhnova Mechanical modulation causes vibration of the instrument.

N. D. Kalinenkov Does noise interfere with the FEU?

I. P. Rozhnova Yes, in the same way as scintillation, etc.

N. N. Mikhelson What is the accuracy of the system?

Z. N. Mamedova It lies between 1' and 3'.

N. N. Mikhelson Are corrections for refraction required?

Z. N. Mamedova Yes, stars were observed up to $z = 80°$ for which refraction was 10'.

P. V. Nikolayev Why is a frequency range from 15 to 100 Hz required?

V. A. Yasevichus This range was established experimentally.

N. D. Kalinenkov Does the background affect the readings?

V. S. Avedisova Very little; the star image is masked by the iris diaphragm.

V. B. Nikonov Were you speaking of internal or external accuracy?

V. S. Avedisova The measurements were made on three independent photographic plates.

V. B. Nikonov How does a different position of the photographic plate affect the accuracy?

V. S. Avedisova The error when the plate is moved does not exceed 1 to 2%.

P. P. Dobronravin We welcome the development of automatic guiding and setting systems for telescopes. The Crimean Astrophysical Observatory thanks the Institute of Electromechanics, and especially to Yu. A. Sabinin, for the work done on automating their instruments, thus reducing the amount of unproductive work by the observer. The continuation of work on automatic techniques is most important for astrophysics. Unfortunately, little use is yet made of semiconductors and magnetic amplifiers. The reliability of operation of all the systems must be improved. I would like to mention the value of the report by V. A. Yasevichus for amateurs.

A. Ye. Solomonovich Between 1953 and 1954 development work was carried out at the Physical Institute of the Academy of Sciences on a drive systems for the 22-m altazimuth radio telescope, with accuracy of $2'$ to $3'$. The work on automation of optical and radio telescopes has thus been duplicated; measures must be taken to see that this does not happen again. The control system of the radio telescope is now being modernized, using digital differential analyzers. I would propose the following resolution: (a) the research program undertaken at the Institute of Precision Mechanics and Computing Techniques concerning digital differential analyzers be supported; (b) work on optical and radio telescopes should be coordinated.

P. N. Kholopov I wish to discuss the problem of photometers. The iris photometer is valuable in stellar photometry, its advantages being: (a) the range of measurement of between 10 and 12 stellar magnitudes with direct calibration; (b) the reduction of the background effect; (c) the high accuracy of the readings; (d) the high speed of measurement and processing up to seventy stars per hour; (e) the stability of the instrument in operation; (f) the possibility of measuring individual stars in highly populated clusters; (g) the simplicity of operation of the instrument and its universality. Photometers of this kind should be mass-produced.

L. A. Ksanfomaliti The system which Yegorov describes for guiding the dome is a complicated one. The system developed at the Abastumani Astronomical Observatory uses radioactive sources placed on the dome, and radiation detectors on the telescope.

N. D. Kalinenkov A large number of observatories in the USSR are not able to use automatic techniques, due to their complexity and high cost. Simpler, cheaper and more accessible systems must also be developed.

P. P. Dobronravin The coronagraph at the Crimean Astrophysical Observatory has a contact system for guiding the dome.

V. P. Yegorov The system we describe will not be expensive if it is mass-produced. The disadvantage of the Abastumani system is that it cannot operate during setting.

N. N. Mikhelson The dome for the 2m-Universal telescope at Tautenberg (GDR)* has a guiding system which uses infrared sources. I would like to comment on the report by Mamedova that with this system there is no need for refraction correction.

P. V. Nikolayev The presence of relay mechanisms and the other units of automatic control systems interfers with photometry of faint light sources.

V. B. Nikonov Working groups should be convened by our committee for the solution of the various problems of astronomical instrumentation.

P. P. Dobronravin I agree with that and suggest that the appropriate measures be taken.

QUESTIONS, ANSWERS AND DISCUSSIONS
1.c FOR CHAPTERS I-15 THROUGH I-20

V. A. Savin What is the size of the rotating plate? Does it limit the field?

Kh. I. Potter The plate is $0.°5 \times 2°$ and the field is $5° \times 5°$.

O. A. Melnikov Does aberration interfere with the observations?

Kh. I. Potter No, it is within tolerable limits.

V. V. Podobed Was there pull-in of air after the tube was raised?

N. N. Pavlov Yes, it was intense at first, and then fell to 0.5 m sec^{-1}.

G. P. Pilnik What is the reason for the presence of a seasonal wave in relation to standard time?

N. N. Pavlov The presence of such a wave has not been firmly established, since the series of observations was not sufficiently long and the observing program within the series was altered.

G. P. Pilnik In what form is the correction obtained?

---
* German Democratic Republic; East Germany. (Ed. English version.)

N. N. Pavlov The device is photoelectric, the observations are averaged over all stars.

G. P. Pilnik Could the errors be refraction errors, and not instrumental?

N. N. Pavlov No, they could not.

V. V. Podobed How is reading of the wedge carried out?

L. A. Sukharev From the reading head of the screw.

G. P. Pilnik What is the effect of lubrication on the reading accuracy?

L. A. Sukharev The loading on the wedge is of the order of 12 kg, and lubrication is not needed so that the readings are not affected.

G. P. Pilnik What is the reliability of the device, and on what instrument is it used?

P. M. Afanaseva It is completely reliable, and will be used on a photoelectric transit instrument.

N. D. Kalinekov Does the device remain horizontal during observations?

Kh. I. Potter It is not important that it should. It is the positioning of the axes which is important, and this is guaranteed to remain unchanged.

P. G. Kulikovsky Why do the visual line have to be horizontal, if you have a mercury level?

Kh. I. Potter We are sure that the visual line is not off the horizontal by more than an allowable amount.

N. D. Kalinenkov What is the limiting magnitude of the device?

Kh. I. Potter It depends on the way it is used. With a 2 m, $f/10$ telescope, it can be used for 8th and 9th magnitude stars.

D. D. Maksutov The disadvantage of the Danjon astrolabe is the use of a glass prism, since deformation of its surfaces impair the quality of the image. In the proposed system, a mirror is to be used instead, which is preferable but it has other drawbacks. Notably a very large flat mirror for large zenith distances.

D. D. Polozhentsev Instead of a printing chronograph an Afanaseyev–Platonov counter may be used. These counters should be put into mass production.

N. D. Kalinenkov It is impossible to make observations on the astrolabe without monitoring.

Kh. I. Potter The zenith distances must be restricted to between 20° and 70°. In this case the dimensions of the flat mirror will be of the order of 600 mm (for a beam of 200 mm). The mirror may be moved. We now have one of 300 mm, and it will not be particularly difficult to make one of 600 mm. There is no reference point for monitoring, and so the system must be sufficiently stable.

L. A. Sukharev Before the instrument is ordered the flat mirror should be manufactured. Serious interference can be originated by variations in the atmospheric pressure, aircraft boom, etc.

QUESTIONS, ANSWERS AND DISCUSSIONS
1.d FOR CHAPTERS I-21 THROUGH I-26

D. Ya. Martynov What is the limiting magnitude of the system?

A. P. Kundzin With the 200-mm reflector, stars of the classes K and M up to the 8th magnitude were observed with accuracy not lower than 1%.

D. Ya. Martynov What is the observation time?

A. P. Kundzin From 16 seconds to 2 minutes.

N. A. Dimov What is the limiting magnitude of the photometer?

L. V. Ksanfomaliti The photometer can be used for stars down to the 11th magnitude.

Ye. M. Neplokhov Why are photoconductors used instead of bolometers?

V. I. Moroz Photoconductors are more sensitive, their long wavelength cut off is 2μ and so radiation from the telescope is not a problem.

L. V. Ksanfomaliti What is the resistance of the photoconductor used?

V. I. Moroz From 2 to 10 MΩ.

N. F. Kuprevich What EOP is used and what is the diameter of the screen?

P. V. Shcheglov It is a single-stage contact EOP, with a 10 mm diammeter screen.

A. A. Kalinyak What is the advantage of your instrument over a monochromator, and what separators are used?

P. V. Shcheglov Our system has higher transmittance; the separators have thickness up to a few tenths of a millimeter.

N. D. Kalinenkov What requirements do the circuit components have to satisfy?

N. A. Dimov The capacitors must have polystyrene as dielectric with leakage resistance not less than 10^{13} ohm.

N. M. Shakhovskoi Recording of integration and the constant component are done simultaneously.

N. D. Kalinenkov How is the angle of the polarization plane determined?

N. M. Shakhovskoi From astronomical observations and by averaging the data.

L. M. Kotlyar An intensity microphotometer has been developed and is being used at the Pulkovo Observatory. The MF-2 type is not automatic. Attempts have been made to improve the manufactured instruments or to develop new systems. The manufactured instruments must be modernized to comply with present-day requirements.

N. N. Mikhelson When a number of different microphotometers exist, it is important to first decide which one is the best, and only then recommend production.

P. V. Slavenas Automatic techniques should be introduced to a great extent in all observatories. Supply of the individual units and electronic components and other equipment should be organized.

V. B. Nikonov The problem of automatic processing of the results has not been discussed at the conference. The astroinstrumentation committee should organize a conference on this question. Industry should produce IPFs of large diameter.

L. V. Ksanfomaliti The polarimeter at the Abastumani Astrophysical Observatory appears to be better than that at the Crimean Astrophysical Observatory.

N. D. Kalinenkov The production of high-quality FEUs for astronomical observations should be recommended.

D. Ya. Martynov Instruments available abroad should be considered in selecting a microphotometer system to be recommended for industrial production. For this purpose a group of astronomers should visit those foreign observatories with original and successful instruments.

Ye. K. Kharadze All the instruments produced by industry should be modernized. The supply of FEUs, EOPs and FSs and other light detectors should be improved.

O. A. Melnikov Yes, this is particularly necessary for the infrared, submillimeter and millimeter regions of the spectrum, which have not been studied despite their indisputable importance (especially in the case of the planets).

V. B. Nikonov Astronomers must work out the technical specifications for the various instruments, and the Committee should form special working groups for this purpose.

QUESTIONS, ANSWERS AND DISCUSSIONS
1.e FOR CHAPTERS I-27 THROUGH I-29

G. V. Borodina Could one make a grating with a hole in the center?

F. M. Gerasimov Yes, this could be done.

G. V. Borodina How wide is the transmission band, and can it be shifted?

S. B. Ioffe The transmission band is 0.5 Å, with a transmittance of about 1%; the shift of the band within small limits is achieved by temperature adjustment.

O. A. Melnikov Is it possible to affect the IPF electrically?

S. B. Ioffe Yes, experiments concerning this point are being conducted at the present time.

N. N. Mikhelson Does decentering of the system when the lens is put in position impair the image?

S. B. Ioffe No.

I. V. Slavenas What are the light losses?

V. A. Savin About 0.1.

I. M. Kopylov What are the characteristics of the grating used?

V. A. Savin 600 grooves per millimeter; at about 4,000 Å the dispersion is 50 Å mm^{-1}.

SECTION II

Preface to Section II

The articles in this book (Section II) are the proceedings of a conference on astronomical instrumentation which took place in May 1964 in Kazan.

During the three years which have elapsed between this conference and the previous one, the proceedings of which have been published*, a great amount of work has been done in industry and in observatories on the construction of new instruments and on analysis and debugging of existing instruments. Research has been carried out on the calculation of new optical systems, the development of modern radiation detectors and the application of electronics and other modern techniques to astronomy.

Further progress has been achieved by the construction of the Shain 2.6-m telescope and its accessory equipment, and new large- and medium-sized telescopes for other Soviet observatories. The development of fast cameras for spectrographs has been of value in increasing the efficiency of the large telescopes. Another important area in which considerable success has been attained is the development of techniques for using image converters. A great amount of attention is being given to the improvement of laboratory instruments and to the introduction of automatic techniques. The large volume of theoretical and experimental work on the planning for new telescopes is also noteworthy. The research carried out in all observations requires the application and mastery of new techniques; the full extent of this was only partially reflected in the reports presented at the conference.

* Section I of this book.

The Commission on Astronomical Instrumentation and the editor of these contributions wish to thank the Rectorate of the Ulyanova-Lenin Kazan State University, the Engelhardt Astronomical Observatory and particularly Sh. T. Khabibullin, A. A. Nefedev and P. S. Konev for their hospitality and their competent organization of the conference.

Main Astronomical Observatory   N. N. MIKHELSON
Pulkovo, USSR

CHAPTER **II-1**

Fast optical systems developed at the Crimean Astrophysical Observatory and their application to astrophysics

G. M. POPOV

Crimean Astrophysical Observatory
Nauchny

---

1 INTRODUCTION

Today, astrophysics and technology are in great need of fast optical systems with large fields of view, of the order of tens of degrees. The widely used Maksutov meniscus systems operates well with small relative apertures, but an increase in the aperture ratio is limited by the residual spherical aberration.

The use of modified Maksutov systems are not recommended, as they are complicated and their cost of manufacture is high.

Although Slevogt–Richter systems (with an afocal two-lens corrector) give a large aperture ratio (up to 1) they have a limited field of view, while the Schmidt system, which can have an aperture ratio near the theoretical limit, is difficult to manufacture. The various modifications of the Schmidt camera (such as the Baker "Super-Schmidt") suffer from the same difficulty.

Research into several concentric optical systems was undertaken at the Crimean Astrophysical Observatory and the results obtained form the basis of this report.

2 CONCENTRIC SYSTEM USING A DOUBLE PASS THROUGH THE MENISCUS

Let us consider the concentric optical system shown in Figure 1-1. It consists of a meniscus and a spherical mirror; the meniscus is used twice over, once in the incoming beam and again in the beam converging from the primary. The radii of the surfaces are chosen to minimize the spherical aberration for an object at infinity. The characteristic feature of this system is the change in sign of the spherical aberration when the meniscus is of a certain thickness, so that the spherical aberration then takes its minimum value. The system can have an entrance aperture up to 200 mm at about $f/0.7$ or $f/0.8$ and a confusion disk not larger than 0.03 mm in diameter. The chromatic aberration can be easily corrected by means of a "chromatic plate", i.e., a plane-parallel plate made of two lenses cemented together, with the same mean refraction indices, but with different dispersion.

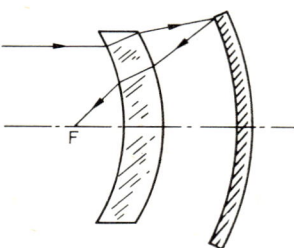

Figure 1-1 Concentric system with a double pass through the meniscus.

The system corrected for chromatic aberration can be recommended for the photography of meteors, artificial earth satellites, and for the study of aurorae and zodiacal light. The system can be used as a fast spectrographic camera (in which case the chromatic plate will obviously be unnecessary). For example, we have used it in a nebular spectrograph mounted at the Cassegrain focus of the 50-in. reflector. The aperture of this camera is 80 mm, $f/0.6$, and has a usable field of $35°$; the spectral lines on the photographic emulsion are not wider than 0.03 mm. Since the field is curved, the photographic emulsion is properly shaped by the cassette.

3 SOLID CONCENTRIC SYSTEM

Let us consider the system shown schematically in Figure 1-2, which consists of a single optical component with three surfaces. The first and second surfaces are concentric; on the third lies the focal surface. Since the spherical aberration can be corrected almost entirely, systems with entrance apertures of about 100 mm up to $f/0.5$ and field of view of up to $20°$ may be used. The system has only small light loss and dispersion, provided that the glass of the optics has a small light absorption.

A variant of the system has a flat field (Figure 1-3), and this has been successfully tested at the Crimean Astrophysical Observatory. It has a 400 mm entrance aperture, $f/0.45$, and a field of view of $7°$; the spectral lines on the emulsion are not wider than 0.03 mm.

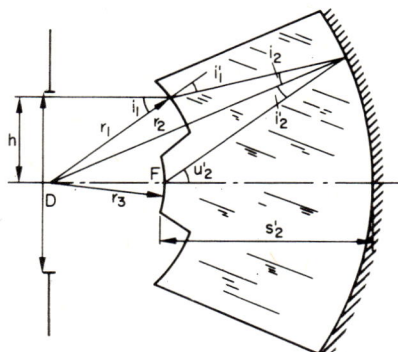

Figure 1-2 "Solid" concentric system.

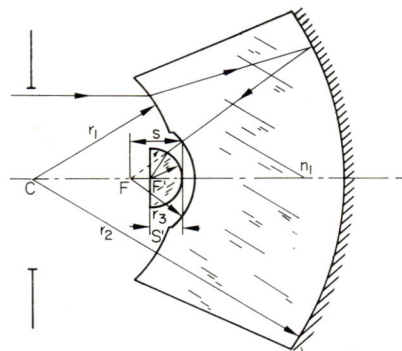

Figure 1-3 "Solid" concentric system with flat field.

3.1 MODIFICATIONS OF THE SOLID SYSTEM

3.1.1 Solid system with a meniscus

The "solid" concentric system may be supplemented by a concentric meniscus (Figure 1-4) to improve the spherical aberration correction. The aperture ratio may then be increased to 0.4 for 100 mm aperture. The obstruction by the cassette section is considerable (more than 30% of the light is blocked). The emulsion is immersed on the exit surface of the system.

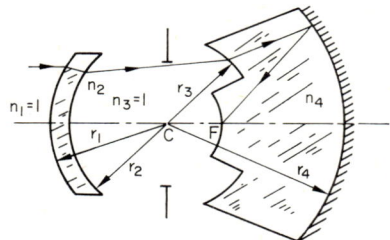

Figure 1-4 "Solid" concentric system with meniscus.

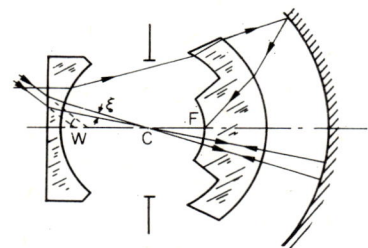

Figure 1-5 "Solid" system with negative lens.

3.2.2 Solid system with a negative lens

The system shown schematically in Figure 1-5 has considerably less obstruction when the aperture is large (100 mm) and the aperture ratio is high (up to 0.4), since the cassette lies in a divergent beam.

While both systems possess certain defects (a curvature of field, uncorrected chromatic aberrations) this does not prevent their use in spectrographs, for photography in monochromatic light, etc.

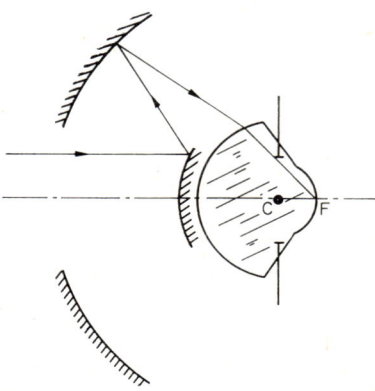

Figure 1-6 Two-mirror concentric system.

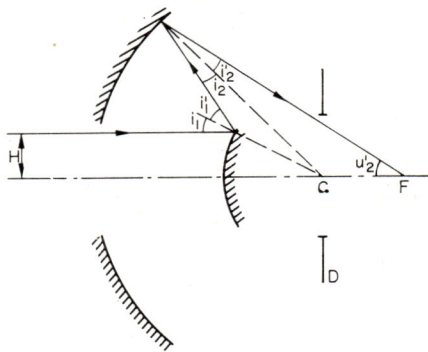

Figure 1-7 Two mirror concentric system with aplanatic lens.

4 A TWO-MIRROR CONCENTRIC SYSTEM

The systems which we have described have an almost inaccessible focal surface, so that image converters, television image pick-up tubes and other bulky equipment cannot be used. Moreover, optical systems with refractive components are not suitable for work in the far ultraviolet and infrared regions of the spectrum. We have investigated the use of an optical system consisting of two concentric spherical mirrors (Figure 1-6) with an accessible focal surface. The field reaches 15° with a 50 mm aperture, $f/0.8$. The system has been used at the Crimean Astrophysical Observatory to lengthen the equivalent focal length of the Slevogt–Richter system with a 640 mm entrance aperture, $f/1.4$. The two-mirror system has 20 mm aperture, $f/1.4$ and 28 mm focal length. The aperture ratio of the two-mirror concentric system can be increased to 0.4 by the use of an aplanatic lens (Figure 1-7).

In conclusion we note that these systems may be used not only in astrophysics (for the study of meteors, nebulae, zodiacal light, luminescence of the night sky) but also in other areas of physics (spectroscopy) and technology (such as the projection of television images on a screen).

REFERENCES

1. G. M. Popov, "A new fast camera with spherical optics", *Izv. KrAO AN SSSR*, **27**, 309 (1962).
2. G. M. Popov, "Some fast catadioptric systems", *Izv. KrAO AN SSSR*, **28**, 341 (1962).
3. G. M. Popov, "A two-mirror concentric system", *Izv. KrAO AN SSSR*, **29**, 318 (1963).
4. G. M. Popov, "A fast mirror–lens system", *Izv. KrAO AN SSSR*, **30**, 320 (1963).

CHAPTER **II-2**

A catadioptric telescope

P. P. ARGUNOV

Odessa Construction-Engineering Institute
Odessa

The development of observational astronomy has put greater and greater demands on the design and quality of telescopes, starting from the small instruments needed by educational institutions and the large groups of amateurs, to the largest instruments used in observatories. Regardless of the difference in size and design, at the present time the most important requirements for practically all telescopes are high optical quality, compactness, accuracy, simplicity of manufacture, convenience in operation and if necessary, universality. Unfortunately, it cannot be said that any of the existing telescopes satisfies all these requirements.

The advantages of refractors include simplicity of maintenance and reliability, but good aberration correction is obtained at the cost of a large f-number, that means long and cumbersome instruments with the detrimental large tube flexure. For a large refractor difficulties arise in the manufacture of the mount and especially of the objective. Further serious disadvantages are the unavoidable residual chromatism and light absorption due to the thickness of the glass, which are particularly important in large instruments, and the unsuitability of refractors for work in the ultraviolet and middle or far infrared regions.

Reflectors are not affected by chromatic aberrations, are suitable for all wavelengths*, and can have small f-numbers, so that they are considerably shorter. Also it is much simpler to manufacture large mirrors than large lenses. However, ordinary fast reflectors using a paraboloid as the main mirror have only a very small usable field, due to coma. Another important

---

* Depends only on the optical properties of the metallic film (aluminum, gold, silver, etc.) deposited on the mirror substrate. (Ed. English version.)

disadvantage is that non-spherical mirrors cannot be manufactured by automatic techniques, and also it is difficult to monitor the surface figure. Thus, in practice, it is difficult to obtain good correction of spherical aberration, particularly for large fast mirrors.

Some authors, including Schwarzschild, Chrétien, and Maksutov have suggested constructing reflectors with two or three mirrors which have nonspherical surfaces, in order to correct for coma. The manufacture of such mirrors is considerably more difficult than that of the paraboloids for classical reflecting instruments; besides, these aplanatic reflectors possess as a rule, large field curvature and astigmatism. Experience in the construction of such instruments for the Washington, Indianapolis and Byurakan Observatories has shown that they are highly sensitive to decentering and that they have low optical qualities in general, this being explained to a great extent by the exceptional difficulty of manufacturing accurate large nonspherical optical surfaces.

Telescopes using a combination of mirrors and lenses could have smaller f-numbers (together with better control of aberrations) than pure refractors or reflectors.

It should be noted, however, that in almost all the catadioptric systems suggested until now, of which the Schmidt and Maksutov systems are the most common, the lenses are placed in the ray path between the object and the mirror. The diameter of the lenses limits the aperture of the telescopes. This greatly complicates the manufacture of large telescopes of this type and also makes it difficult for use outside the visible region. With the lenses placed in this way, the quality of the glass and the accuracy of the surfaces must be much higher than those of a refracting objective. Also note that in the case of the Schmidt system the chromatism introduced by the correcting plate is fairly substantial; for example, in the instrument mounted at the Abastumani Observatory, the chromatic difference attains 5×10^{-4} of the focal length for a narrow interval of wavelengths from 4,000 Å to 5,000 Å, which is greater than in a photographic refractor. Additional inconveniences are the considerable curvature of the field and the position of the focal plane inside the telescope tube.

Similar systems have been proposed by Baker, Linfoot, Hawkins, Richter and Slevogt, Penning and many others, where attempts have been made to improve the chromatism correction and other aberrations by means of refractive components of various types placed in front of the main mirror. Sometimes lenses are placed in the beam reflected by the mirror. But all heset systems have the defects mentioned above and many of them, in addition, are characterized by their exceptional complexity. The majority of these suggestions have not been put into practice.

This group of catadioptric telescopes includes instruments in which there are lenses only in the reflected beam from the main mirror. Such systems have been proposed by Ross, Sonnefeld and Slyusarev. In the Ross system, the main mirror is a paraboloid; limited coma correction by means of two cemented lenses is possible. For example, in an ordinary reflecting telescope ($f/4$), the meridional coma at the edge of a 3′ field already exceeds 1″. The same system with the use of the Ross corrector increases the usable field to only 15′ to 20′. The Sonnefeld is a similar system, but the main mirror has a complex aspherical surface, which gives better coma correction at the cost of an increase in the astigmatism. In both systems, the focal plane is inside the telescope tube. In the Slyusarev system, the main mirror is an $f/5.6$ spherical, but aberration correction is achieved by means of five lenses, one of which has a reflecting surface. Data about the degree of aberration correction in this system are not available.

It is our purpose to design an aplanatic system as simple and as fast as possible consisting of a minimum number of optical components and with spherical surface only. Undoubtedly the catadioptric instruments in the last of the groups we have considered have a number of important advantages, due to the comparatively small dimensions of the lenses, but the Slyusarev system still seems to be too complex and not fast enough.

Preliminary analysis showed that good correction of the basic aberrations of the telescope could be obtained with a correction element consisting of only two lenses placed in the convergent beam reflected on the main mirror. Figure 2-1 shows the layout of the telescope; the lenses 2 and 3 are separated by an air gap. Lens 3 has a reflecting back surface; as in the

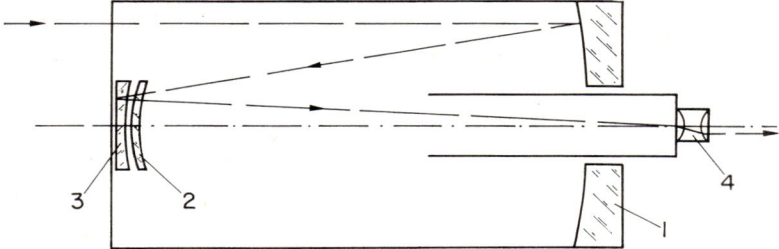

Figure 2-1 Optical layout of the catadioptric telescope developed at the Odessa Construction-Engineering Institute: 1, main spherical mirror; 2, positive lens; 3, mirror; 4, eyepiece.

Cassegrain system the beam reflected by it passes through the central hole in the main mirror to the eyepiece 4. Calculations show that with this system it is possible to correct simultaneously up to third-order spherical aberration and coma, as well as chromatic aberration for two wavelengths, as opposed to the general case when a minimum of four refractive surfaces with independent curvature are needed for similar correction.

In order to establish the best basic design features, such as the type of glass to be used, the position of the correction element and secondary focus, subsequent computations were carried out using Gaussian formulas and third-order theory of aberrations. The effect of these factors on residual chromatism and on the curvature of the lens surfaces, which is a fairly good indicator of the effect of high-order aberrations, was investigated.

We first determined that the lens 2 should be manufactured of a glass with large dispersion (Flint), since the surfaces have less curvature and can be better positioned. Figure 2-2 gives the parameter z (which characterizes the diameter of the confusion disk due to chromatic aberration) versus the distance d_2 from the vertex of the main mirror to the secondary focus; the distances d_1 from the vertex of the main mirror to the correction ele-

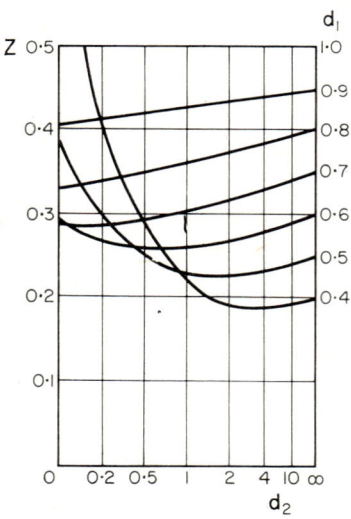

Figure 2-2 The values of z, which characterizes the diameter of the confusion disk due to chromatic aberration, versus d_z, which is the distance from the vertex of the main mirror to the secondary focus; d_1 is taken as a parameter. These curves have been normalized for $f_0 = 1$.

ment is taken as parameter for the family of curves. These curves have been normalized for a focal length f_0 of the main mirror, $f_0 = 1$.

Similarly, Figure 2-3 gives the values of t (which are proportional to the curvature of the lens surfaces) versus d_2; d_1 is taken also as a parameter.

The family of curves given in Figures 2-2 and 2-3 are useful to establish design criteria since it shows the effects of d_1 and d_2 on the image quality. The distance d_2 may not be chosen arbitrarily; for a Cassegrain or Nasmyth system, this distance is about 0.1–0.3, and for a Coudé system between 1 and 2. If we wish to reduce aberrations or increase the aperture ratio with a given degree of correction, it would be most advantageous to take d_1 between 0.6 for the Cassegrain system and 0.4 to 0.5 for the Coudé system. However, in this case the lenses would be too large, so that the best compromise is to take the values $d_1 = 0.7$ and $d_2 = 0.2$ for the Cassegrain system, and this will give relatively little increase in the aberrations. In this case the equivalent focal length f is $3f_0$.

The formulas obtained show that the curvature of the inner lens surfaces (which is greater than that of their outer surfaces and in general depends little on the factors considered above) is directly proportional to the quantity

$$C = \left(\frac{n_3 v_3}{n_3 - 1} - v_2 \right) : \left(\frac{n_3 + 1}{n_3} v_3 - \frac{n_2 + 1}{n_2} v_2 \right), \qquad (2\text{-}1)$$

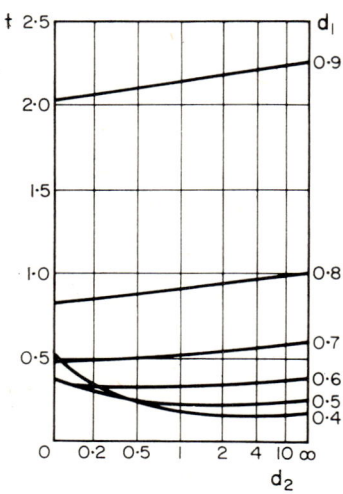

Figure 2-3 The values of t, which are proportional to the curvature of the lens surfaces, versus d_2; d_1 is taken as a parameter. These curves have been normalized for $f_0 = 1$.

Table 2-1 Longitudinal chromatic aberration of a catadioptric system for different characteristics of the correcting elements and using different glass-type combinations

Variant	Glass-type lens				Surface curvature of lenses					Longitudinal chromatic aberration $\times 10^4/f$	
	No. 2	No. 3	No. 4		No. 2		No. 3		No. 4		
					ϱ_1	ϱ_2	ϱ_3	ϱ_4	ϱ_5	ΔS_D	$\Delta S_{G'}$
1	TF 5	TK 5			−2.34	−3.46	−2.80	0.15	—	2.56	−10.60
2	TF 3	TK 4			−2.15	−3.65	−3.18	0.15	—	2.39	−10.45
3	TF 3	BK 2			−2.65	−3.87	−3.30	0.19	—	2.16	−9.30
4	OF 3	BK 2			−2.33	−6.28	−6.24	0.21	—	0.86	−4.94
5	OF 3	CaF$_2$			−3.05	−1.30	−3.68	0.21	—	0.48	−2.17
6	TF 3	BK 2	K 7		−2.47	−4.34	−4.15	0.21	19.33	0.00	−8.59
7	OF 3	BK 2	K 7		−2.36	−6.99	−7.04	0.24	6.04	0.00	−4.05
(1)	TF 5	TK 5			−1.91	−2.87	−2.26	0.27	—	—	—
(2)	TF 3	TK 4			−1.73	−3.06	−2.66	0.27	—	—	—

where n_2 and n_3 are the mean refraction indices (n_D) of the lenses 2 and 3 and ν_2 and ν_3 the respective dispersion coefficients. At the same time, the residual chromatic aberration is directly proportional to the quantity

$$B = (P_3 - P_2) : \left(\frac{n_3 + 1}{n_3} \nu_3 - \frac{n_2 + 1}{n_2} \nu_2 \right), \tag{2-2}$$

where P_2 and P_3 are the respective dispersions. The need to reduce the quantities B and C simultaneously is partly self-defeating since is necessary to select glass with as high index of refraction as possible, especially for lens 2, and, on the other hand, with very different dispersion coefficients. At the same time, the difference in particular dispersions must be minimal.

Table 2-1 gives the results of the calculation for certain combinations of glass. The values of the surface curvature of the lens 2 are given as ϱ_1 and ϱ_2, and for the lens (3) as ϱ_3 and ϱ_4. It can be seen, for example, that replacing the heavy crown TK 4 of lens 3 by the barium crown BK 2 gives an increase in curvature of only 4–6% in ϱ_2 and ϱ_3, while if the heavy flint TF 3 of lens 2 is replaced by the special short flint OF 3 the curvature is increase by 60–90%. If we use a still lighter flint, such as type OF 1, then ϱ_2 and ϱ_3 can be increased as much as 2.5–3 times (as this combination is not suitable it is not given in Table 2-1). At the same time, the use of BK 2 or especially of OF 3, with reduced spectral transmittance, reduces residual chromatism, as shown by the values ΔS_D and $\Delta S_{G'}$ given in the Table 2-1. The curvature of the outer surfaces, ϱ_1 and ϱ_4, depends very little on the type of glass used. Calculations were also made for variant No. 5 using fluorite (CaF_2) and short flint OF 3, which is especially good for reducing residual chromatism. In this case, too, the lens curvature is small.

Further reduction of chromatism is possible if a three-lens correction element is used. Calculations have been done for combinations consisting of a positive lens made of flint and a two-element negative lens made of two different kinds of crown glass cemented together and separated from the flint lens by an air gap. It can be seen from Table 2-1 that the best combination of glasses is OF 3, BK 2 and K 7, for which the residual chromatism is almost zero, and the surface curvature ϱ_5 of the interface of the two-element lens is moderate. There will be practically no reflection or refraction at this surface, since the mean refraction indices of the two kinds of crown are almost identical and the surface of the two lenses are in contact.

Residual chromatism of different telescopes given as the ratio of longitudinal chromatic aberration ΔS_h to the focal length f of the system versus wavelength, are given in Figure 2-4. The curves are for the Fraunhofer achromate class E, the Sonnefeld AS semiapochromate, the Zeiss apochromate class A and for the catadioptric telescope using corrector variants 2, 4 and 7 (see Table 2-1). It can be seen that even the simplest combination, No. 2, with an ordinary kind of glass, approaches the apochromate with respect to the quality of correction, while the two others are considerably better.

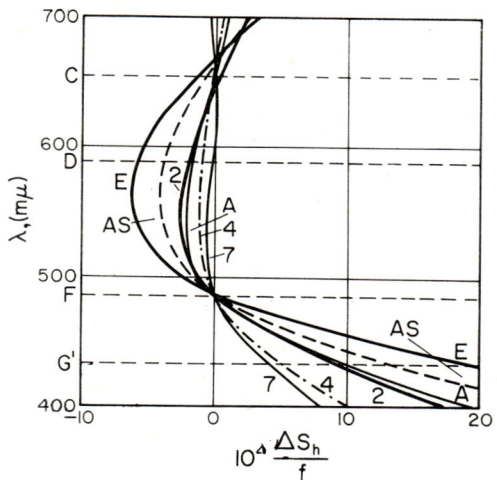

Figure 2-4 Residual chromatism of different telescopes given as the ratio of longitudinal chromatic aberration ΔS_h to the focal length f of the system versus wavelength. The curves are for the Fraunhofer achromate (E), Sonnefeld semiapochromate (AS), the Zeiss apochromate (A) and for the catadioptric telescope using corrector variants 2, 4, 7 given in Table 2-1.

Note that in our instrument the residual chromatism is of the opposite sign to that of a refractor, which is advantageous if the residual chromatism of the eyepiece is used for compensation. This fact is not allowed for in the graphs given in Figure 2-4 as it would complicate the picture.

After we had determined approximate values of the surface curvature using third-order theory of aberrations we made accurate and detailed trigonometric calculations to obtain the relationship between the higher-order aberrations and the design data of the lenses, and to select the best values. We used a trial and error method in which the calculations had to be satisfied to seven decimal places. A large number of variants were considered, with different values of the surface curvature for the first two glass combinations given in Table 2-1. In this table the final values of the curvature, rounded to two decimal places, are given. If we compare them with the values obtained from the third-order theory, we see how inaccurate these values were. Clearly the accuracy of the latter is limited in a preliminary analysis of the correction capabilities of the given optical system when its design data are chosen only approximately.

Starting from actual possibilities of manufacturing the experimental model, we selected the TF 3 and TK 4 glass combination. For this combination, Figure 2-5 gives the ratio of the distance h between the optical axis and the point of entrance of the ray to the primary mirror focal length f_0 versus the ratio of longitudinal spherical aberration ΔS_c to the system equivalent focal length f. All the curves are for variants a, b, c and d of the catadioptric system corrector with corrected coma and visually corrected chromatism. The exact values of the surface curvatures of lens 2 for these variants are given in Table 2-2. The surface curvature of lens 3 is the same for all variants: $\varrho_3 = -2.6615$, $\varrho_4 = 0.2679$. Positive curvature corresponds to the surface being concave to the left in the system of Figure 2-1. The axial thickness of lens 2 is equal to 0.012, of lens 3 to 0.008, the distance between them is 0.001, the distance between the vertices of lens 2 and of the main mirror is 0.688. The equivalent focal length f is 3.

Table 2-2 Performance of a catadioptric system for different characteristics of the correcting element No. 2 and using a TF 3 and TK 4 glass-type combination

Variant	Surface curvature of lens No. 2*		Relative aperture	Diameter confusion circle	Max. useful magnification
	ϱ_1	ϱ_2			
a	−1.7486	−3.0806	0.18	0″.04	3000
b	−1.7412	−3.0732	0.21	0.10	1200
c	−1.7338	−3.0658	0.24	0.19	630
d	−1.7190	−3.0510	0.30	0.50	240

* The surface curvature of the lens No. 3 is fixed for all variants: $\varrho_3 = -2.6615$ and $\varrho_4 = 0.2679$.

An investigation based on caustic surface showed that the smallest confusion circle due to spherical aberration will be obtained if the entrance pupil radius is taken equal to $0.89h_0$, where h_0 is the height of the ray at the entrance corresponding to $\Delta S_c = 0$. These heights are marked for each variant by the circles on the curves given in Figure 2-5. Using these

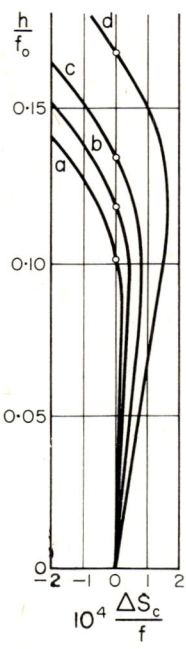

Figure 2-5 Ratio of the distance h between the optical axis and the point of entrance of the ray to the primary mirror focal length f_0 versus the ratio of longitudinal spherical aberration ΔS_c to the system equivalent focal length f. These curves are for variants a, b, c, and d of the catadioptric system corrector (see Table 2-II).

values we give in Table 2-2 the value of the best relative aperture of the main mirror and the diameter of the confusion disk at the center of the field, expressed in seconds of arc. The last column gives the angular magnification for which the visible diameter of the confusion disk is 120″, i.e. is equal to the normal resolution of the eye. These values characterize the maximum useful magnification.

The values in the last column show that even for the largest instruments intended for visual earth-bound observations, variant c (Figure 2-5) may be used, since atmospheric turbulence practically never allows the use of a magnification greater than 600. But for photography, the limit resolution usually does not exceed 0″.5, and so we can use variant d with relative aperture 1 : 3.3. This variant is also suitable for small instruments. Variants a and b, therefore, are of value only for observations from outside the Earth's atmosphere.

The residual meridional coma for rays with entry height $\pm 0.67 h_0$ is 1″ at the edge of 1° field.

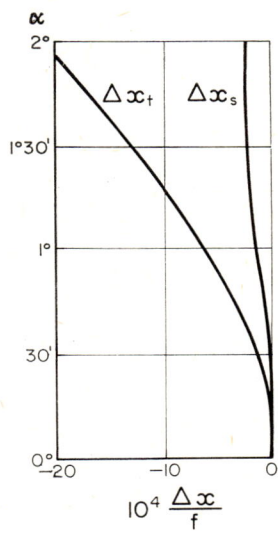

Figure 2-6 Astigmatic difference versus angular distance α from the optical axis. $\Delta x/f$ is the relative distance from the intersection of the surface of least confusion with the optical axis and Δx_t and Δx_S indicates the tangential and sagittal image surfaces respectively.

Astigmatism is not corrected in this system. The astigmatic difference is given in Figure 2-6; the diameter of the smallest confusion disk is 5″ at the edge of 1° field, i.e. of the same order as for a refractor with a field of 1.5 to 2°. Astigmatism and curvature of the field can be corrected with two correction elements separated by a large gap. This system seems very promising but is beyond the scope of this contribution.

The magnification of the radial scale at the edge of 1° field and due to distortion reaches 0.01% compared to the scale at the center. This corresponds to only 0″.21 so that the image can be taken to be fully orthoscopic. Reflexes from the lens surfaces do not fall in the image field.

Thus, this catadioptric telescope is as good as the better apochromate refractors with respect to image quality, and has a number of important advantages over them and also over the reflectors. The tube is 5 to 6 times shorter than that of a refractor, and 1.5 to 2 times shorter than that of a fast reflector; all the optical surfaces are spherical and the lenses are relatively small. Since the mirror is spherical the instrument can be centered easily, and as the lenses are placed in the convergent beam coming from the main mirror, the quality of the glass and accuracy of the finished optical surfaces do not require to satisfy such high requirements as a refractor or catadioptric telescopes of other designs.

When the lenses are made of other materials, such as clear fused silica, fluorite, rock salt, etc., the telescope can perform in different spectral regions and the instrument need only have removable correction elements. It is also possible to change the mode of operation of the telescope to other configurations such as the Schmidt or Maksutov. It can also operate as a Nasmyth or Coudé without difficulty. A telescope thus equipped is completely universal, as well as being the simplest and least expensive between many telescopes of other designs.

Using these calculations, the author designed and constructed a prototype telescope with a 225-mm mirror diameter (Figure 2-7). Variant d of glass combinations (see Table 2-2) was chosen for this instrument. All the mechanical parts and the accessories were made by the author, while the optics was manufactured by V. G. Shreiber based on the calculations by the author. Experimental operation over a year has demonstrated the convenience in use and the excellent optical performance of the instrument.

Hartmann tests gave a value of 0.25 for the constant which is close to the theoretical value for variant d. No retouching of the surfaces was required in order to compensate for residual aberrations or small manufacturing defects in the glass. This result shows that it is not difficult to produce high-quality optics for a telescope with this optical system. For large instruments, clearly, it will not be difficult to obtain still lower values of the Hartmann constant; its theoretical value for variant c is 0.08.

The residual chromatism has practically no effect on the sharpness of the image either in visual observations or in photographic work. Sharp photographs of the lunar surface details, planets, and also of stellar photographs have been obtained with visual focusing on frosted glass and without the use of a yellow filter.

It has also been confirmed that the instrument is quite free from stray reflections: when a partial lunar eclipse was photographed, the umbra

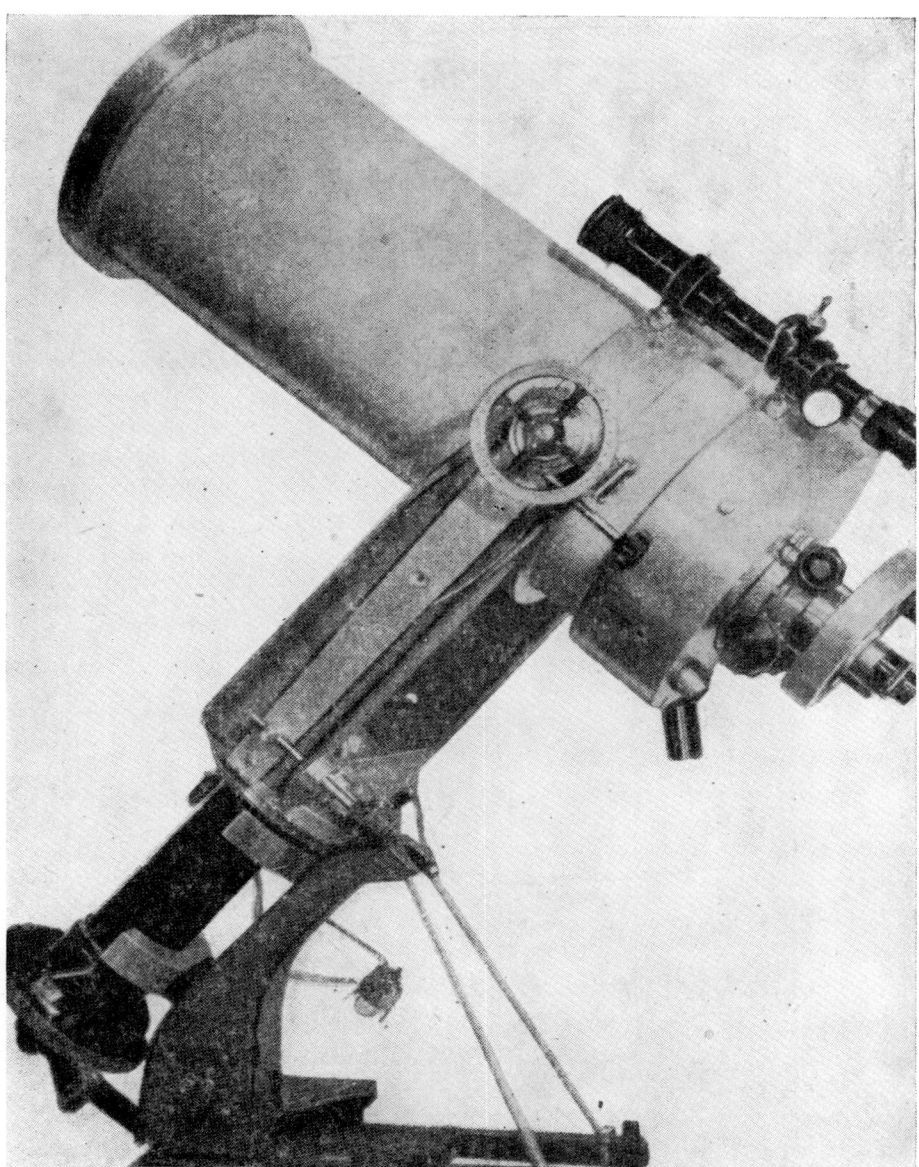

Figure 2-7 General view of the catadioptric telescope developed at the Odessa Construction-Engineering Institute.

portion of the image was not affected by light scattering or reflection on the optical surface, even though the penumbra portion was more than 1000-times brighter than the part in umbra.

A second prototype, with a 425-mm diameter mirror was constructed by Professor V. P. Tsesevich, and has been installed at a subsidiary of the Odessa Astronomical Observatory, in the village of Mayaki.

The telescope system has been given certificate number 158697 with priority from 15 December, 1962.

CHAPTER **II-3**

An astrometric study of the 70-cm meniscus telescope at the Abastumani Observatory

A. SH. KHATISOV

Abastumani Astrophysical Observatory
Mount Kanobili

1 INTRODUCTION

In photographic astrometry, refractors have been used exclusively in the past. It was considered that reflectors are less accurate, having a smaller field free from aberrations and being more subject to mechanical deformation, air turbulence inside the tube, deterioration of the reflecting film of the mirror, etc. However, refractors are not perfect instruments either for accurate work in photographic astrometry. They possess large aberrations which depend on the brightness and color brightness of the stars. It is very difficult to make allowance for these aberrations. Large light losses are caused by absorption in the multielement objective and by reflection from their surfaces which reduce the limiting magnitude of the instrument. For technical reasons it is impossible to manufacture fairly fast refractors with an objective of large diameter. This is particularly noticeable now when there is a great demand for observational data on the positions and proper motions of faint stars.

With the appearance of the Schmidt cameras and Maksutov meniscus systems which have wide fields with high quality stellar images, it seemed worthwhile using such telescopes in photographic astrometry, and the first experiments conducted abroad have confirmed this [1–3].

We have been studying the field of the Maksutov meniscus telescope at Mount Kanobili (Abastumani Observatory) which has a 700-mm objective, a 995-mm spherical mirror and a 2,100-mm focal length.

Similar studies were made by Omarov in 1961 [4], but were not completed. In 1962 the primary mirror of the telescope needed recoating, but since it had a crack it was replaced. We therefore had to repeat the determination of certain characteristics of the telescope which had already been done by Kiladze [5].

The optical layout and general description of the telescope and control console have been described in detail in [5], and need not be repeated.

Let us describe the observational material used for the study (Table 3-1).

Table 3-1 Plates taken with the 70-cm meniscus telescope for position accuracy determination

Plate serial No.	Date	Hour angle (h, m)	Zenith distance (degrees)	Exposure time (sec)	Temperature (°C)
3487	15–16 VII 1963	−0 33	16	40	+16
3488		−0 26	15	30	+16
3489		−0 11	11	30	+16
3490		−0 02	11	40	+16
3491		+0 03	11	30	+16
3494		0 00	38	9×30	+14
3496		0 00	0	9×30	+14
3497		0 00	38	9×30	+14
3498	1 I–1 II 1964	−0 10	8	60	−04
3900		+0 03	7	60	−04
3901		+0 09	7	60	−04
3902		+0 15	8	60	−04

2 POSITION ACCURACY DETERMINATION

To study the accuracy with which the equatorial coordinates of star images and other star-like objects could be determined, we used the plates numbered from 3487 to 3491. An area with the center at $\alpha = 18^h 10.8^m$, $\delta = +52°18'$ was photographed on these plates.

The reference stars chosen were of similar brightness and color and had small proper motions. The field was divided into three concentric zones with a star at the center and with radii $r_1 \leq 1°0$, $r_2 \leq 1°5$, $r_3 \leq 2°0$ respectively. The part of the field outside the third zone (the radius of the whole field is $2°5$) was not studied because the star images further than $2°$ from the center of the plate are seriously affected by various aberrations. In the first zone, were the stellar images with ideal circular shape; 20 reference stars were selected in this zone. In the second we chose another 20 with slightly elongated images which could not be labeled as good quality images. The second zone had 40 stars in total. The third zone contained 60 stars, whose images could still be said to be quite satisfactory.

We first examined the accuracy of measurement of the plates by comparing the measured coordinates on each plate with the coordinates of the "average" plate, using the following systems of linear equations:

$$\bar{x} - x_i = ax_i + by_i + c,$$
$$\bar{y} - y_i = dx_i + ey_i + f, \quad (3\text{-}1)$$

where

$$\bar{x} = \frac{1}{5}\sum_{i=1}^{5} x_i, \quad \bar{y} = \frac{1}{5}\sum_{i=1}^{5} y_i. \quad (3\text{-}2)$$

and $\bar{x}_i$, $\bar{y}_i$ are the measured coordinates on the ith plate.

We then had 10 systems of simultaneous equation for each zone, and used the Doolittle method [6] to obtain the least squares solution. The resulting residuals in the simultaneous equations are the errors of the measurements. The mean square error of the measurement of a plate (for one star) is obtained from the formula [6]

$$\varepsilon = \pm \sqrt{\frac{[(O - C)]}{n - m}}, \quad (3\text{-}3)$$

where $[(O - C)]$ is the sum of the squares of the residuals; n is the number of equations in the system and m is the number of unknowns.

The following values of ε were obtained for each zone separately:

$$\left.\begin{array}{ll} \varepsilon_{1x} = \pm 0.0012 \text{ mm}, & \varepsilon_{1y} = \pm 0.0010 \text{ mm}, \\ \varepsilon_{2x} = \pm 0.0015 \text{ mm}, & \varepsilon_{2y} = \pm 0.0015 \text{ mm}, \\ \varepsilon_{3x} = \pm 0.0018 \text{ mm}, & \varepsilon_{3y} = \pm 0.0018 \text{ mm}, \end{array}\right\} \quad (3\text{-}4)$$

where the first suffix denotes the zone and the second the coordinate axis.

The accuracy to which the position of an object is determined is obtained by comparing the measured coordinates with the absolute coordinates calculated from the star catalog. The equatorial coordinates of the reference stars were taken from the Yale catalog [7]. The equatorial coordinates were translated into absolute coordinates using the formulas [8]

$$X = F \tan (\alpha - A) \cos d \sec (d - D),$$
$$Y = F \tan (d - D), \quad (3\text{-}5)$$

where $\tan d = \tan \delta \sec (x - A)$; F is the focal length of the telescope, and A, D are the equatorial coordinates of the center of the plate.

To reduce the measured coordinates to absolute coordinates we used Turner's linear equations [8]

$$\left. \begin{aligned} X - \bar{x}_i &= ax_i + by_i + c, \\ Y - \bar{y}_i &= dx_i + ey_i + f, \\ i &= 1, 2, 3, \ldots, n, \end{aligned} \right\} \quad (3\text{-}6)$$

where $\bar{x}, \bar{y}$ were obtained from Eq. (3-2) and n is the number of reference stars.

The residuals obtained after the least squares solution of Eqs. (3-6) give the difference between the measured and absolute coordinates of corresponding stars.

The results of the solution are given in Table 3-2.

Table 3-2

Zone	$a \cdot 10^6$	$b \cdot 10^6$	$c \cdot 10^4$	$d \cdot 10^6$	$e \cdot 10^6$	$f \cdot 10^4$	ε_x (mm)	ε_y (mm)
I	−931 ± 22	−489 ± 22	+78 ± 4	+636 ± 22	−896 ± 22	+26 ± 4	±0.0020	±0.0019
II	−930 ± 13	−511 ± 12	+70 ± 3	+618 ± 12	−870 ± 12	+19 ± 4	±0.0022	±0.0021
III	−955 ± 12	−499 ± 12	+67 ± 3	+605 ± 11	−900 ± 11	+14 ± 4	±0.0032	±0.0031

The values a, b, c, d, e and f in the table are the plate constants; $\varepsilon_x, \varepsilon_y$ are the mean square errors calculated from Eq. (3-3) for each zone separately.

It is clear that the mean square errors result from the errors ε_{cat} in the star positions obtained from the catalog, the errors ε_{pl} in plate measurements and errors ε_T introduced by the optics of the telescope, which were not taken into account in any way. Thus, we have

$$\varepsilon_T = \pm \sqrt{\varepsilon^2 - (\varepsilon_{\text{cat}}^2 + \varepsilon_{\text{pl}}^2)}. \quad (3\text{-}7)$$

The mean square error for right ascension and declination of the star is equal to $\pm 0\rlap{.}''135$ and for the proper motion is $\pm 0\rlap{.}''0075$. The difference in epochs of the catalog observations and ours is 16.16 years. Hence the mean square error of a single catalog position of a star at the epoch of our observations is $\varepsilon_{cat} = \pm 0''\!.18$. The plate measurement errors are given by Eqs. (3-4).

If the errors given above are introduced into Eq. (3-7) we obtain

$$\text{I} - \varepsilon_{Tx} = \pm 0.0001 \text{ mm}, \quad \varepsilon_{Ty} = \pm 0.0000 \text{ mm},$$

$$\text{II} - \varepsilon_{Tx} = \pm 0.0003 \text{ mm}, \quad \varepsilon_{Ty} = \pm 0.0002 \text{ mm},$$

$$\text{III} - \varepsilon_{Tx} = \pm 0.0012 \text{ mm}, \quad \varepsilon_{Ty} = \pm 0.0011 \text{ mm}.$$

Thus, inside the second zone there is practically no perceptible additional systematic error. Within the third zone there is a systematic error of up to 0.0012 mm. A further study showed that this error varies and sometimes even vanishes, depending on the position of the reference stars. We may therefore assume that the source of error is in the field correcting lens.

BRIGHTNESS AND COLOR EQUATIONS

We know that the main source of systematic error in determining the coordinates of celestial objects is the brightness and color equations, i.e. the dependence between the relative position of the object on the plate and its brightness and spectral class. The first is caused by coma and residual misadjustment of the telescope, after all possible adjustments have been made, and the second by chromatic aberration of the objective. In our case chromatic aberration is caused by the meniscus and the Piazzi-Smyth correcting plate. The effect of these errors can be minimized by selecting reference stars which are as similar in brightness and color to the given object as possible. However, it is not always possible to do this, so we first need to establish the "degree of freedom" with which these conditions may be satisfied, by determing permissible deviations in the stellar magnitudes and spectral classes of the reference stars for which the effects of the brightness and color equations are negligible.

The method is a follows. Stars of identical color are grouped according to brightness, and the plate constants are determined for each group separately. By comparing corresponding constants with one another, the relationship between the brightness of the reference stars and the scale, orientation and zero point of the plate can be evaluated. Another method would be to take the plate constants for any single group as the basic constants and substitute these values in the simultaneous equations for the other groups. By graphical or analytical means we can establish the form of the brightness equation from the way in which the residuals change with the change in the coordinates. For the color equation, we have to

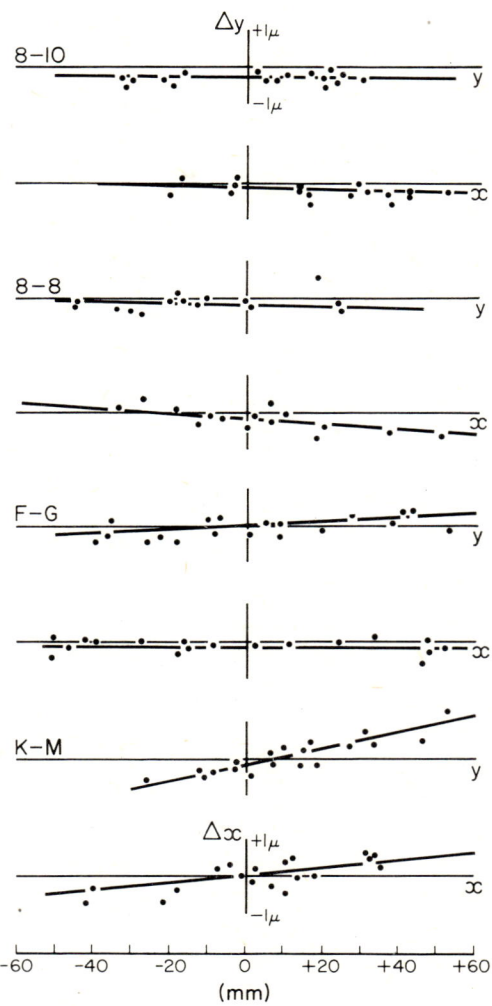

Figure 3-1 Residuals Δx versus coordinate x and Δy versus coordinate y for each group of stars.

group stars of identical brightness according to their spectral classes and then proceed as before.

Corrections for differential refraction and aberration need not be introduced if the plates are obtained at small zenith distances and, more important, if the stars of all the groups in the given area are uniformly distributed.

We used four photographic plates, serial numbers 3898, 3900, 3901 and 3902 of the region near α Perseus, which is used as an astrometric standard (Table 3-1). The exact positions and proper motions of the stars of this region, from the brightest down to 12 th magnitude stars, are given in [9].

We considered only the field $r \leq 1°.5$, where there are no systematic errors of any other kind, and selected 132 stars between 6^m and 12^m, grouping them as shown in Table 3-3.

Table 3-3

Group No.	Group designation	Number of stars	Magnitude	Spectral class
1	A	26	11–12	AO-A 9
2	6-8	15	6–8	B-A 9
3	7-9	20	7–9	B-A 9
4	8-10	20	8–10	B-A 9
5	9-11	16	9–11	B-A 9
6	F-G	29	11–12	FO-G 9
7	K-M	19	11–12	KO-Ma

For each group we obtained simultaneous equations of the type in Eqs. (3-6). The solutions of all the systems are given in Table 3-4.

The first group was taken as the basic group and the plate constants obtained from it were substituted into the simultaneous equations corresponding to the other groups. Graphs were drawn for the residuals obtained for each group, with x or y plotted along the abscissa and the residuals along the ordinate axis (Figure 3-1).

Analytic representation of all the graphs was done using the linear relationships

$$\Delta x = ax + b,$$
$$\Delta y = cy + d,$$
(3-8)

Table 3-4

Group	Group designation	$a \cdot 10^6$	$b \cdot 10^6$	$c \cdot 10^4$	$d \cdot 10^6$	$e \cdot 10^6$	$f \cdot 10^4$	ε_x (mm)	ε_y (mm)
1	A	-273 ± 11	-4586 ± 13	-25 ± 4	$+4498 \pm 13$	-306 ± 15	$+191 \pm 4$	± 0.0021	± 0.0021
2	6–8	-371 ± 21	-4624 ± 24	-47 ± 6	$+4483 \pm 22$	-313 ± 25	$+168 \pm 6$	± 0.0019	± 0.0020
3	7–9	-357 ± 17	-4613 ± 18	-40 ± 5	$+4487 \pm 15$	-303 ± 16	$+172 \pm 4$	± 0.0019	± 0.0017
4	8–10	-318 ± 16	-4593 ± 17	-38 ± 4	$+4488 \pm 15$	-291 ± 16	$+164 \pm 4$	± 0.0016	± 0.0016
5	9–11	-296 ± 13	-4588 ± 15	-37 ± 4	$+4463 \pm 19$	-281 ± 21	$+175 \pm 5$	± 0.0016	± 0.0021
6	F-G	-279 ± 13	-4588 ± 17	-32 ± 5	$+4486 \pm 16$	-249 ± 20	$+185 \pm 6$	± 0.0020	± 0.0024
7	K-M	-152 ± 25	-4588 ± 28	-25 ± 6	$+4493 \pm 20$	-107 ± 24	$+117 \pm 5$	± 0.0023	± 0.0019

The 70-cm meniscus telescope at the Abastumani Observatory 339

where Δx and Δy were the corresponding residuals. A least squares solution of these systems gave the following values of the constants a, b, c and d:

$$
\begin{aligned}
&\text{6-8} & \Delta x &= -0.000091x - 0.0017 \\
& & &\ \pm 21 \pm 5 \\
& & \Delta y &= -0.000017y - 0.0021 \\
& & &\ \pm 18 \pm 6 \\
&\text{7-9} & \Delta x &= -0.000083x - 0.0014 \\
& & &\ \pm 20 \pm 4 \\
& & \Delta y &= -0.000000y - 0.0020 \\
& & &\ \pm 18 \pm 5 \\
&\text{8-10} & \Delta x &= -0.000014x - 0.0018 \\
& & &\ \pm 14 \pm 3 \\
& & \Delta y &= -0.000000y - 0.0027 \\
& & &\ \pm 18 \pm 6 \\
&\text{9-11} & \Delta x &= -0.000036x - 0.0012 \\
& & &\ \pm 17 \pm 5 \\
& & \Delta y &= +0.000033y - 0.0018 \\
& & &\ \pm 16 \pm 5
\end{aligned}
\tag{3-9}
$$

$$
\begin{aligned}
&\text{F-G} & \Delta x &= -0.000005x - 0.0010 \\
& & &\ \pm 12 \pm 5 \\
& & \Delta y &= 0.000054y - 0.0005 \\
& & &\ \pm 19 \pm 5 \\
&\text{K-M} & \Delta x &= 0.000122x - 0.0001 \\
& & &\ \pm 20 \pm 5 \\
& & \Delta y &= 0.000193y - 0.0015 \\
& & &\ \pm 20 \pm 6
\end{aligned}
\tag{3-10}
$$

We calculated the mean square errors after corrections had been made to the measured coordinates according to Eqs. (3-9) and (3-10)

6-8 $\varepsilon_x = \pm 0.0020$ mm,   9-11 $\varepsilon_x = \pm 0.0018$ mm,

       $\varepsilon_y = \pm 0.0017$ mm, $\varepsilon_y = \pm 0.0023$ mm,

7-9 $\varepsilon_x = \pm 0.0020$ mm,   F-G $\varepsilon_x = \pm 0.0024$ mm,

       $\varepsilon_y = \pm 0.0016$ mm, $\varepsilon_y = \pm 0.0022$ mm,

8-10 $\varepsilon_x = \pm 0.0016$ mm,   K-M $\varepsilon_x = \pm 0.0022$ mm,

       $\varepsilon_y = \pm 0.0015$ mm, $\varepsilon_y = \pm 0.0022$ mm.

The following conclusions may be drawn from the data of Table 3-4 and Eqs. (3-9) and (3-10).

(1) Examination of the brightness equation has shown that the system of bright stars is shifted relative to the faint stars by roughly $2\,\mu$, i.e. $0''\!.2$ (see the coefficients c and f in the first five rows of Table 3-4 or the constant terms in Eqs. (3-9)). This shift can be explained by the effect of residual misadjustment of the optical components of the telescope. The fact that the coefficients of x and y in Eqs. (3-9) are small means that there is no noticeable coma in this part of the field ($r \leq 1°\!.5$).

(2) The color equation reveals the presence of chromatic enlargement; the scale for red stars is 1.00015 times larger than for white [see the corresponding scale coefficients in Table 3-4 and the coefficient of x and y in the last two equations of (3-10)]. A shift of the red star population relative to the white by approximately $1\,\mu$ is also observed, and this corresponds to the magnitude of the atmospheric dispersion [10].

(3) In choosing reference stars, the permissible difference in stellar magnitudes is $\pm 1^{m}\!.0$ and in spectral classes ± 1 class.

4 DISTORTION

The distortion coefficient and factors of the inclination of the plate relative to the optical axis can be determined by a method using pairs of stars separated wide apart; this method was applied in the study of the Hamburg AG astrograph [11].

The pairs of stars chosen were either near the equator, had identical declinations and differed in right ascension by roughly half the field of the plate, or had identical right ascensions and different declinations, but were in the zenith region. The paired stars had to be of approximately the same brightness. The pairs were photographed at the meridian, each in the following three positions (see Figure 3-2): (a) with star No. 1 at the center of the plate and star No. 2 at the edge; (b) with stars Nos. 1 and 2

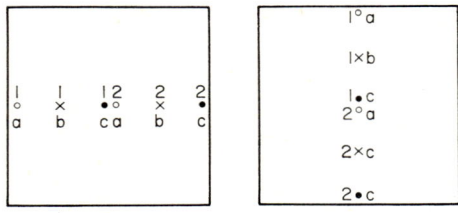

Figure 3-2 Schematic showing the relative location of the stellar images on the photographic plate used to determine the distortion coefficient and the plate inclination factor.

symmetric with respect to the center; (c) with star No. 2 at the center and star No. 1 at the edge. For V_x (the coefficient for α) the telescope is shifted along α, and for V_y (the coefficient for δ) along δ. If the system of measured coordinates coincides with that of the ideal coordinates, we easily derive the following formulas [11]

$$V_x = \frac{x_{1b} - x_{2b} - \frac{1}{2}(x_{1a} - x_{2a} + x_{1c} - x_{2c})}{x_{1b}^3 - x_{2b}^3 - \frac{1}{2}(x_{1a}^3 - x_{2a}^3 + x_{1c}^3 - x_{2c}^3)},$$

$$V_y = \frac{y_{1b} - y_{2b} - \frac{1}{2}(y_{1a} - y_{2a} + y_{1c} - y_{2c})}{y_{1b}^3 - y_{2b}^3 - \frac{1}{2}(y_{1a}^3 - y_{2a}^3 + y_{1c}^3 - y_{2c}^3)},$$

(3-11)

where x and y are the measured coordinates and the first suffix refers to the star and the second to the corresponding position.

To determine V_x and V_y we used two plates each (Nos. 3494, 3496, 3497, 3498) taking three exposures in each position. It proved possible to choose 10 pairs of stars on each plate.

After averaging the results over all pairs we obtained the following values of the distortion coefficient

$$V_x = (8.22 \pm 0.20) \times 10^{-8} \text{ mm}^{-2},$$
$$V_y = (8.23 \pm 0.20) \times 10^{-8} \text{ mm}^{-2}.$$

The mean distortion coefficient was

$$V = (8.2 \pm 0.2) \times 10^{-8} \text{ mm}^{-2}.$$

We note that for the old mirror Kiladze obtained a distortion coefficient of 10.3×10^{-8} mm^{-2}.

5 PLATE INCLINATION

To calculate the factors of inclination of the plate relative to the optical axis, p and q, we used the coordinate measurements which corresponded to positions a and c (Figure 3-2). These were used for the distortion coefficient. The equations to calculate p and q [11] are

$$p = \frac{(x_{1a} - x_{2a}) - (x_{1c} - x_{2c})}{(x_{1a}^2 - x_{2a}^2) - (x_{1c}^2 - x_{2c}^2)},$$

$$q = \frac{(y_{1a} - y_{2a}) - (y_{1c} - y_{2c})}{(y_{1a}^2 - y_{2a}^2) - (y_{1c}^2 - y_{2c}^2)}.$$

(3-12)

By averaging, we obtained $p = 0.1 \pm 0.5$ mm and $q = 0.1 \pm 0.3$ mm.

Clearly, the optical center of the plate coincides with the geometric center to sufficient accuracy for it to be unnecessary to introduce terms, allowing for inclination of the plate in the reduction formula.

6 PLATE SCALE

The scale of the plates was determined for the two temperatures $+16°C$ and $-4°C$. For this purpose we used the plate measurements which had been used in the study of the position accuracy and the brightness and color equations (Nos. 3490, 3491, 3901, 3902). We selected 10 pairs of stars on each plate. The scale was obtained from the formula

$$M = \sqrt{\frac{(X_2 - X_1)^2 + (Y_2 - Y_1)^2}{(x_2 - x_1)^2 + (y_2 - y_1)^2}}, \qquad (3\text{-}13)$$

where X and Y are the absolute coordinates of the stars in seconds of arc and x and y the measured coordinates in millimeters.

Before comparing the absolute coordinates and the measured coordinates we corrected the latter for distortion, differential refraction and first-order aberration [7]. Second-order corrections were negligibly small, since all the plates were taken almost at the meridian (Table 3-1).

We obtained the following plate scale values:

$$M = 98\rlap{.}''115 \pm 0\rlap{.}''003 \text{ mm}^{-1}, \quad \text{at} \quad t = +16°C;$$
$$M = 98\rlap{.}''147 \pm 0\rlap{.}''003 \text{ mm}^{-1}, \quad \text{at} \quad t = -4°C.$$

The corresponding focal lengths of the telescope were

$$F = 2102.28 \pm 0.04 \text{ mm}, \quad \text{at} \quad t = +16°C;$$
$$F = 2101.59 \pm 0.04 \text{ mm}, \quad \text{at} \quad t = -4°C.$$

Thus, when the temperature changes by $1°C$ the plate scale changes by $0\rlap{.}''0016 \text{ mm}^{-1}$.

7 HARTMANN CONSTANT

The constant is obtained by carrying out the Hartmann test as described in [12]. The screen used in our case was described by Kiladze [5]. The value that we obtained is $T = 0.13$, which in accordance with the Hartmann criterion is excellent. It should be noted that for the old mirror it was $T = 0.41$ [5].

8 CONCLUSION

The results of this examination show that the central part of the field ($d \leq 3°$, area 7 square degrees) of the 70-cm meniscus telescope at the Abastumani Astrophysical Observatory is entirely suitable for astrometric observations. To obtain the equatorial coordinates of point sources to the accuracy required today (with a mean error of $\pm 0''.15$), two plates have to be measured, but no corrections need to be made to the measurements and only first order terms need to be used in the reduction formulas. When the whole field is used, the number of measured plates must be increased to four to give similar accuracy.

I would like to express my gratitude to Prof. A. N. Deutsch and to my former colleague at Pulkovo, A. A. Kiselev, for their valuable comments.

REFERENCES

1. I. Barney, *Mitt. Hamburg Sternw., Bergedorf,* **23**, 114 (1955).
2. W. Dieckvoss, "The Bergedorf 32-inch conventional Schmidt telescope as an astrometric instrument", *Astron. J.,* **65**, No. 4, 214 (1960).
3. M. N. Dixon, *Monthly Notices Astron. Soc. South Africa,* **22**, No. 3, 32 (1963).
4. S. Z. Omarov, *Izv. AN AzSSR, Ser. Fiz-Mat. Tekhn. Nauk,* No. 1, 181 (1961).
5. R. I. Kiladze, "On the experimental determinations of stellar radial velocities by means of the objective prism attached to the 70-cm meniscus type telescope", *Byull. Abastumanskoi Astrofiz. Obs.,* No. 24, 35 (1959).
6. N. I. Idelson, *The Least Squares Method and the Mathematical Theory of Observation Processing* (Sposob naimenshikh kvadratov i teoriya matematicheskoi obrabotki nablyudenii), Geodezizdat, Moscow (1947).
7. I. Barney, D. Hoffleit and R. B. Jones, "Catalogue of 8380 stars $+50°$ to $+55°$", *Trans. Astron. Obs., Yale Univ.,* **26**, 11 (1959).
8. Ye. Ya. Bugoslavskaya, *Photographic Astrometry* (Fotograficheskaya astrometriya), OGIZ-Gostekhizdat, Moscow and Leningrad (1947).
9. O. Heckmann, W. Dieckvoss and H. Kox "Eigenbewegungen in der Umgebung von α Persei", *Mitt. Hamburg. Sternw., Bergedorf,* **10**, No. 101, 109 (1956).
10. O. A. Melnikov, "Corrections for Refraction to be Used for Stars of Different Colors", *Astron. Zh.,* **33**, No. 2, 266 (1956).
11. R. Schorr and A. Kohlschutter, *Zweiter Katalog der Astronomischen Gesellschaft,* Vol. 1, Hamburg-Bergedorf (1951).
12. A. A. Mikhailov (Ed.), *Course in Astrophysics and Stellar Astronomy* (Kurs astrofiziki i zvezdnoi astronomii), Vol. 1, Gostekhizdat, Moscow and Leningrad (1951).

CHAPTER II-4

A study of the optical systems of the PM-700 telescope

G. I. BOLSHAKOVA and A. V. SHUMAKHER

Main Astronomical Observatory
Pulkovo

1 INTRODUCTION

The PM-700 telescope is designed to have three optical systems which allows operation at the prime focus ($f = 2,100$ mm, $f/3$), the Cassegrain focus ($f = 8,400$ mm, $f/12$) and the Coudè focus ($f = 21,000$ mm, $f/30$). Layout of the optical systems is shown in Figure 4-1.

Studies were made at the prime focus, with and without corrector and at the Cassegrain focus. After coarse adjustment of the main mirror, the prime focus optics was collimated. We know that off-axis images will be affected by coma, and this property in fact was used as the basis for the collimation of the optics. The Pleiades constellation was photographed at the focus and out of focus. The comatic stellar images are oriented along radial directions from the optical center of the plate (intersection of the optical axis with the photographic plate).

The position of the optical center was determined from the comatic images and marked on the photographic plate. The geometric center of the plate was located at the same time as the optical center. The two centers should coincide in a perfect collimated system, but if they did not the following procedure was used.

The plate with the marked centers was placed in a camera attached to the corrector of the prime focus. A stellar image was first set on the optical center, then the image was moved toward the geometric center by an amount equal to the distance between the two centers; this was done by readjusting the primary mirror. By adjustment of the telescope tube coincidence of the two centers was obtained. After this, control plates were taken and the coincidence was checked.

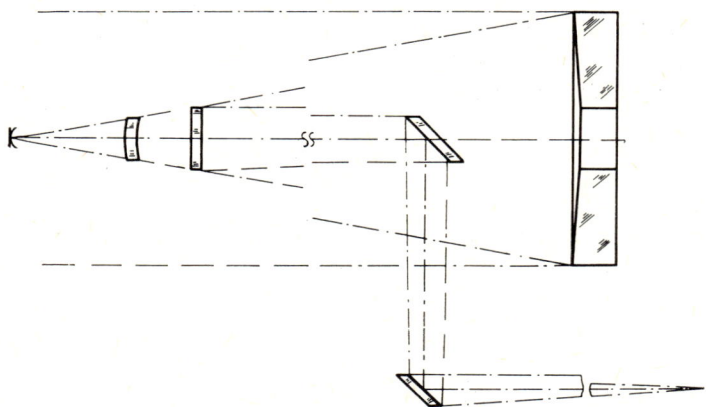

Figure 4-1 Layout of the PM-700 telescope optical systems.

The Cassegrain system was collimated visually so that the clear aperture of the telescope tube, the reflection of the primary mirror on the secondary mirror and the hole in the primary were concentric. Then the correct shape of the in and out of focus images was obtained by proper tilting of the secondary mirror.

2 QUALITATIVE METHOD

The Foucault test was the qualitative method used to study the optical systems of the PM-700. It uses a point source (star) and a knife edge at the focal plane of the telescope. For this purpose we designed the device shown in Figure 4-2 which has a glass plate with a knife edge; this device was attached to the front of the objective of a photographic camera.

The camera at the prime focus was focused on the primary mirror; with the camera at the Cassegrain it was focused on the image of the primary mirror on the secondary mirror. The focused image of a star was set on the knife edge and the shadow pattern was then photographed.

The shadowgram of the primary mirror is shown in Figure 4-3. The zonal errors and edge are clearly seen.

A study of the optical systems of the PM-700 telescope 347

Figure 4-2 Knife-edge attachment to carry out the Foucault test of the PM-700 telescope.

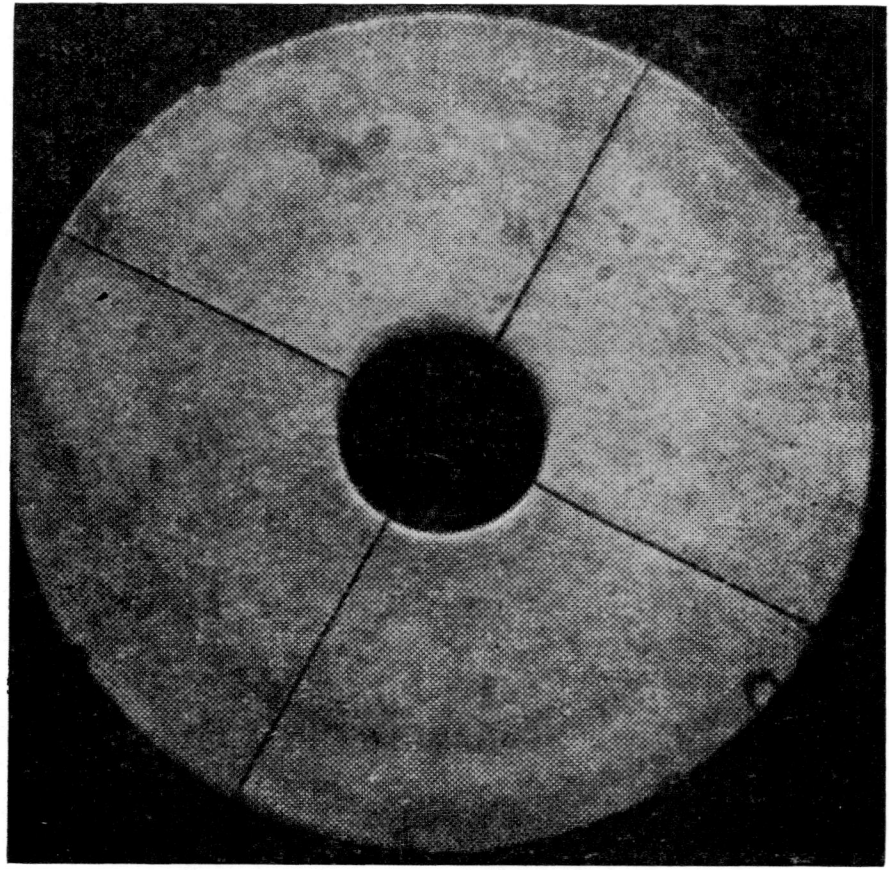

Figure 4-3 Shadowgram of the PM-700 telescope primary mirror.

3 QUANTITATIVE METHOD

The quantitative method was a modified Hartmann test. We used an aluminium screen with 41 apertures, one of which was for monitoring purposes only. The distribution of the apertures on the screen is shown in Figure 4-4; the 40 apertures have a diameter of 30 mm ± 0.1 mm.

The screen was mounted on the upper part on the telescope tube and centered visually. After the screen was secured in place the in and out of focus stellar images were then photographed. The plates were measured on the coordinate measuring machine KIM 3 to an accuracy of 1 μ (fractions of a micron were estimated by eye).

A study of the optical systems of the PM-700 telescope 349

The next step was to find the best focus plane. If an imaginary plane perpendicular to the optical axis is moved along the axis so that the diameter of the field of points converges to a minimum, this will then be the plane of best focus. The path of each light bundle through the screen apertures was examined separately, i.e. it was assumed that light bundles passing through diametrically opposite apertures in the screen did not intersect in the optical axis, and their point of intersection in the best focus plane was determined. A field of points was thus obtained.

The coordinates of the intersections in the best focus plane were obtained by geometric construction (Figure 4-5)

$$X_0 = \frac{\varepsilon}{d}(X - X') + X', \tag{4-1}$$

$$Y_0 = \frac{\varepsilon}{d}(Y - Y') + Y', \tag{4-2}$$

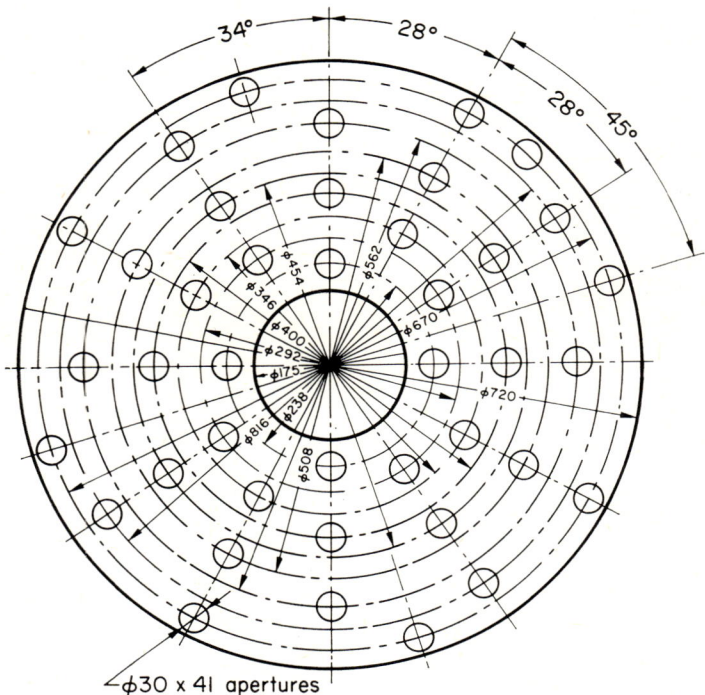

Figure 4-4 Schematic of the screen used to carry out the Hartmann test of the PM-700 telescope.

where X, Y, X', Y' are the coordinates of the intersections on the plates by the rays corresponding to the in and out of focus plate respectively; d is the distance between the in and out of focus plates; ε is the distance between the best focus plane and out of focus plate and X_0, Y_0 are the coordinates in the best focus plane.

By changing the value of a (Figure 4-5), we found X_0 and Y_0 for many planes and selected the plane with the best concentration of points, i.e. the smallest confusion disk. For this disk we obtained its diameter $2x\varrho$ for the different zones, and then using the following formulas we compute from the diameter the angular and front wave aberrations (η_y, h_y):

$$\eta_y = -\frac{\varrho_y}{f} \dots, \qquad (4\text{-}3)$$

$$h_y = \int_0^y \eta_y \, dy \cong \sum_0^y \eta_y \, \Delta y \dots \qquad (4\text{-}4)$$

The graphical representation of the front wave aberrations contribute by different portions of the primary mirror is given in Figure 4-6. The maximum value of the wave aberration basically characterizes the quality of the image given by the optical system. To give a complete representation of the optical systems of the telescope we calculated the Hartmann constant.

Table 4-1 Results of the Hartmann tests of the PM-700 telescope

Test at the focus	diameter confusion disk $\varrho \times 2$ (μ)	Front wave aberration h_{max} (μ)	Hartmann constant
prime	34	0.65	0.95–0.80
prime with corrector	46	0.68	1.00–0.800
Cassegrain	60	0.81	0.80–0.60

The diameters of the smallest confusion disks, the maximum fronte wave aberrations and the Hartmann constant for the prime focus, the prime focus with a corrector and the Cassegrain focus are given in Table 4-1.

In our case the wave aberrations exceed permissible values by approximately 5 to 6 times. The primary mirror has been removed from the telescope for laboratory tests and final retouching.

A study of the optical systems of the PM-700 telescope   351

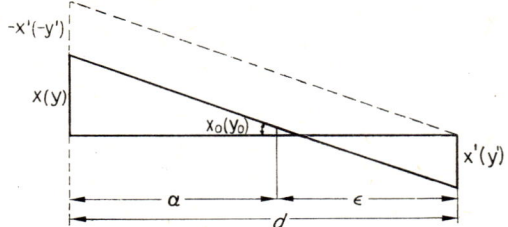

Figure 4-5 Geometry used to determine the coordinates of the intersection of the light bundles with the best focus plane.

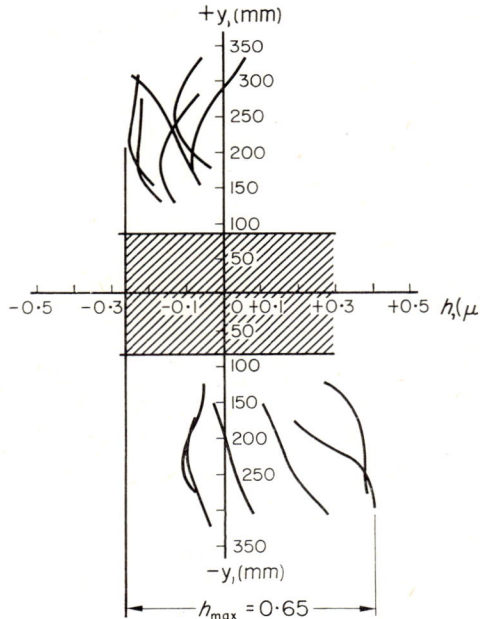

Figure 4-6 Front wave aberrations contributed by different portions of the PM-700 telescope primary mirror. The cross-hatched region corresponds to the central hole of the primary mirror.

CHAPTER **II-5**

A study of the deformation of the PM-700 telescope mount

G. I. BOLSHAKOVA and N. N. MIKHELSON

Main Astronomical Observatory
Pulkovo

A study of the deformation of the mount of the PM-700 telescope was made at the prime focus ($f = 2,100$ mm, $f/3$) using photographs taken with the telescope set at $\delta = +90°00'$ and driving at sidereal rate.

Besides determining the amount of deformation, we obtained the deviation of the instrumental pole from the apparent celestial pole (which is displaced by atmospheric refraction from the celestial pole defined by spherical astronomy), and the optical center* of the plate.

When the telescope is rotated around the hour axis, its optical axis describes a cone (for $\delta \neq 90°$). As a result, each star and the apparent pole will generate arcs of a circle on the photographic plate with center corresponding to the instrumental pole. Since the mount of the telescope experiences variable deformations due to unequal rigidity on different planes containing the polar axis and passing through the intersection of the polar and declination axes, the instrumental pole does not remain fixed when the tube changes attitude.

Therefore, when we speak of the rotation of the stellar field around the instrumental pole, we mean the rotation of the star images around the instantaneous instrumental pole. As a result the traces of each star on the photographic plate are closed curves which are not perfectly circular.

In order to determine the instantaneous positions of the instrumental pole of the PM-700 telescope, we obtained plates of the polar zone at the hour angles 18^h00^m, 19^h00^m, 21^h00^m, 0^h00^m, 2^h00^m, 4^h00^m, 6^h00^m, 8^h00^m,

---

* By optical center, it is understood the intersection of the optical axis and the photographic plate. (Ed. English version.)

20^h00^m, 12^h00^m, 16^h00^m. In each of the positions the shutter was opened for three minutes during which time the drive mechanism was stopped.

To determine the position of the apparent pole, we obtained seven exposures on the same plate of the polar region of the sky with the drive mechanism stopped at ten-minute intervals. The length of each exposure was one minute. As a result, each star gave 7 images along the diurnal parallels. The intersection of the normals to these sections of the diurnal parallels gives the position of the apparent pole.

The photographic plates used were 40 mm × 40 mm in size. The several exposure photograph of the polar region of the sky that we obtained is shown in Figure 5-1.

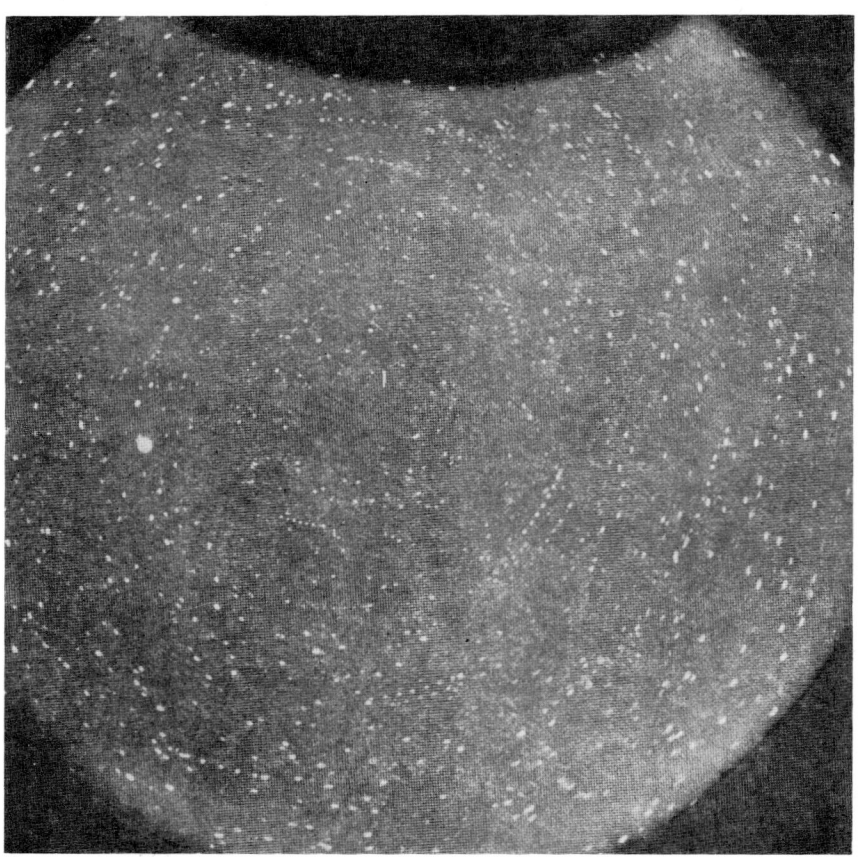

Figure 5-1 Several-exposure photograph of the polar region of the sky.

The processed plate was measured and the positions of the optical center of the plate, the apparent pole and the instantaneous positions of the instrumental pole were located. All the measurements were made on the UM-21 to an accuracy of 1 µ (tenths of a micron were estimated by eye).

The problem of determining the position of the optical center is reduced to the determination of the geometric center of the plate, since when the telescope was collimated the optical center coincided with the geometric center of the plate.

To determine the position of the apparent pole, we selected the traces of 7 stars which were distributed approximately uniformly over the plate. The coordinates of the beginning and end of the traces were measured and the coordinates of the centers were calculated. Perpendiculars were drawn through the centers to the straight lines connecting the ends of the traces. The system of equations of the perpendiculars was solved by least squares to give the position of the apparent pole.

The instantaneous positions of the instrumental pole were found in a similar way. The coordinates of the stellar images were measured at various hour angles and for each pair of successive images the instantaneous center

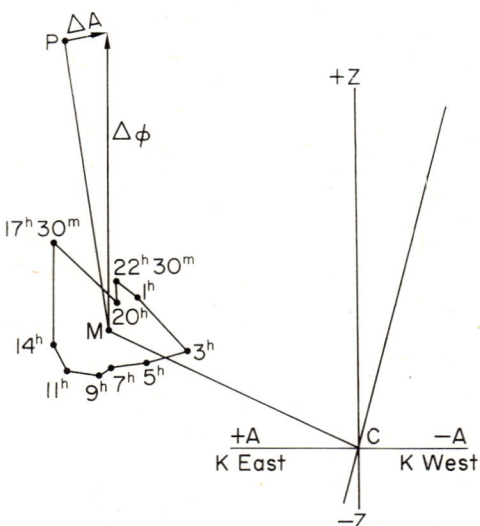

Figure 5-2 Graph showing the optical center C of the plate, the apparent pole P, the instrumental pole for different hour angles and the mean position M of the instrumental pole; $\Delta\varphi$ is the error component in the setting of the polar axis in the meridian and ΔA in azimuth.

was found as the point of intersection of the perpendicular drawn through the centers of the segments connecting successive images. We thus found the instantaneous centers of the instrumental pole for hour angles $1^h.00^m$, $3^h.00^m$, $5^h.00^m$, $7^h.00^m$, $9^h.00^m$, $11^h.00^m$, $14^h.00^m$, $17^h.30^m$, $20^h.00^m$, $22^h.30^m$.

Figure 5-2 shows the optical center C of the plate, the instantaneous centers of the instrumental pole which describes an irregular polygon, the mean position M of the instrumental pole and the apparent pole P on the scale of 40 : 1 relative to the plate scale.

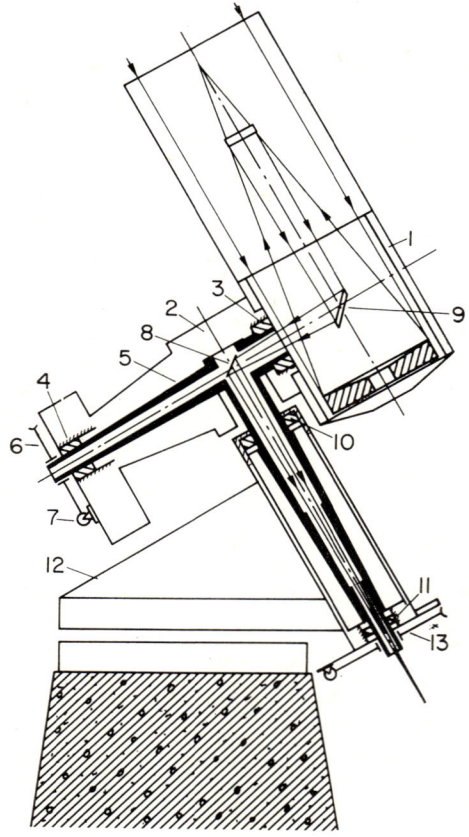

Figure 5-3 Schematic of the PM-700 telescope: 1, telescope tube; 2, declination trunnion housing; 3, main bearing of the declination trunnion; 4, end bearing of the declination trunnion; 5, declination trunnion; 6, declination gear drive; 7, declination worm; 8, 9, diagonal mirrors; 10, main polar axis bearing; 11, end polar axis bearing; 12, bedplate; 13, hour angle gear drive.

The error in setting the telescope by the coordinates produces a deviation of the optical center C from the mean instrumental pole M. The error in setting the telescope mount causes a deviation of the mean instrumental pole M from the apparent pole P; this deviation can be resolved into the two components $\Delta\varphi$ and ΔA:

$$\Delta\varphi = 2.85 \text{ mm},$$

$$\Delta A = 0.45 \text{ mm},$$

corresponding to

$$\Delta\varphi = 4'40'',$$

$$\Delta A = 1'25''.$$

The adjustment screws in the base of the telescope can be used to correct the errors $\Delta\varphi$ and ΔA. The maximum scatter in the positions of the instantaneous instrumental pole is equal to the maximum deformation (X_{max}) of the telescope mount.

In our case X_{max} is the distance between the instrumental poles at 3^h and 17^h30^m hour angles which on the photographic plate, is equal to 1.65 mm or equivalent to $2'40''$.

In the mount of the PM-700 telescope (designed by B. K. Ioannisiani and Yu. S. Streletskii) the polar axis and the housing for the declination trunnion are rigidly attached to one another (Figure 5-3). The telescope tube is fixed to the housing of the declination trunnion, which can turn around the declination axis together with the counterweight. A long cut is made in the housing so that the polar axis may pass through it, allowing the telescope to be rotated in declination by 160°. Although the central part of the housing is also very rigid, its rigidity cannot be constant for different cross-sections. This is the probable cause of the mount deformation as a function of hour angle.

Deformations of the mount have an adverse effect not only on the accuracy of setting and guiding, but also on the quality of the stellar images for long exposures in the polar region.

CHAPTER **II-6a**

Investigation of models of the large telescope primary mirror by photoelastic methods*†

YE. G. GROSSVALD and K. S. TAVASTSHERNA

Main Astronomical Observatory
Pulkovo

To analyze theoretically the elastic deformations of astronomical mirrors, it is assumed that they are thin infinite plates, which can be treated as a two-dimensional problem. Using this assumption, it is difficult to compute local stresses which causes large inaccuracies and for this reason experimental data is required. Since all calculations are done on the assumption that the reactions at the support points (adjustable supports) are equal, so that the problem is statically defined, the experimental results may differ from the theoretical to some extent. This is due to the technical difficulties of equalizing the reactions by the adjustable supports.

Since 1961 studies have been carried out at the Main Astronomical Observatory on the elastic deformation of the large telescope primary mirror. Three designs were analyzed:

(1) A ribbed mirror with a flat back: (a) with triangular ribbing lattice (Figure 6a-1(a)); (b) with a mixed ribbing lattice.

---

* In the title "... the large telescope ...", refers to the 6-meter altazimuth telescope; see second footnote page 367. (Ed. English version.)

† The contribution No. 6 "Experimental methods for study of telescope components and assemblies" by Ye. G. Grossvald has been replaced by the contribution Chapters II-6a and II-6b in the English version. These two contributions appeared in the *Izv. GAO AN SSSR*, **24**, No. 177 (1964) and **24**, No. 180 (1966). The substitution was made in order to give a clearer description of the work presented in contribution No. 6 of the Proceedings of the Second Conference on *New Techniques in Astronomy*.

(2) A ribbed mirror with spherical back surface and triangular ribbing lattice (Marksutov variant, Figure 6a-1(b)).

(3) Solid mirror with support sockets (Ioannisiani variant, Figure 6a-1(c)). Two variants were made with relative thicknesses 1 : 7.5 and 1 : 9.

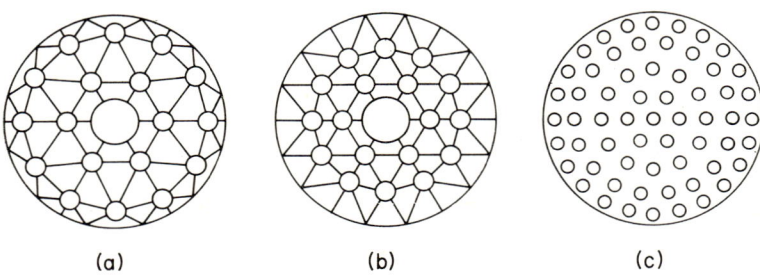

Figure 6a-1 Designs used for the mirror models analyzed by photo-elastic techniques: (a) ribbed structure with triangular ribbing lattice and spherical back; (b) structure with another triangular ribbing attice; (c) solid mirror with sockets for the support units

Two methods of analysis were used: (1) the "frozen-in" method using models made of epoxy resin ED-6-type and (2) the stereometric method using models made of plyable materials ("valtzmassa")*.

The frozen-in method has been described in detail [1]. The models used were cast at the Leningrad State University (LGU) and processed at the Leningrad Association of Experimental Machine Establishments (LOOMP). They had modulus of elasticity between 91 and 179 kg cm^{-2}.

The model was mounted on a support unit made textolite, and was "frozen" in a centrifuge. The textolite support was chosen as its coefficient of linear expansion ($\alpha = 3.3 - 4 \times 10^{-5} °C^{-1}$) is close to that of the model ($\alpha = 6 \times 10^{-5} °C^{-1}$) and this minimized the effect of the supports on the deformed model.

The centrifuge was designed in the instrumentation section of the observatory, and was made in the mechanical workshops. It consists of rocking shaft with an arm one meter long, which rotates around a vertical

---

* Valtzmassa is made from a mixture of gelatin, glycerin and water. Animal-base gelatin combined with glycerin and water in varying proportion is one of the oldest photoelastic materials used for determination of stresses (or strain) produced by the weight of the structure. Since 1940, it has been used for modeling of dams, tunnels and other civil engineering structures; also it has been used for modeling of geological cross sections. (Ed. English version.)

axis. The load developed by it reaches 200 g at 400 rev. min.$^{-1}$ The arm of the shaft is balance at the ends within 50 g. Thermostats are also incorporated in the centrifuge and the temperature conditions during "freezing" are monitored externally on a pen-recorder.

We made detailed temperature calculations of the heaters in the thermally controlled compartment and also calculated the mechanical characteristics of the centrifuge.

The setting of the model in the centrifuge gives the corresponding load conditions on the actual mirror when the telescope points to the zenith.

The freezing was done twice. The first time the model was supported in a plane or sphere, depending on the shape of the back surface. This corresponded to deformation of the upper surface of the mirror before it was mounted in the cell. The second time the model was supported to correspond to the deformation of the mirror when it is mounted in the telescope.

The flexures of the deformed models were measured using the universal measuring microscope UIM-21-type by the contact method with an interval of 5 mm along the different radii under analysis.

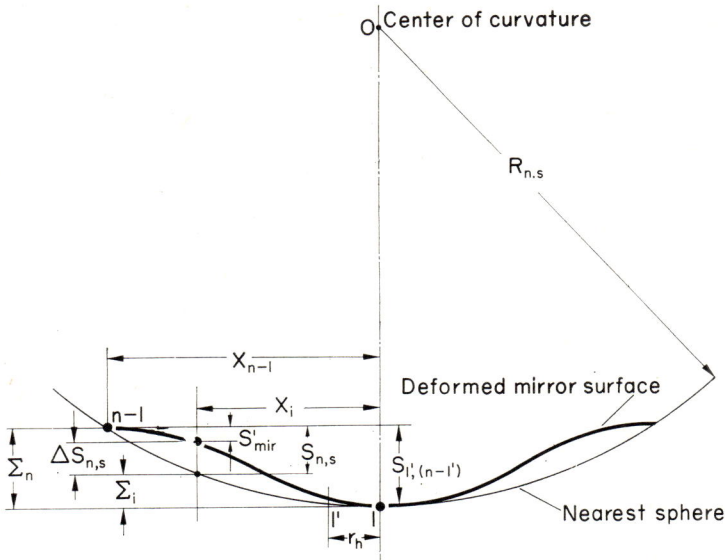

Figure 6a-2 Diagram showing the relationship between the deformed mirror surface and the nearest sphere. The radius of the central hole of the mirror, if existent, is r_h. (This diagram has been added to the English version to make the text clearer; Ed. English version.)

Using these measurements, we scaled the deformations of the model to correspond to the actual mirror size. As the final result, we took deviations $\Delta S_{n,s}$ from the actual mirror to the nearest sphere (see Figure 6a-2). This sphere is characterized by the fact that the whole deformed surface lies on one side of it and the cross section of the two surfaces has three points in common [2].

The deviation $\Delta S_{n,s}$ is expressed by the following relationship:

$$\Delta S_{n,s} = S_{n,s} - S'_{\text{mir}}, \qquad (6\text{a-}1)$$

where $S_{n,s}$ is given by

$$S_{n,s} = \Sigma_n - \Sigma_i \qquad (6\text{a-}2)$$

and S'_{mir} by

$$S'_{\text{mir}} = S_{\text{mir}} - \Delta S_{\text{mir}}, \qquad (6\text{a-}3)$$

where ΔS_{mir} is the deflection at each point of the deformed mirror surface.

By expanding in series Σ_i, we obtain

$$\Sigma_i = \frac{X_i^2}{2R_{n,s}} + \frac{X_i^4}{8R_{n,s}^2} + \frac{X_i^6}{16R_{n,s}^5} + \cdots, \qquad (6\text{a-}4)$$

where $R_{n,s}$ is the radius of the nearest sphere for a mirror without a central hole. The value of $R_{n,s}$ can be obtained from the following relationship:

$$R_{n,s} = \frac{X_{n-1}^2 + S_{1,(n-1)}^2}{2S_{1,(n-1)}}, \qquad (6\text{a-}5)$$

where $S_{1,(n-1)} = \Sigma_n$.

For a mirror with a central hole, the radius $R_{n,s}$ is expressed by

$$R_{n,s}^2 = \frac{-X_{1'}^2 \cdot X_{n-1'}^2 + \left(\dfrac{X_{1'}^2 + X_{n-1'}^2 + S_{1',(n-1')}^2}{2}\right)^2}{S_{1',(n-1')}^2}. \qquad (6\text{a-}6)$$

The deformations ΔS_{mir} at each point of the actual mirror surface can be obtained by multiplying the deformation ΔS_{mod} of the model at the homologue point by the scaling coefficient K [3], that is

$$\Delta S_{\text{mir}} = \Delta S_{\text{mod}} K. \qquad (6\text{a-}7)$$

The ΔS_{mod} is the difference between the initial model deflection S_{mod} at each point, and the S'_{mod} obtained at the same points but after freezing, that is:

$$\Delta S_{\text{mod}} = S_{\text{mod}} - S'_{\text{mod}}. \qquad (6\text{a-}8)$$

The values of S_{mod} and S'_{mod} are obtained experimentally.

Investigation of models of the large telescope primary mirror 363

The deviations from the nearest sphere $\Delta S_{n,s}$ were calculated to an accuracy of 0.001 μ.

The following conclusions may be drawn from the experiments and calculations:

(1) From the models with a ribbed structure those with a flat back gave the best results.

The maximum flexures of the model with a spherical back surface (Figure 6a-1(b)) when scaled to the full size mirror are 0.8 μ. The maximum deviation from the nearest sphere is 0.34 μ, which greatly exceeds the established tolerable limit of 0.037 μ [2] (see Figure 6a-3). The average difference between the flexure measured along different radii for given distances from the optical axis of the mirror is 0.1 μ. This is due to the

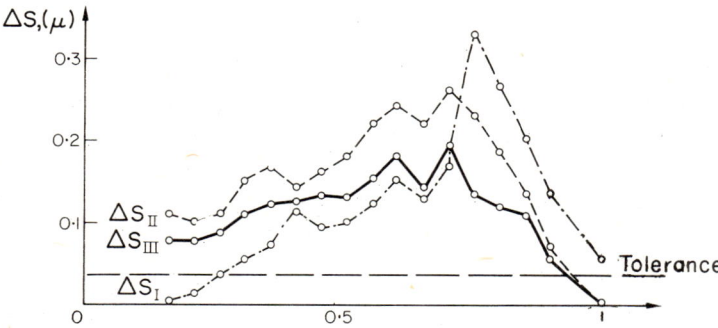

Figure 6a-3 Deviations ΔS of the actual mirror surface from the nearest sphere versus distance from the center of the mirror for the model given in Figure 6a-1(b). The values ΔS_I have been measured at points along radii 1, 5, 9, ...; ΔS_{II} along 3, 7, 11, ...; ΔS_{III} along even number radii. The deviations have been scaled to actual mirror size.

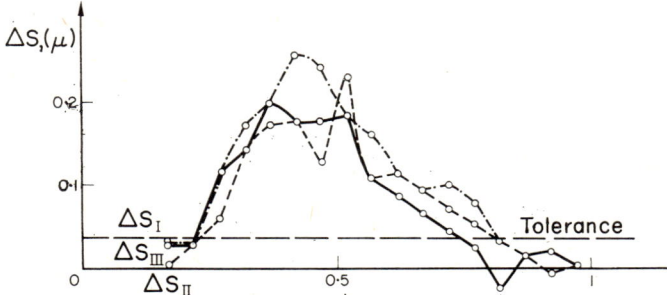

Figure 6a-4 Similar to Figure 6a-3 but for the model given in Figure 6a-1(a).

fact the main weight falls on the unsupported center. One can say with certainty that mirrors of this configuration are unsuitable, despite the large number of support points (36).

For the model with a flat surface, also resting on 36 support points (Figure 6a-1(a)), the deviations from the nearest sphere have a maximum of 0.24 μ. Although this also exceeds the tolerable limit, it is considerably less (see Figure 6a-4) than that for the model with a configuration shown schematically in Figure 6a-1(b).

(2) Of the models analyzed, solid models are better than ribbed models. For ribbed models with a small number of supports (Figure 6a-1(a) and 6a-1(b)) the deviations from the nearest sphere, as we have said, correspond to 0.3 μ in the actual mirror.

The solid model (Figure 6a-1(c)) is a solid disk with 61 support sockets. The upper and back surfaces have the same radii. The maximum deviation from the nearest sphere for these models was 0.03 μ.

(3) As indicated, the deviations from the nearest sphere for the solid models lie within the tolerable limits. For models with relative thickness 1 : 9 larger deviations from the nearest sphere were obtained over the whole surface than for the models with relative thickness 1 : 7.5 as the plotted data in Figures 6a-5 and 6a-6 show.

Besides using the frozen-in method, we applied the stereometric method for the study of solid models [4]. This is an experimental method used to solve problems in thin elastic plates. Models made of materials with low modulus of elasticity (1 kg cm^{-2}) and relative high proportionality limit were used. One such material is valtzmassa. The basic ingredients of valtzmassa are gelatine and glycerine, and its characteristics, such as

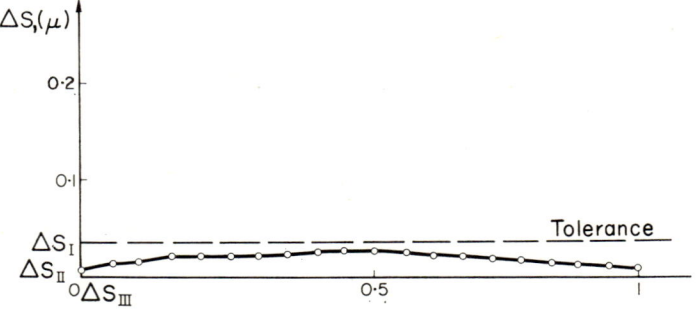

Figure 6a-5 Similar to Figure 6a-3 but for the model given in Figure 6a-1(c) (thickness-to-diameter ratio 1 : 9).

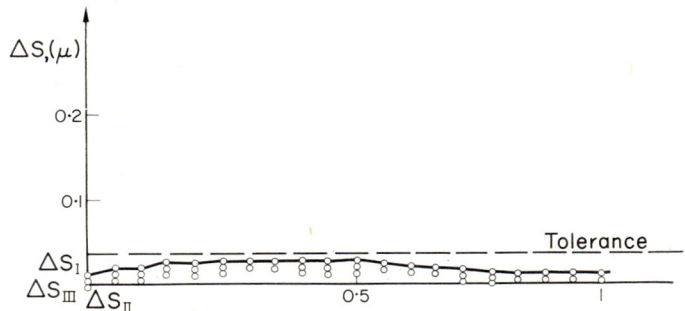

Figure 6a-6 Similar to Figure 6a-3 but for the model given in Figure 6a-1(c) (thickness-to-diameter ratio 1 : 7.5).

strength, elasticity and optical activity depend on the purity and ratio of the two.

The models were obtained by casting valtzmassa in a Plexiglas mold. The prepared models were deformed under their own weight on a special support unit; the measurements were made on the UIM-21 by an electric contact method.

The measurements showed that models with structures given in Figure 6-1(c) are deformed significantly less than models with structures given in Figure 6-1(a) and 6-1(b) for equal thickness. The flexures in the former did not exceed 0.2 mm, while those in the latter reached a maximum of 1 mm.

The authors are indebted to D. D. Maksutov and N. N. Mikhelson for their valuable advice, and to G. A. Batranina for the measurements.

REFERENCES

1. Ye. G. Grossvald, "An investigation of the details of the telescope by the method of photoelasticity", *Izv. GAO AN SSSR*, **22**, No. 4 (No. 169), 162 (1961).
2. D. D. Maksutov, *The Manufacture and Analysis of Astronomical Optics* (Izgotovleniye i issledovaniye astronomicheskoi optiki) GITTL, Moscow (1948).
3. S. G. Lekhnitskii, *Proceedings of the Conference on the Optical Method of Studying Stresses* (Tr. konf. po opticheskomu methodu izucheniya napryazhenii), NII Matematiki i mekhaniki LGU i NII mekhaniki MGU, Mathematics and Mechanics Research Institute of Leningrad State University and Mechanics Research Institute of Moscow State University (1937).
4. N. S. Rozanov, *The Sterometric Method for the Analysis of Stresses in Plates* (Stereometricheskii metod issledovaniya napryazhenii v plitakh), Moscow and Leningrad (1954).

CHAPTER **II-6b**

Experimental model investigations of the large telescope hyperbolic mirror*

YE. G. GROSSVALD and K. S. TAVASTSHERNA

Main Astronomical Observatory
Pulkovo

In 1964 the instrumentation section of the Main Astronomical Observatory began experimental studies using models of the secondary mirror for the BTA† telescope. In 1962 Maksutov had suggested certain modifications to the structural configuration of the secondary mirrors to improve their performance.

The project was divided into the following stages:

(1) Deformation analysis of three telescope mirrors, 800 mm in diameter which were supported at six points uniformly spaced around the edge. Flat, flat-convex and flat-concave models of identical weight were studied (see Figure 6b-1(a)).

(2) Analysis of the same models with the addition of a seventh support at the center and central hole. Determination of the reaction at the central support for minimal elastic deformation.

(3) Analysis of similar models with a central hole and different relative thickness, i.e. with a different ratio of the thickness to diameter (see Figure 6b-1(b)).

(4) For comparison between the computed deformations and the experimental results, models similar to those used in the first stage but supported along the edge were studied.

---

* See first footnote page 359. (Ed. English version.)
† BTA is the acronym for большой телескоп азимутальный (Large Altazimuth Telescope) which is the 6–M telescope. (Ed. English version.)

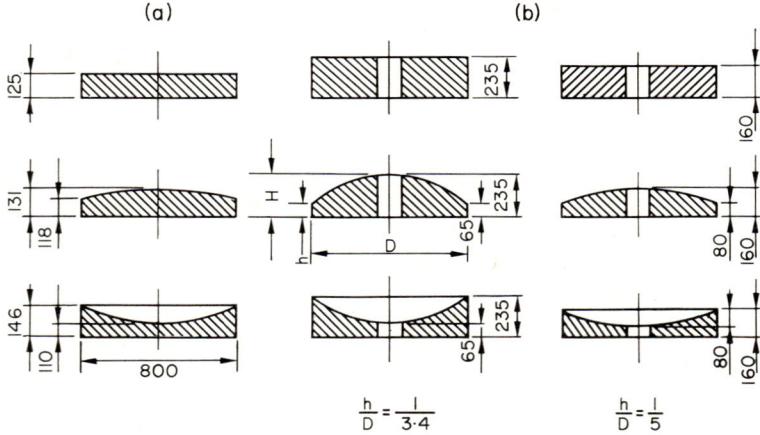

Figure 6b-1 Schematic of different mirror models used at the Main Astronomical Observatory for the experimental study of the elastic deformations in a large telescope secondary mirror. The dimensions are given in millimeters and are for the actual mirror, not the model.

The models of the mirror were made of "igdantin", an elastic material with a small modulus of elasticity. The process used to make the models is described in [1, 2], but in our studies certain modifications were introduced.

The main constituents of igdantin are gelatin, glycerin and water—the same as for the optically active material "valtzmassa" [3] with which we worked previously. The difference between them consists in the purity of the ingredients used and their manufacturing processes, which greatly affect their mechanical and birefringent properties.

It is not possible to discuss here in detail the physical and mechanical properties of igdantin but the following observations should be made:

(1) It is not possible to work with models of igdantin for a few hours after they have been cooled, since during this time their properties alter rapidly. Measurements are best made not less than thirty to forty hours after the models have been cast.

(2) The value of the modulus of elasticity largely depends on the "thermal history" of the models, i.e. on the different temperature conditions of their melting, cooling and subsequent molding time. Thus, models casted from one composition and used after the same elapsed time from casting, exhibits similar values of the modulus of elasticity (5 to 10% difference) while models casted from different batches may give very different values of the modulus of elasticity (20% to 40% difference).

(3) It has been stated by a number of authors that a gelatinous mass can be melted again and used several times: our research has not shown this to be so. When igdantin was remelted, i.e. under lengthy heating, the gelatin apparently undergoes modification, and the material loses its elastic properties.

We have arrived at the conclusion, after lengthy experiments independently of a number of authors [1, 2, 4], that igdantin has great advantages as a material with extremely stable properties.

It was established after a series of preliminary mixtures that for our purposes the optimal composition of igdantin (in % of weight) is the following: gelatin—11%; glycerin—45%; water—44%. Both photogelatin and pure glycerin were used. Other authors [1, 2, 4] advise adding β-naphthol in the amount of 0.01% (in weight) when the models have to be preserved for several months or longer. We did not use β-naphthol, since our models were not kept longer than two to three days; deterioration could not take place, and any additions would only have complicated the production process of the material.

The models were made in three stages:

(1) *Mixing*: the gelatin was placed in an aluminum container, and while thoroughly mixing cold water was poured into the container. After completion of the mixing, the container was tightly covered and the gelatin kept in the container for 20 to 24 hours.

(2) *Boiling*: after this time had elapsed, the container with the gelatin was set in a hot water bath and the cover was removed. After one hour the gelatin had completely melted and its temperature reached 50°C. During heating, the mixture was stirred two or three times and the foam was removed. The required amount of glycerin was then stirred into the melted mixture. An hour after the addition of the glycerin, the temperature of the mixture reached 75–78°C, and the mixture was maintained at this temperature for about an hour. The resulting igdantin was cooled in the bath for two hours.

(3) *Molding*: the igdantin at a temperature of 50–55°C was poured into Plexiglas molds to which a thin layer of machine grease had been applied as a mold releaser; the molds were placed in a thermally controlled chamber and were held at 60°C for 20 to 30 minutes. The molds were wrapped in rags to reduce heat losses to enable the substance to be cooled without bubbles; the molds could be left in that condition for days.

At the same time as the models, other castings were made to determine the modulus of elasticity (E) and the specific weight (γ) of the material.

The E was determined as follows. The igdantin was poured into a rectangular mold ($2 \times 10 \times 25$ cm^3), and wooden splints were inserted in it with their rough sides towards the igdantin for better adhesion. The sample which was obtained in this way was stretched on a special unit at the same time as the model was tested under the action of its own weight.

When it was in the mold, marks were made in printer's ink on the surface of the sample, using a zinc block. These marks were transferred onto tracing paper before and after the sample was loaded. The distance between the marks, before and after loading, were measure with an MMI-type microscope. Then E was obtained using the following relationship [5]:

$$E = \frac{Pl_0}{td\,\Delta l} \text{ (kg cm}^{-2}\text{)}, \tag{6b-1}$$

where P is the applied load in kg, l_0 is the distance in cm between marks before loading, Δl_0 is the increase in cm of the distance l_0 due to the load P, td is the area of the cross-section in cm^2.

The Poisson coefficient (μ) is the ratio of the relative transversal deformation ($\varepsilon_{\text{trans}}$) to the relative logitudinal deformation ($\varepsilon_{\text{long}}$) [5]:

$$\mu = \frac{\varepsilon_{\text{trans}}}{\varepsilon_{\text{long}}}. \tag{6b-2}$$

The $\varepsilon_{\text{trans}}$ and $\varepsilon_{\text{long}}$ in this case are measured using the same grid as for the determination of E. The method used to determine γ is not discussed in this paper.

The first stage of the studies consisted of the following: the models shown in Figure 6b-1(a) were supported at six points; each support pad was 12 mm in diameter. The center of the supports lay on a circle of 74 mm radius. The dimensions of the supports were calculated to avoid the possibility of local deformation in the model. To measure the elastic deformations due to its own weight, the model on its six support pads were mounted on a coordinate measuring engine (KIP) (designed by Shkutov and made at the observatory workshops). The use of the following measuring method was chosen due to the electrical conductivity of the igdantin. One terminal of a flashlight battery was connected to the model and the other to the tip of an indicator in series with a sensitive galvanometer. The indicator was used to measure the elastic deformations. During the measuring process, the tip of the indicator was brought into contact with the surface of the model; the contact was determined by the galvanometer deflections. At this point the indicator was read.

The results of the analysis of three models (Figure 6b-1(a)) scaled up to the actual mirror size using the formulas for modeling from the theory of elasticity [6], are presented in Table 6b-1, and in Figure 6b-2. The

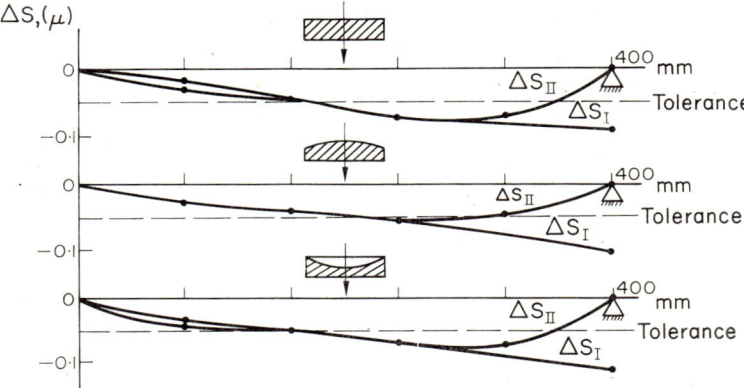

Figure 6b-2 Deviations ΔS of the actual mirror model surface from the nearest sphere versus distance from the center of the mirror. The values ΔS_{II} has been measured at points along radii passing through the support points and ΔS_I at points along radii passing between the support points. The deviations have been scaled to actual mirror size.

Table 6b-1 Experimental data scaled to actual mirror size from models using six support points and for $h_M/D_M = 1/6.4$ (Figure 6b-1(a))

Dist. from mirror center X (mm)	Flat mirror		Flat-convex mirror		Flat-concave mirror	
	$\Delta S_I(\mu)$	$\Delta S_{II}(\mu)$	$\Delta S_I(\mu)$	$\Delta S_{II}(\mu)$	$\Delta S_I(\mu)$	$\Delta S_{II}(\mu)$
400	−0.09	0.00	0.09	0.00	−0.09	0.00
320	—	−0.05	—	−0.04	—	−0.05
240	−0.05	−0.05	−0.04	−0.03	−0.05	−0.05
160	−0.03	−0.03	−0.03	−0.03	−0.03	−0.03
80	−0.01	−0.02	−0.01	−0.01	−0.01	−0.02
0	0.00	0.00	0.00	0.00	0.00	0.00

ΔS_{II} is the deviation of the actual mirror surface from the nearest sphere and measured at points along radii passing through the support points; ΔS_I is similar to S_{II} but measured at points along radii between the support points.

igdantin used for the models to obtain the data in Table 6b-1 had $E_M = 0.40$ kg cm^{-2}, $\gamma_M = 1.14$ g cm^{-3} and $\mu_M = 0.5$; the diameter of the model was $D_M = 160$ mm and the thickness $h_M = 25$ mm which gave a ratio $h_M/D_M = 1/6.4$. To scale up the results from the models to the actual mirror, it was assumed a glass with $E_g = 10^6$ kg cm^{-2} and $\gamma_g = 2.5$ g cm^{-3}. It is clear from the data that the largest deformations appeared in the flat-concave mirror (0.33 μ) and the least in the flat-convex (0.30 μ). The tolerable limit for the deformed surface of the secondary mirror is $\lambda/16$, or 0.037 μ.

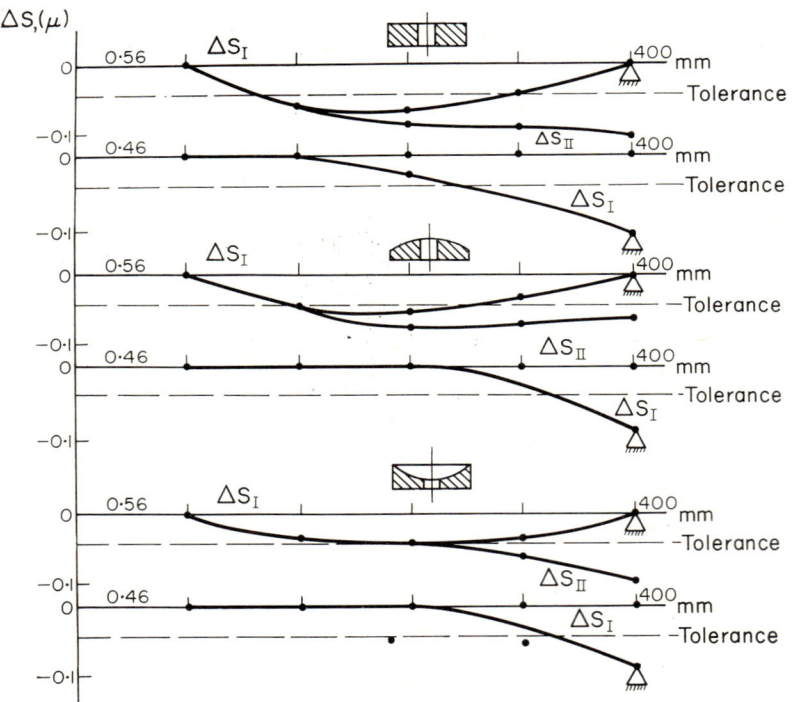

Figure 6b-3 Deviation ΔS of the actual mirror model surface from the nearest sphere versus distance from the center of the mirror. The values ΔS_I have been measured at points along radii passing through the support points, and ΔS_{II} has been measured at points along radii passing between support points. The mirrors have seven supports (one in the center and six evenly spaced along the edge). The deviations have been scaled to actual mirror size. Per mirror type there are two sets of measurements: for $0.5G$ and $0.4G$ reaction at the central support, where G is the weight of the mirror.

Thus, when the mirror is supported at the six points the resulting deformations greatly exceed the permissible limits. It was therefore proposed to support the mirror at seven points and to determine the optimal reaction at the seventh central support. The reaction at the central support was determined by using weights from a pharmaceutical scale. For a $0.4\,G$ reaction at the center, where G is the weight of the model, the center and the edge between the supports lie on the same level, below the supports. It is clear from the results given in Figure 6 b-3 that the flexure occurs in the extreme edge region. For a $0.5\,G$ reaction at the central support the center lies on the same level as the supports, but considerable sag of the edge is observed; apparently, a reaction equal to $0.4\,\text{G}$ is the most acceptable choice. This comprised the second stage of the work. In the third stage we considered models with equal relative rigidity but of different weights (Figure 6 b-1(b)). The smallest deviations from the nearest sphere were observed in the models with the smallest rigidity— in the flat-convex model $(0.08\,\mu)$ and in the flat model $(0.06\,\mu)$. The results are given in Tables 6 b-2 and 6b-3.

Table 6 b-2 Experimental data scaled to actual mirror size from models with a central hole, using seven support points and for $h_M/D_M = 1/5$ and $h_M/H_M = 1/2$ (Figure 6 b-1(b))

Dist. from mirror center X_i (mm)	Flat mirror		Flat-convex mirror		Flat-concave mirror	
	$\Delta S_{\mathrm{I}}(\mu)$	$\Delta S_{\mathrm{II}}(\mu)$	$\Delta S_{\mathrm{I}}(\mu)$	$\Delta S_{\mathrm{II}}(\mu)$	$\Delta S_{\mathrm{I}}(\mu)$	$\Delta S_{\mathrm{II}}(\mu)$
400	−0.08	0.00	−0.08	0.00	−0.07	0.00
320	—	−0.02	—	−0.02	—	−0.04
240	−0.03	−0.02	−0.03	−0.02	−0.05	−0.04
160	−0.02	−0.02	−0.01	−0.01	−0.04	−0.04
80	0.00	0.00	0.00	0.00	0.00	0.00

Note: ΔS_{I} and ΔS_{II} are deviations measured as in Table 6b-1.

The igdantin used for the models to obtain the data of Table 6 b-2 had $E_M = 0.32$ kg cm^{-2}, $\gamma_M = 1.14$ g cm^{-3} and $\mu_M = 0.5$; the model had also the ratios $h_M/D_M = 32$ mm/160 mm $= 1/5$ and $h_M/H_M = 1/2$. For the data given in Table 6 b-3 the material had $E_M = 0.6$ kg cm^{-2}, $\gamma_M = 1.14$ g cm^{-2} and $\mu_M = 0.5$; the model had also the ratios $h_M/D_M = 47$ mm/160 mm $= 1/3.4$ and $h_M/H_M = 1/3$.

Thus, the best variant of those considered is the flat-convex model with seven-point support. The deflections of the flat model are somewhat larger.

Table 6b-3 Experimental data scaled to actual mirror size from models with central hole, using seven support points and for $h_M/D_M = 1/3.4$ and $h_M/H_M = 1/3$ (Figure 6b-1(b))

Dist. from mirror center X_i (mm)	Flat Mirror		Flat-convex mirror		Flat-concave mirror	
	$\Delta S_I(\mu)$	$\Delta S_{II}(\mu)$	$\Delta S_I(\mu)$	$\Delta S_{II}(\mu)$	$\Delta S_I(\mu)$	$\Delta S_{II}(\mu)$
395	−0.06	0.00	−0.09	0.00	−0.07	0.00
325	—	−0.03	—	−0.03	—	−0.03
200	−0.06	−0.05	−0.07	−0.05	−0.06	−0.05
75	0.00	0.00	0.00	0.00	0.00	0.00

Note: ΔS_I and ΔS_{II} are deviations measured as in Table 6b-1.

Finally, the fourth stage of the work covers the comparison between theoretical and experimental results for a model simply supported along a circle ($r = 74$ mm). As was to be expected, the deflections in this case were somewhat less than for six-point support. The worst case was that of the flat-concave mirror (0.24 μ). The deflections for the flat and the flat-convex mirrors were the same—0.22 μ. The deflections ΔS_i are given in Table 6b-4.

Table 6b-4 Deflections ΔS_i of models simply supported along the edge scaled to actual mirror size (Figure 6b-1(a))

Dist. from mirror center X_i (mm)	Flat mirror $\Delta S_i(\mu)$	Flat-convex mirror $\Delta S_i(\mu)$	Flat-concave mirror $\Delta S_i(\mu)$
400	0.00	0.00	0.00
320	−0.04	−0.04	−0.04
240	−0.11	−0.11	−0.11
160	−0.16	−0.16	−0.18
80	−0.21	−0.20	−0.22
0	−0.22	−0.22	−0.24

For comparison with the computed deflections S_{comp} (simply supported at the edge, $r_{comp} = 80$ mm), the maximun deflection of the flat mirror $S_{exp} = 0.24$ μ, determined experimentally, was corrected for the location of the supports along a circle $r_{exp} = 74$ mm in radius:

$$S_{comp} = S_{exp}\left(\frac{r_{comp}}{r_{exp}}\right)^4 = 0.22\left(\frac{80}{74}\right)^4 = 0.30 \ \mu. \tag{6b-3}$$

The maximum flexure S_{max} of a solid circular plate of equal thickness and simply supported at the edge is given by the following approximate expression [7]:

$$S_{max} = \frac{0.7 q r^4_{comp}}{E h^3}, \qquad (6b\text{-}4)$$

where q is the distributed load*. For the actual mirror ($h = 125$ mm, $D = 800$ mm) made of glass ($E_g = 10^6$ kg cm^{-2}, $\gamma_g = 2.5$ g cm^{-3}), $S_{max} = 0.29\,\mu$.

CONCLUSION

The smallest deviations from the nearest sphere among all the variants of mirrors analized and supported at six points is for the flat with relative thickness 1 : 3.4 and the flat-concave with relative thickness 1 : 5. If a weight reduction is required, the best results are obtained with a flat-convex mirror and a seven-point support.

REFERENCES

1. V. F. Trumbachev and N. A. Suvorov, *Tr. IGD, I. Izd. AN SSSR*, Moscow (1954).
2. V. F. Trumbachev and L. S. Molodtsova, *The Application of an Optical Method to the Study of the Stress State of Rocks Near Mines* (Primeneniye opticheskogo methoda dlya issledovaniya napryazhennogo sostoyaniya porod vokrug gornykh vyrabotok) Izd. AN SSSR, Moscow (1963).
3. Ye. G. Grossvald and K. S. Tavastsherna, "Investigations of Models of the Principal Mirror of the Large Telescope by Photoelastic Methods", *Izv. GAO AN SSSR*, **24**, Issue 1, No. 177, 114 (1964).
4. D. N. Osokina, *Plastic and Elastic Low-Modulus Optically Active Materials* (Plastichnyye i uprugiye nizkomodulnyye opticheski aktivnyye materialy), Izd. AN SSSR, Moscow (1963).
5. N. M. Belyaev, *Strength of Materials* (Soprotivleniye materialov), GITTL, Moscow and Leningrad (1949).

---

* The equation for S_{max} and for the boundary conditions given above, is

$$S_{max} = \frac{5 + \mu}{64(1 + \mu) D} q r^4,$$

where D is the flexural rigidity of the plate and is given by

$$D = \frac{E h^3}{12(1 - \mu^2)}.$$

For $\mu = 0.30$ (value for most common glasses) the expression for S_{max} is reduced to (6b-4). (Ed. English version.)

6. S. G. Lekhnitskii, *Proceedings of the Conference on the Optical Method of Studying Stresses* (Tr. konf. po opticheskomu methodu izucheniya napryazhenii), NII matematiki i mekhaniki LGU i NII mekhaniki MGU, Mathematics and Mechanics Research Institute of Leningrad State University and Mechanics Research Institute of Moscow State University (1937).
7. *Machine Construction Manual*, Vol. 3 (Spravochnik mashinostroeniya) Moscow (1951).

CHAPTER II-7

On the design selection for the support system of an astronomical mirror

N. N. MIKHELSON

Main Astronomical Observatory
Pulkovo

The proper design of the mirror support system for a large modern astronomical telescope is essential to preserve the figure of the mirror surface, regardless of the attitude of the telescope. At the present time solid mirrors with flat and convex backs are used, and also with back sockets for the support system. We shall consider here these types of mirrors in conjuction only with the Lassel-type support system. We will analyze first the back support of a solid mirror with a flat back. Let us consider one support unit as indicated in Figure 7-1.

The departure of lever ABC from a straight line by any angle ε originates an unbalance moment Δ expressed by

$$\Delta = QL \cos (z - \varepsilon) - Pl \cos z. \tag{7-1}$$

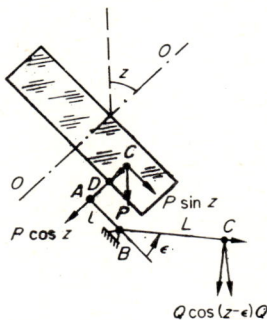

Figure 7-1 Vector diagram of the main forces acting on a back support unit (Lassel type) of a solid mirror to compute the unbalance moment Δ due to the angular misalignment ε.

This unbalance can be corrected by the proper adjustment of the load Q depending of the z value.

We set the condition that the unbalance Δ shall not exceed a given percentage (k) of the moments originated by loads at the ends of the lever for $z = 0$. We then obtain the condition

$$\tan \varepsilon \sin z \leq \frac{k}{100}. \tag{7-2}$$

If the linkage DA (Figure 7-2) is not perpendicular to the back of the mirror and the lever ABC is not parallel to it (forming the angles $90° + \gamma$ and β respectively) then the unbalance will be

$$\Delta = Pl \frac{\cos(\beta - \gamma)}{\cos \gamma} \cos z - QL \cos(\beta + z). \tag{7-3}$$

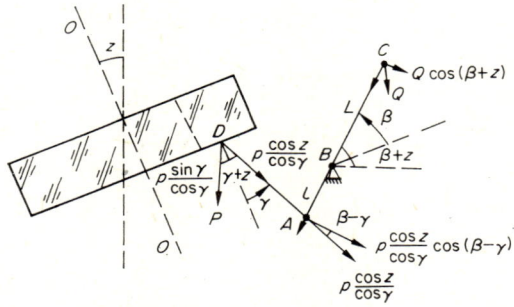

Figure 7-2 Vector diagram of the main forces acting on a back support unit (Lassel type) of a solid mirror to compute the unbalance moment Δ due to the angular misalignment β and γ.

Obviously, if $\beta = 0$, Δ is only a function of the angle z. For the support system to operate properly, it is important that $\beta = 0$. The limit for the tolerable value of β is given by

$$\tan \beta \sin z \leq \frac{k}{100}, \tag{7-4}$$

where k is the tolerable unbalance expressed as a percentage of the moment produced by lever ABC, the given load conditions and for $z = 0$.

In a lateral support mechanism for a solid mirror with back sockets (see Figure 7-3), the unbalance Δ caused by the deviation ϑ of the linkage HE with respect to HC (where C is in the axis of the socket and on the neutral plane of the mirror) and the deviation ψ of the lever EFG from the normal to the mirror back will be

$$\Delta = Pm \frac{\sin z}{\cos \vartheta} \cos (\vartheta + \psi) - RM \sin (z + \psi). \qquad (7\text{-}5)$$

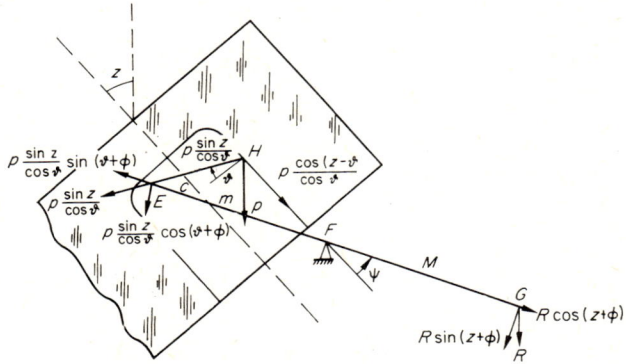

Figure 7-3 Vector diagram of the main forces acting on the lateral support unit (Lassel type) and in the mirror socket to compute the unbalance moment Δ due to the angular misalignment ϑ and ψ.

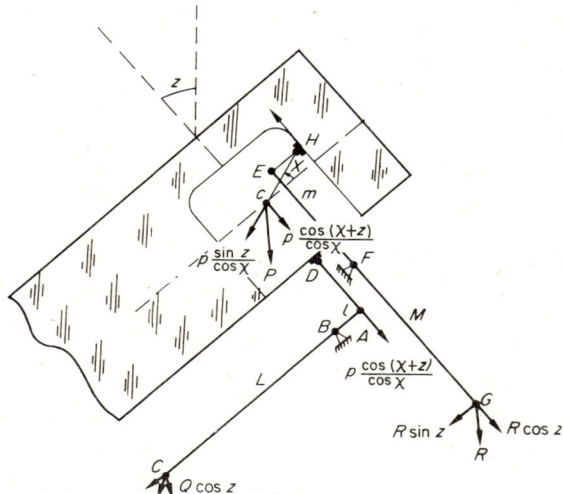

Figure 7-4 Vector diagram of the main forces acting on the lateral and back support unit (Lassel type) in the mirror socket to compute the unbalance moment Δ and reactions due to the angular misalignment χ.

The limits for the tolerable θ and ψ is expressed by the relationship

$$\frac{\sin \psi}{\cos \vartheta} \cos (z - \vartheta) \leqq \frac{k}{100}. \qquad (7\text{-}6)$$

If $\vartheta \neq 0$, the component $P \dfrac{\cos (z - \vartheta)}{\cos \vartheta}$ causes an unbalance of the back support

$$\varDelta = Pl \sin z \tan \vartheta. \qquad (7\text{-}7)$$

If the lateral support mechanism does not support the mirror in the neutral plane (Figure 7-4) the additional moment causes the unbalance

$$\varDelta = Pl \sin z \tan \chi \qquad (7\text{-}8)$$

of the back support mechanism.

The following conclusions may be drawn from the previous analysis: (1) the points A and B and the center of gravity of the load C on the lever ABC for a back support unit (see Figure 7-2) must lie in a straight line; (2) similarly, for the lateral support lever; (3) the lever ABC must be parallel to the back surface of the mirror (see Figure 7-2); (4) the lateral support point must lie in the neutral plane of the mirror; (5) the linkage which transmits the load to the lateral support unit must lie also in the

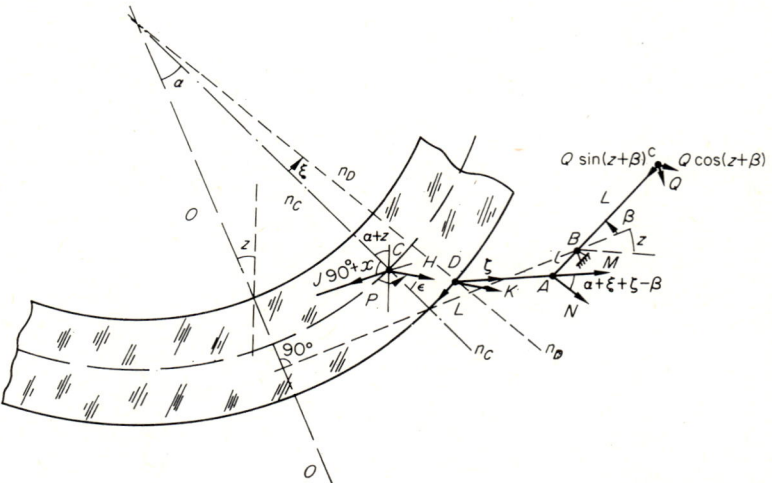

Figure 7-5 Vector diagram of the main forces acting on the support unit of a solid mirror with a convex back to compute the unbalance moment $\varDelta$ due to angular misalignment of the lever mechanism.

neutral plane; (6) the lateral support lever EFG (see Figure 7-4) must be parallel to the optical axis of the mirror.

We will consider now the back support for a solid mirror with a convex back. Figure 7-5 shows a schematic diagram of the mirror and a support unit with misalignment in the lever system; the line 0-0 is the optical axis of the mirror. The main forces for the load conditions given in Figure 7-5 are given by the following expressions:

$$J = P \frac{\sin(z + \alpha + \varepsilon)}{\cos(\chi + \varepsilon)}, \qquad (7\text{-}9)$$

$$K = P \frac{\cos(\chi - z - \alpha)}{\cos(\chi + \varepsilon)}, \qquad (7\text{-}10)$$

$$L = P \frac{\cos(\chi - z - \alpha)\sin\zeta}{\cos(\chi + \varepsilon)\cos(\varepsilon - \xi + \eta)}, \qquad (7\text{-}11)$$

$$M = P \frac{\cos(\chi - z - \alpha)\cos(\varepsilon - \xi)}{\cos(\chi + \varepsilon)\cos(\varepsilon - \xi + \zeta)}, \qquad (7\text{-}12)$$

$$N = P \frac{\cos(\chi - z - \alpha)\cos(\varepsilon - \xi)}{\cos(\chi + \varepsilon)\cos(\varepsilon - \xi + \zeta)} \cos(\alpha + \varepsilon + \zeta - \beta). \qquad (7\text{-}13)$$

In the case described in Figure 7-5 the misalignment of the support unit will originate an unbalance moment expressed by

$$\varDelta = Pl \frac{\cos(\chi - z - \alpha)\cos(\varepsilon - \xi)}{\cos(\chi + \varepsilon)\cos(\varepsilon - \xi + \zeta)} \cos(\alpha + \varepsilon + \zeta - \beta) - QL\cos(z + \beta). \qquad (7\text{-}14)$$

Taking K as a tolerable unbalance of the system, we obtain the relation

$$\frac{\sin(\chi - \alpha + \beta)}{\cos(\chi - \alpha)\cos\beta} \sin z \leq \frac{k}{100}. \qquad (7\text{-}15)$$

The unbalance expressed by Eq. (7-14) becomes $\varDelta = 0$ only if the conditions $\varepsilon = \xi = -\chi$ and $\beta = \alpha - \chi = \alpha + \varepsilon$ are satisfied.

Of the set of possible values of the angle β only two are of practical interest:

(A) $\beta = 0$ and $\chi = \alpha$, i.e. all the levers (ABC, see Figure 7-5) of the back support mechanism must be perpendicular to the optical axis of the mirror, and the direction determined by C (center of mass of the element supported by the unit) and the lateral support point H must also be perpendicular to the optical axis.

(B) $\beta = \alpha$ and $\chi = 0$, $\varepsilon = 0$, $\zeta = 0$, i.e. each back support lever (*ABC*, see Figure 7-5) must be perpendicular to the normal n_c to the mirror surface passing through the center of mass C of the supported mirror element.

We will consider now the lateral support for a solid mirror with a convex back and sockets. Figure 7-6 shows a schematic diagram of the mirror and a support unit with misalignment in the lever system. The main forces for the load conditions given in Figure 7-6 are given by the following expressions:

$$S = P \frac{\cos(\chi - \alpha - z)}{\cos(\chi + \varepsilon)}, \qquad (7\text{-}16)$$

$$T = P \frac{\sin(z + \alpha + \varepsilon)}{\cos(\chi + \varepsilon)}, \qquad (7\text{-}17)$$

$$U = P \frac{\sin(z + \alpha + \varepsilon) \sin \psi}{\cos(\chi + \varepsilon) \cos(\varphi - \psi)}, \qquad (7\text{-}18)$$

$$V = P \frac{\sin(z + \alpha + \varepsilon) \cos \varphi}{\cos(\chi + \varepsilon) \cos(\varphi - \psi)}, \qquad (7\text{-}19)$$

$$X = P \frac{\sin(z + \alpha + \varepsilon) \cos \varphi}{\cos(\chi + \varepsilon) \cos(\varphi - \psi)} \cos(\alpha - \chi + \psi - \tau). \qquad (7\text{-}20)$$

Similarly, the unbalanced moment of the lateral support mechanism, due to misalignments shown schematically in Figure 7-6, will be

$$\varDelta = Pm \frac{\sin(z + \alpha + \varepsilon) \cos \varphi \cos(\alpha - \chi + \psi - \tau)}{\cos(\chi + \varepsilon) \cos(\varphi - \chi)} - RM \sin(z + \tau), \qquad (7\text{-}21)$$

and the limit values are obtained from the condition

$$\frac{\sin(\alpha + \varepsilon - \tau)}{\cos(\alpha + \varepsilon) \cos \tau} \cos z \leqq \frac{k}{100}. \qquad (7\text{-}22)$$

For proper operation of this lateral support system the conditions $W = -\varepsilon$ and $\alpha + \varepsilon = \tau$ must be satisfied.

The errors φ and ψ can be compensated by adjustment of the counterweight R at G.

The following conclusions may be drawn from this analysis:

(1) the directions *CH* and *CD* of the back and lateral support unit should be perpendicular (see Figure 7-5);

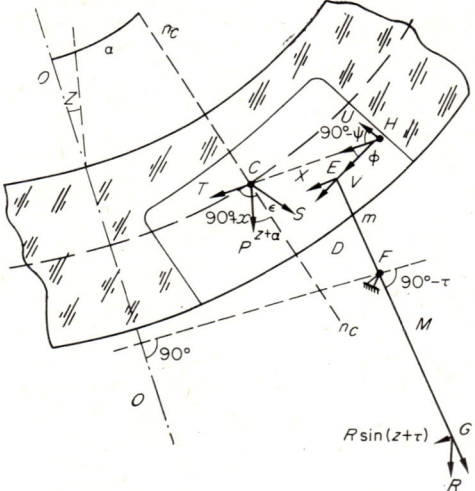

Figure 7-6 Vector diagram of the main forces acting on the support unit in the socket of a mirror with a convex back to compute the unbalance moment due to angular misalignment of the lever mechanism.

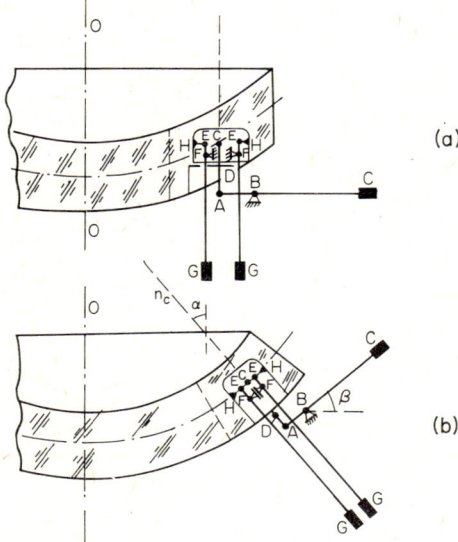

Figure 7-7 Schematic of two support unit designs for mirrors with a convex back and sockets: (a) unit with levers parallel and perpendicular to the optical axis; (b) unit with levers parallel to the surface tangent to the back of the mirror and perpendicular to it.

(2) there are two possible designs for the lateral support mechanism (see Figure 7-6): (a) with CH perpendicular to the optical axis of the telescope, and the lever EFG parallel to this axis (see Figure 7-7(a)); (b) with CH on the neutral plane of the mirror and the lever EFG parallel to the normal to the mirror which passes through the point C (see Figure 7-7(b)).

Variant (a) is simpler to adjust and less sensitive to errors in the back support mechanism than variant (b), but is less satisfactory from the point of view of tangential forces.

For high-aperture ratio mirrors, variant (b) is to be preferred.

CHAPTER **II-8**

Electrical control for the telescope mount APSh-6

U. K. VEISMAN and T. E. KYUBAR

Institute of Physics and Astronomy
Tartu

The automatic control of telescopes of small aperture usually involves the following operations:

(1) Setting the telescope by the electric drive controls from a hand block or small portable console.

(2) Hour angle guiding at different speeds without a gear mechanism.

(3) Photoelectric guiding.

(4) Automatic setting from the control console.

These operations have been studied separately to a certain extent, but until recently no telescope mount was available with universal electric drives or workable solutions to adapt existing instruments to automatic control. A number of automatic control systems have been developed for certain (mostly large aperture) telescopes. Many designs have not reached further than the modeling stage, and the realization of some projects has encountered technical problems (the use of amplidynes, power supply at 400 Hz, etc.). The automatic control of moderate-sized telescopes must satisfy the requirements of simplicity, universality, low cost and reliability.

Starting from these considerations, an attempt has been made at the Estonian Institute of Physics and Astronomy to construct universal electric drives for the widely-used equatorial (parallactic) mount, APSh-6 type. (The APSh-6 is suitable for reflectors of up to 50 cm and refractors of up to 25 cm in diameter.) For the hour angle and the declination drives, a kinematic circuit with two motors was selected and a worm-gear was mounted in the declination axis similar to the main worm in the hour axis. Figure 8-1 shows the block diagram of the control system. The blocks left of the dashed vertical line correspond to the declination drive, and the blocks to the right of the second line correspond to the hour angle drive. The other blocks represent the equatorial mount APSh-6, the cabinet with ten relays of the RMUG type and the hand block RP-1.

The slew motors M_g, UL-062 type, operate at 3,000 rev min^{-1} and are coupled to the respective telescope axis through a 14,400 : 1 worm-gear reduction which gives a slew motion of approximately 1° sec^{-1}. The slow motion motor M_t, D-32 type, operates at 1,080 rev min^{-1}. Through a gear reduction built in the motor and another external gear reduction unit, the motor M_b is coupled to the M_g by means of an electromagnetic clutch (see Figure 8-1). The slow motion has three different rates available*: 8, 12 and 24″ sec^{-1}. When the motor M_g for slew motion is switched on, the slow motions are disconnected by an electromagnetic clutch.

The only difference between the hour angle and declination drive is the addition of a unit to drive in the hour angle at guide rate (see Figure 8-1). This unit consist of a standard drive mechanism ChM-1 in which the gears have been replaced by the motor M_{Ch}, D-32 type (the centrifugal governor

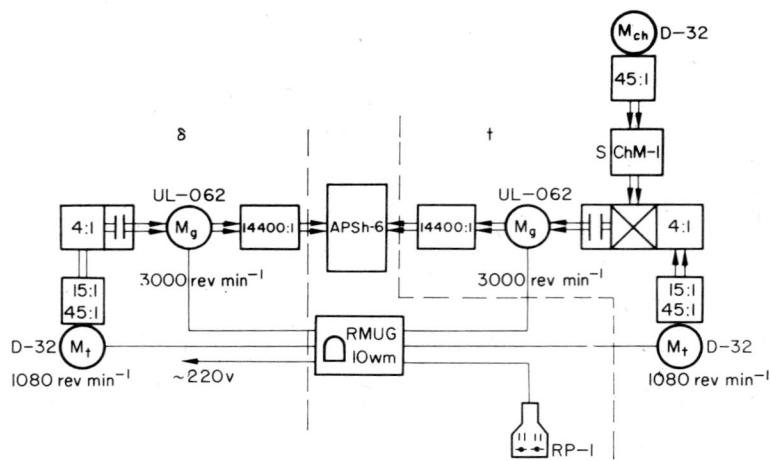

Figure 8-1 Block diagram of the electrical control developed at the Estonian Institute of Physics and Astronomy for the telescope mount APSh-6.

* The block diagram given in Figure 8-1 only indicates the speed reduction 15 : 1 and 45 : 1 which gives 8 and 24″ sec^{-1}. (Ed. English version.)

and seconds control are kept); it is coupled to the hour angle drive via a differential between the clutch and the 4 : 1 gear reduction.

For a simple two-speed drive operation of the telescope, the motors are switched on by push buttons on the hand block RP-1 through a relay cabinet (ten relays of the type RMUG) as shown in Figure 8-1.

For more flexible electrical control and for automatic setting and photoelectric guiding, the drives use servoamplifiers. The block diagram of this control system is shown in Figure 8-2†. In this case the hand block is connected to the terminals RP-2. This hand block has two potentiometers, which allow the selection of a continuously adjustable speed rate from -8 to $+8''$ sec^{-1} (or -16 to $+16''$ sec^{-1}, -24 to $+24''$ sec^{-1}); at the extreme settings of the potentiometer knobs, the relays R switch on the coarse motors. For automatic control, we plan to use the VT input to connect rotary transformers VTs. The accuracy of this control will depend mainly on the accuracy of the VTs, but our practical experience has confirmed the accuracies given in the literature of a few minutes of arc, or even higher. The input FG is to be used for photoelectric guiding by a dc signal. The drives were constructed by E. Vakkur, R. Koppel and O. Tammemyagi.

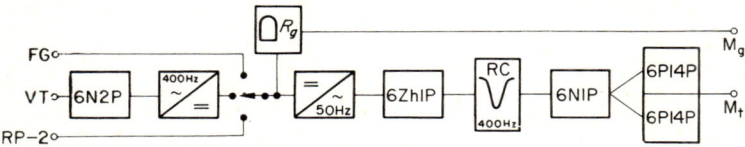

Figure 8-2 Block diagram of the additional circuitry to be used in connection with the system shown in Figure 8-1 for more flexible electrical control, automatic setting and photoelectric guiding.

† The block diagram shown in Figure 8-2 is not fully described in the original Russian text. It is apparent that this diagram is from another publication on the same subject, but since no bibliographical references are given it was difficult to make it clear in the English version. (Ed. English version.)

CHAPTER II-9

A photoelectric photometer with digital printout

U. K. VEISMAN

Tartu Observatory
Tartu

The increased accuracy and speed required in modern scientific measurements, including astrophotometric observations, cannot be provided by analog techniques. The potentiometer type recorders EPPs widely used in photoelectric photometry have only limited reading accuracy (0.5%), plus the fact that the data from the recordings require subsequent manual reduction. Improved performance of measuring instruments can be achieved by combining automatic compensation with digital output and automation of the recording process.

Digital measuring instruments have the following advantages: (1) high accuracy (0.1–0.01%); (2) convenient display and the elimination of subjective reading errors; (3) automation of the measurement process; (4) possibility of performing computation; (5) possibility of printed or punched output; (6) high speed of operation.

Since the minimum detectable signal in photometers with photomultiplier tubes (FEUs) is of the order of 10^{-10} A and the input resistances of modern digital voltmeters do not usually exceed 10 MΩ for a range of 0.1 to 1 V. Full scale, it is necessary to amplify the FEU output current.

Figure 9-1 shows the block diagram of the photoelectric photometer with digital output of the Tartu Observatory (Estonia). The front optics for the photometer is a Cassegrain reflector AZT-14 type (480 mm aperature and 7.7 m focal length). The photometer head is of standard design with filters, diaphragms, a shutter, standard luminous source, photomultiplier tube, range selector and a cathode follower. The anode dark current of the FEUs EMI 6094 B or EMI 9502 B is of the order of 10^{-10} A, and the anode responsivity reaches 2,000 Alm^{-1}. The load resistors of the FEUs are of high values (up to $1 \times 10^{10}\,\Omega$) which give the required output voltage of up to 0.2 V. The cathode follower uses the 6 ZhlB vacuum tube connected as an electrometer tube which drives the digital voltmeter Solartron LM-902, with an input resistance of 10 KΩ for the range of 0.15 V.

Electrometer cathode followers and differential amplifiers using double triodes of the type 6N16B also performed well when used as current preamplifiers. The cathode follower is fed from the stabilized power supply U 1136 (stability 0.01%), but stabilized power supplies using semiconductors are to be preferred for reasons of economy and reliability. In order to reduce the noise level we used an RC-filter between the cathode follower and the digital voltmeter, giving time constants which can be selected from 0.1 to 5.0 sec.

The Solartron LM-902 digital voltmeter (which may be substituted equally successfully by the Soviet digital voltmeter V 2-8) operates on the principle of automatic compensation using electromechanical relays. For 280 msec integration time the accuracy is 0.1%. The output of the digital voltmeter is in parallel digital code and is fed to the digital printer which was taken from the digital voltmeter ETsPV-1; the maximum printing

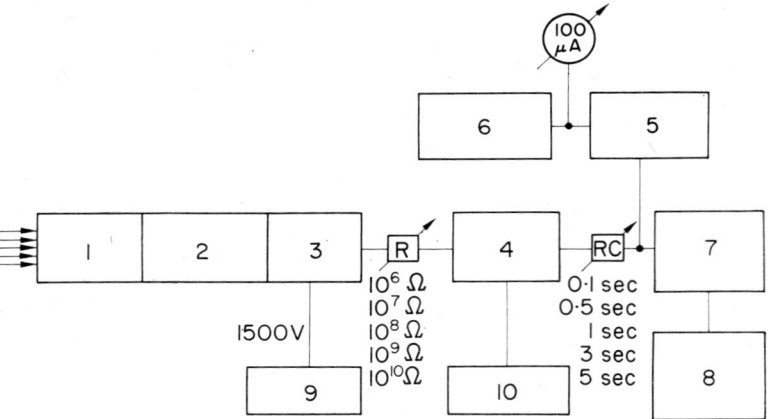

Figure 9-1 Block diagram of the photoelectric photometer with digital printout developed at the Tartu Observatory: 1, telescope AZT-14 ($D = 480$ mm, $f = 7715$ mm); 2, photometer head; 3, photomultiplier tube EMI 6094 B or EMI 9502 B; 4, cathode follower 6ZhlB; 5, dc amplifier (Ye 6-3); 6, recorder EPP-09; 7, digital voltmeter—SOLARTRON LM 902; 8, digital printer (ETsPV); 9, high-voltage power supply VS-23; 10, power supply U-1136; R, range sector (voltage divider); RC, time constant.

rate is two measurements per second. When the output of the digital voltmeter and the input of the digital printer are matched, the projection type display unit should be replaced by a new unit which uses the digital display tubes IN-1. For monitoring purposes the signal is recorded by the EPP-09 in parallel with the digital voltmeter. The recorder is driven by a two-stage dc balanced amplifier, Ye 6-3 type.

Part of the digital photoelectric photometer and its main control is at the telescope: it allows for setting of the photometer head, selection of the measuring range of the microammeter and a push-button which triggers the measurement cycle of the digital voltmeter. The rest of the system is located in a heated laboratory 30 m away from the telescope. When the measurement cycle ends, the voltmeter prints the output signal in digital form; operation of the voltmeter was also tested under conditions in which the digital printer was triggered on by a push-button. For subsequent identification of the photometer printouts, two-digit numbers to be printed between the readings may be selected by the observer on the twenty-key keyboard. The digital photoelectric photometer was successfully tested.

CHAPTER II-10

Image converters in astronomical research

P. V. SHCHEGLOV

Shternberg Astronomical Institute
Moscow

Image converters have been used in astronomical work for approximately fifteen years, long enough for their performance and applications to be assessed*.

1 OPTIMUM GAIN

In astronomy, images usually have a high information content, while in nuclear physics, for example, ten small images are sufficient to define a track which can be measured. When the gain of a cascaded image converter is very high, each output electron (or group of electrons) gives an image of about 0.1 mm in diameter on the photographic emulsion. It is difficult for the image converters, to transmit a complicated pattern by using only a black and white mosaic, since finer details and half-tones cannot be easily handled.

The size of the mosaic elements should be reduced (i.e. the resolution increased) and an attempt should be made to record halftones by reducing the gain. The optimum gain of the converter cannot greatly exceed 10^3. This problem has yet to be solved.

2 DARK BACKGROUND

The lowest dark background is obtained in simple sealed-off image converters. The cascaded converters used in nuclear physics have large irreducible backgrounds of the order of 10^3–10^4 electrons cm^{-2} sec^{-1}. This has been

---

* The conclusions on the state-of-the-art presented in this section refers mainly to the work carried out in the USSR. (Ed. English version.)

measured by Ye. K. Zavoisky and his colleagues. The difficulty in nuclear physics is circumvented by switching on the converter for the duration of the particle flight which is only a few microseconds. This period of time is too short for the photocathode to emit a single background electron. Unfortunately, in astronomy, exposures of a few microseconds are not feasible, and thus the converter must have a small dark background. Studies made at the Crimean Astrophysical Observatory have shown that cascaded image converters must have a gain of only a few hundreds, so that the converter background is comparable to the sky background. In this case, normal operation of the converter is handicaped, as sensitivity to stray magnetic fields is increased. In single-stage image converters the coefficient of amplification reaches 10^2 and less information is lost, since their resolution is much greater. As yet there has not been any published spectrum of a faint object recorded with a cascaded image converter.

The improvement of photocathodes is of importance. Many completely new astronomical problems can be solved with image converters using the new multi-alkali cathodes, although practical difficulties have arisen. For instance, precautions must be taken to eliminate the optical feedback, which have otherwise insignificant effects in low-sensitivity cathodes.

3 ELECTRONIC CAMERA

As yet, there are many difficulties for the effective use of the "electronic camera". Its dark background (10^3–10^4 electrons cm^{-2} sec^{-1}) is much larger than for simple sealed-off tubes (1–10 electrons cm^{-2} sec^{-1}). The focusing problem has yet to be solved. For focusing, the camera has to be opened and the casette removed; after the operation is completed and the camera is closed again, there is no assurance that the focusing is maintained. Thus, although the electronic camera is ideal in principle, it is still far from perfection.

CHAPTER II-11

A low-order Fabry–Pérot etalon used in astronomical research

P. V. SHCHEGLOV

Shternberg Astronomical Institute
Moscow

A Fabry-Pérot etalon with multilayer coatings has proved to be a very efficient monochromator for astronomical research. In the classical configuration, it gives excellent results in the study of the emission from the night sky and the Galaxy. Spectrographs used in conjunction with the Fabry–Pérot etalon allows operation of the spectrograph with a wider slit while maintaing a given resolution; this is important to increase the efficiency of the spectrograph. Also it allows to obtain high and extra-high resolution from existing spectrometers provided, of course, that sufficient light is available.

The design and construction of such systems is determined by the properties of the available etalons, i.e. by the quality of their plate surfaces, coatings and the separators. The available coatings are good: high-quality layers with low absorption and with a reflection coefficient of about 95% have been produced. The best surfaces produced for etalons at the present time have flatness of $\lambda/40$, as can be seen from the small increase in resolution which is observed upon increasing the reflectance of the coating on a given substrate from 0.91 to 0.95. For a sufficiently flat substrate, this would have the effect of nearly doubling the number of interfering rays, and hence the resolution.

Industry is not able to manufacture separators thinner than 300 μ, although sometimes thicknesses of 50–100 μ is advantageous. This forces us to use small lengths of steel wire and mica as Fabry did.

Plate surfaces with accuracies of the order of $\lambda/100$ have been reported in foreign journals.

The realization of a high-quality low-order etalon will have great potential. With separators of the order of 100 μ the separation of adjacent passbands of such an etalon in the red region is about 20 Å, and with an ordinary interference filter as a monochromator, transmission of the combination can be restricted to one order alone. The etalon can operate as a spectrograph: for instance, it can be used to study absorption lines. By regulating such a system appropriately and by combining the interference pattern with the output pupil of the telescope, we obtain a monochromator for protuberances, for example, with a bandpass of the order

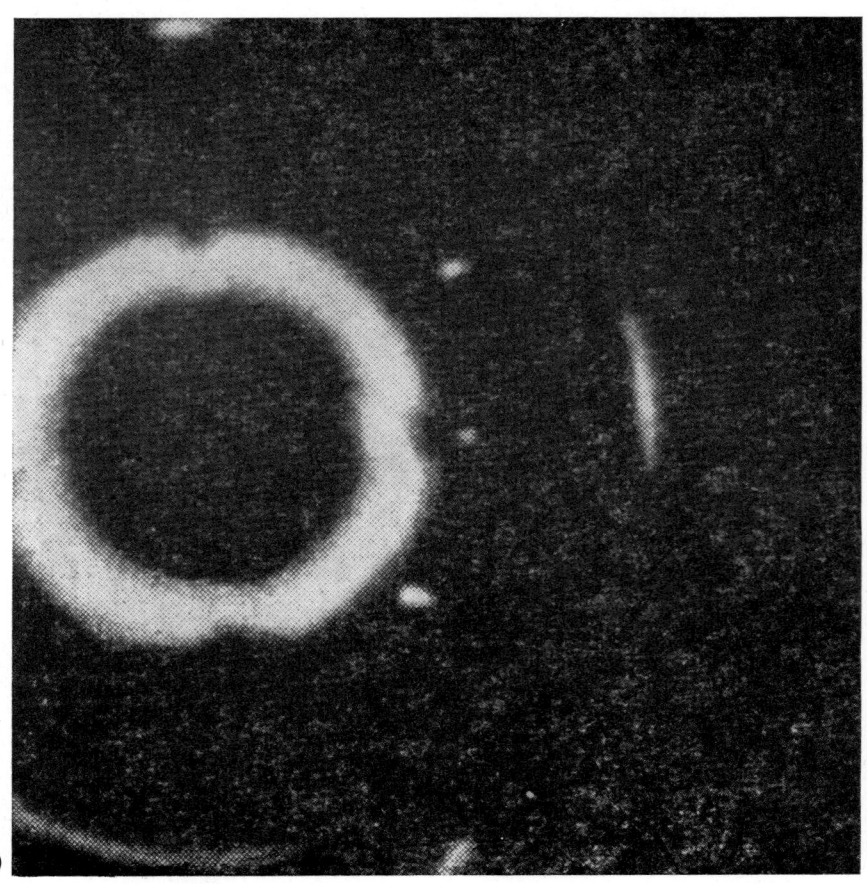

Figure 11-1 Photographs $(H_\alpha, \Delta\lambda = 0.3 \text{ Å})$ of nebulae obtained, with a Fabry-Pérot etalon in conjunction with the 13-in. reflector at Simeiz Station: (a) nebula NGC 7000; (b) nebula NGC 6618.

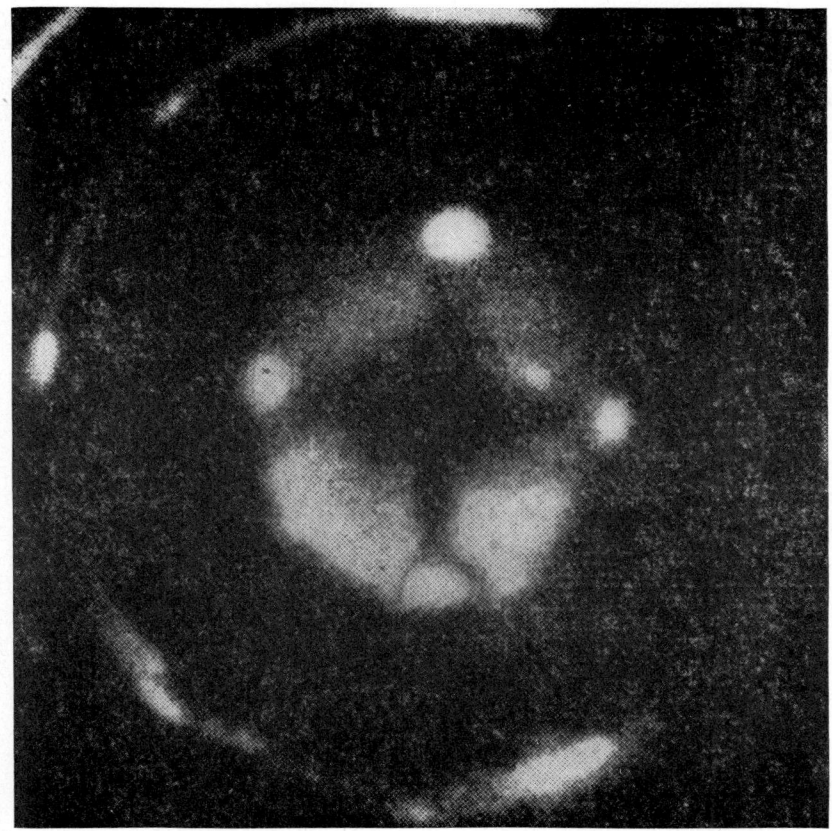
b)

of 1 Å and transmittance of about 50%. For comparison, the transmittance of IPFs* with a bandpass of about 1 Å is not greater than a few per cent. This is of great importance in geophysics.

In 1963 a low-order etalon was used to study the profiles of the H_α emission line in the gas nebulae. The etalon was mounted on the 13-in. reflector at the Simeiz Station†; the resolution of the etalon was 0.3 Å. In Figure 11-1(a) and (b) photographs of the nebulae NGC 7000 and NGC 6618 are shown. In the first case the emission line is narrow, and in the second is wide, corresponding to velocities of ± 40 km sec^{-1}. This could not be seen if less efficient etalons or slit spectrographs were used.

* See "An interference-polarization filter for astrophysical studies of the sun in the K-line of ionized calcium", Chapter I-28 of this book. (Ed. English version.)
† Branch of the Crimean Astrophysical Observatory. (Ed. English version.)

CHAPTER II-12

Rowland ghosts in double pass diffraction monochromators and their compensation in photoelectric spectrophotometry

P. P. KOZAK

Lvov Astronomical Observatory
Lvov

There are several papers devoted to the study and theoretical interpretation of Rowland ghosts*. The majority of them consider Rowland ghosts on the diffraction gratings used in ordinary single pass monochromators [1]. Rowland ghosts in double pass monochromators are discussed by Karpinskii [2]. One of the monochromators described by Karpinskii [3] was used as a photoelectric solar spectrophotometer [4], in which a flat mirror was added for scanning the spectrum. This also produced periodic changes in the intensity of the ghosts; making it necessary to carry out a detailed study of the behavior of ghosts.

Consider the double pass monochromator with diffraction grating as shown schematically in Figure 12-1. The solar image is projected by means of the coelostat mirrors Z_1Z_2 and the vertical solar telescope, Z_3 and Z_4 on the plane of the entrance slit S_1 and rays passing through the slit and the objective O_1 reaches the diffraction grating G. The diffracted monochromatic parallel beams are reflected from the flat aluminized mirror Z_5 to the grating G. The double diffracted rays are imaged by the objective O_1 and the beam splitter Z_6 on the exit slit S_2; the photomultiplier F is located behind the exit slit. In the monochromator the mount of the mirror

---

* In the actual grating ruling the grooves will deviate to some extent from the ideal equal spacing. These deviations could be random, periodic or continousy increasing in one direction. The periodic deviations give rise to what are called Rowland ghosts (one period) and Lyman ghosts (two periods). (Ed. English version.)

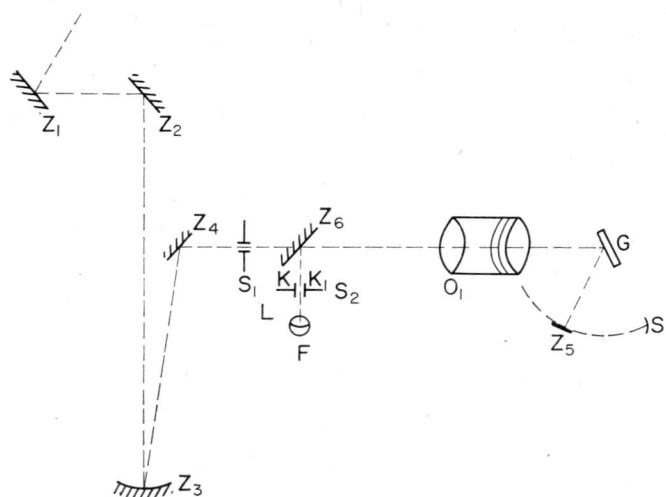

Figure 12-1 Optical scheme of the coelestat, vertical solar telescope and double pass monochromator with diffraction grating.

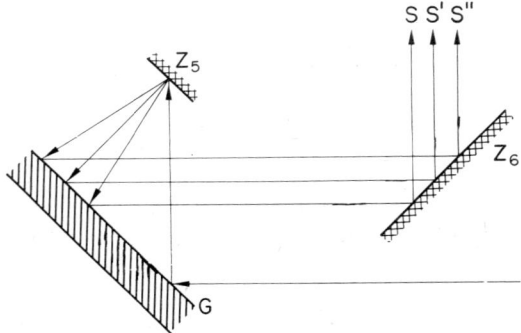

Figure 12-2 Ray paths in the photoelectric section of the spectrophotometer for different settings of the scanning mirror Z_5; G is the diffraction grating and Z_6 beam splitter.

Z_5 is driven by a worm gear mechanism providing continuous motion of the mirror along an arc S thus moving the spectrum from S to S' to S'' (Figure 12-2) in the photoelectric section L of the spectrophotometer. The singly diffracted beam is not affected by the mirror Z_5 but only the one diffracted twice. The spectrum obtained from the first diffraction is in the lower portion of the photoelectric section and it is possible to set the wavelength of this spectrum by only rotating the diffraction grating.

We studied the behavior of Rowland ghosts in the photoelectric section for continuous displacement of the spectral lines from the position K to position K_1, and vice versa. The light sources used were neon and mercury gas discharge tubes, which have intense lines so that Rowland ghosts could be easily observed. It was noticed that when the spectral lines were displaced from K towards K_1 the intensity of the ghosts altered periodically, giving a series of successive maxima and minima. The distance (period) between two maxima of the blue ghost, for example, was 10 mm or equivalent to 20 Å (2 Å mm^{-1} dispersion). The half-periods of ghost intensity changes were strictly symmetrical, and the periods were equal. The distance between the two ghosts (maximum or minimum) associated with the parent line is 4 Å. The ghosts were symmetrical about the parent line. When the spectrum was scanned by means of Z_5, decrease in intensity of both ghosts ocurred in the photoelectric section at the same setting of Z_5.

We checked all the marked minima using the 5,852 Å neon line, and then using all possible lines of the neon and mercury gas discharge tubes. It was typical that, regardless of the spectral line wavelength, the monochromator could be set to minimize the ghost effect. When certain lines fell simultaneously in the positions of several minima, they too were free from intense ghosts. If two lines were 10 Å apart, for example, and on of them was at minimum, it was free of ghosts, and the second, which was in the middle of the period, was affected by intense ghosts.

It should be remembered that for single diffraction the intensity of the Rowland ghosts remained constant, and no periodicity was observed. The distance between minima (or maxima) in the K–K_1 plane were constant, and did not change by displacement of the spectrum, or of the diffraction grating, or by a shifht of Z_6.

Measurements of the doubly diffracted beam showed that the relationship between the intensity of the ghosts, $I_{gh,\max}$, and the maximum intensity, $I_{l,\max}$, of the parent line is expressed by

$$I_{gh,\max} = 0.075 I_{l,\max}. \tag{12-1}$$

At a minimum, the intensity of the ghost is

$$I_{gh,\min} = 0.$$

Photoelectric measurements showed that

$$I_{l,\max} = I_{l,\min} + 0.075 I_{l,\max}, \tag{12-2}$$

i.e. at a minimum intensity of the ghosts, the intensity of the parent line increased by 7.5% due to ghost compensation. Correspondingly, at a maximum of the ghost, the intensity of the parent lines was reduced due to ghost formation.

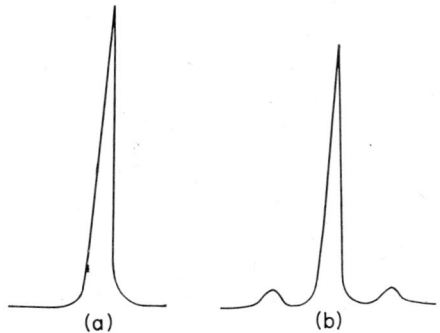

Figure 12-3 Photoelectric record of the 5,852 Å neon line: (a) with minimum effects of ghosts; (b) with maximum effects of ghosts.

In Figure 12-3 is shown the photoelectric record of the 5,852 Å neon line at a minimum (a) and at a maximum (b) of the ghosts. Superimposition of the two records demonstrates the symmetry of the parent line and shows also a line intensity increase due to ghost compensation. This increase is observed over the entire spectrum.

CONCLUSION

The effect of the periodic intensity change of the Rowland ghosts can be reduced to a minimum in a photoelectric spectrophotometer, but only if the spectrum is scanned with an additional mirror or by driving the diffraction grating. By setting the exist slit to a position for a minimum, we can obtain profile lines unaffected by Rowland ghosts.

When compensated, the effect of the ghosts on the intensity of the recorded Fraunhofer lines is negligible compared with the uncompensated intensity using better quality diffraction gratings.

The phenomenon we have been studying is of great importance in practical work with double pass monochromators using diffraction grating. Theoretical interpretation will be considered elsewhere.

REFERENCES

1. W. F. Meggers and C. C. Kies, "False spectra from diffraction gratings, Part I, secondary spectra", *J. Opt. Soc. Am.*, **6**, No. 5, 417 (1922).
2. V. N. Karpinskii, "On the compensation of Rowland's 'ghosts' in spectrographs with double diffraction on the grating", *Opt. Spektrosk.*, **8**, No. 3, 401 (1960).
3. V. E. Stepanov and A. A. Kopytyanskii, *Byull. Kom. po Issled., Solntsa*, Akad. Nauk SSSR, No. 10, 24 (1954).
4. P. P. Kozak, "A high-speed solar photoelectric spectrophotometer", *Astron. Zh.*, **38**, No. 3, 549 (1961).

CHAPTER II-13

An iris microphotometer based on the MF-2 microphotometer

L. A. URASIN

Crimean Astrophysical Observatory
Nauchny

The use of iris microphotometers [1, 2] to measure stellar magnitudes from photographic plates is becoming widespread.

In an iris microphotometer star brightness measurements are obtained by means of the adjustable iris diaphragm. The principle of operation of this instrument is based on the relationship between the diameter of the stellar image and its magnitude. The operation is based on the balancing of two light beams generated by the same lamp, one of which passes through the photographic plate (measuring beam) and the other through a wedge (balancing beam). The MF-2 microphotometer shown schematically in Figure 13-1 can be modified into an iris microphotometer without dif-

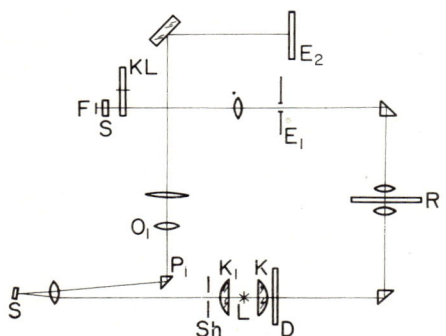

Figure 13-1 Schematic of the optical system in the MF-2 microphotometer.

ficulties by making simple changes to the photometer scheme. The alterations required include a small change in the reading section of the optical system to transform it into the balancing beam (Figure 13-2). The objective O_1 with prism P_1 must be turned by 180° around the optical axis and the prism has to be lowered. The light beam passing through O_1 is directed by the flat mirror Z_1 to a photocell or photomultiplier tube F; also the light beam which passes through the photographic plate R is fed to F.

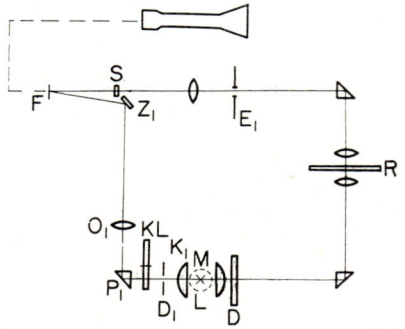

Figure 13-2 Schematic of the modified optical system of the MF-2 for operation as an iris microphotometer.

The small scale Sh in the MF-2 is removed and is replaced by the wedge KL; in the MF-2 the KL is in front of F (photometric beam path).

In the modified MF-2 an iris diaphragm or a transparent disk D is used instead of the entrance slit. This disk has a superimposed opaque plate with a set of apertures of different diameters. The calibration of the apertures (measured with the KIM-3) are given in the table 13-1. The output slit E_1 of the MF-2 is replaced by a screen E_1 with circular aperture which can be closed by means of an electromagnet. This aperture is somewhat larger than the aperture on D projected on the screen E_1, but still small enough not to interfere with the field of the plate projected on the screen.

The incandescent lamp L is covered by the cylindrical chopper M with three apertures 120° apart. The chopper is driven by the motor RD-09 in such a way that each light beam is modulated at a rate of 20 times per second.

The projection screen E_2 in the modified MF-2 is replaced by an ENO-1 (Sl-4) oscilloscope.

The beam passing through the photographic plate R and the balancing beam are nulled by selecting the proper aperture of the disk D. The null is monitored on the oscilloscope screen. The aperture number which can be seen on the screen E_1, serves as a measure of the light intensity passing through the stellar image. Since the diameters of the apertures differ by discrete amounts, a grid is placed on the oscillograph screen to enable fairly accurate interpolation to determine the aperture for which both beams will reach the null condition.

In the iris microphotometer, due to the switching between the two beams, there is no need to monitor the conditions of the incandescent lamp. Since a considerable amount of measurements were made and are being made [3] using the first version of the MF-2 and MF-4 [4], a study was made, showing that it takes 2.5 hours to stabilize the operating conditions of the lamp. Figure 13-3 gives the reading A in the MF-2 and MF-4 versus the time elapsed, t, from the time that the lamp was switched on. The

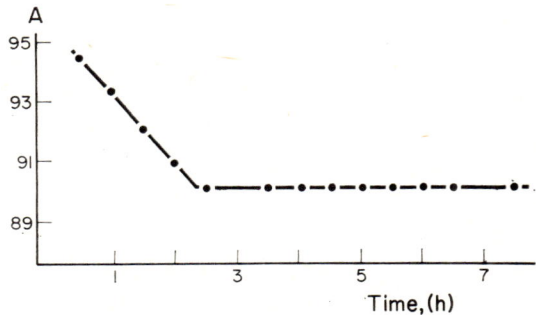

Figure 13-3 MF-2 or MF-4 reading A versus time elapsed from the time that the lamp was switched on.

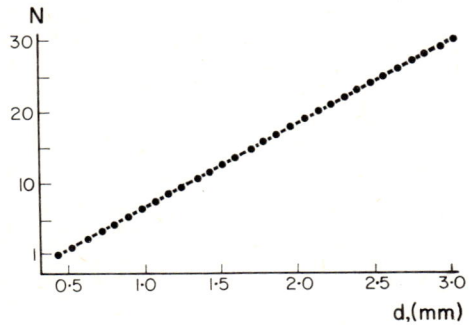

Figure 13-4 Aperture number N versus diameter d for the iris microphotometer.

heating of the disk D due to its proximity to the lamp is of importance only during the stabilization period.

The measurements of normally exposed stars (the rectilinear portion of the characteristic curve) are mainly the ones affected by unstabilized conditions of the lamp during 2.5 hours. In this case the reading for any one star lies within the range of 0.4, corresponding to roughly 0.2 stellar magnitudes. For bright or faint stars this effect is undetected.

For microphotometers which have a disk with a set of apertures instead of an iris diaphragm, the accuracy of the aperture diameters is of great importance. We found it difficult, to make a disk with apertures of linearly increasing diameters. Figure 13-4 gives the aperture numbers N versus the diameter d; this relationship can be expressed analytically by the formula

$$N = 11.18d - 3.97. \tag{13-1}$$

The diameter d of the disk apertures versus B magnitudes in the UBV system are given in Figure 13-5. This relationship has been obtained from a photographic plate (Sh 4162) of the cluster region NGC 1664 [5]. The plate has been taken using the Crimean Astrophysical Observatory Schmidt Telescope and Agfa-Astro plates (40 minutes exposure time). This relationship can be expressed by the formula

$$d = 50.8 \, m^{-1.37}. \tag{13-2}$$

For the magnitude $m = 11.^m0$ (corresponding to the middle of the rectilinear portion of the characteristic curve of photograph Sh 4162, see Figure 13-5) and when the root mean square error in the aperture diameter

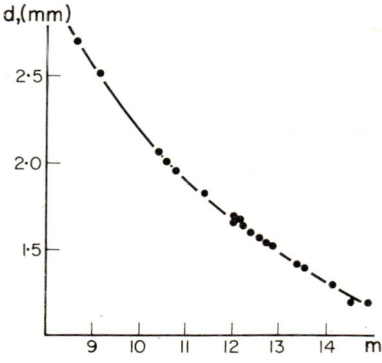

Figure 13-5 Diameter d of stellar images (measured with the iris microphotometer) versus stellar B magnitudes (UBV system).

is 0.01 mm (see Table 13-1 and Figure 13-4) the error in determination of the stellar magnitudes is equal to $0.^m04$. For brighter stars it is a little less, and for faint stars a little more.

Table 13-1 Calibration of the apertures on the disk D

Number	Diameter (mm)	Number	Diameter (mm)
1	0.438	16	1,787
2	0.523	17	1.857
3	0.625	18	1.965
4	0.726	19	2.067
5	0.802	20	2.153
6	0.908	21	2.236
7	0.983	22	2.329
8	1.088	23	2.400
9	1.163	24	2.490
10	1.244	25	2.582
11	1.359	26	2.668
12	1.424	27	2.767
13	1.511	28	2.849
14	1.594	29	2.950
15	1.718	30	3.024

Apertures to an accuracy of 2–3 μ will make it possible to reduce the error in determining stellar magnitudes to $0.^m01$.

The simple conversion of the MF-2 into an iris microphotometer will be of great value to photometric research.

REFERENCES

1. H. Haffner, *Nachr. Akad. Wiss., Göttingen, Math-Phys. Kl.*, **11a**, No. 4 (1953).
2. D. Ya. Martynov, *Course in Practical Astrophysics* (Kurs prakticheskoi astrofiziki), Fizmatgiz, Moscow (1964).
3. L. A. Urasin and A. I. Urnatsky, "Iris microphotometer", *Byull. AOE*, No. 38, 101 (1965).
4. L. A. Urasin, "The stellar microphotometer of the Engelhardt Observatory", *Astron. Tsirk.*, No. 232, 16 (1962).
5. A. A. Hoag, H. L. Johnson, *et al.* "Photometry of stars in galactic cluster fields", *Publ. U. S. Naval Obs., Second Ser.*, **17**, Part 7 (1961).

CHAPTER II-14

Time marking in high-speed solar cinematography

U. I. ILYASOV

Main Astronomical Observatory
Pulkovo

1 INTRODUCTION

High-speed cinecameras can be effectively used to record solar features. A long sequence of very sharp photographs can be obtained at times when the atmospheric turbulence is minimum and thus has a least effect on the image quality, i.e. in those short intervals of time when the quality of the photograph is determined only by the resolution of the astronomical optics.

Although the cinecamera was invented some seventy years ago, it was only in 1937 that it was first used in solar research by McMath.

At present cinematography is used in solar research at the following observatories: McMath–Hulbert (Michigan, USA) [1, 2], Mount Wilson (California, USA), Meudon (France), High Altitude, (Colorado, USA), the Crimean Astrophysical Observatory, (Crimea, USSR), Sacramento Peak (New Mexico, USA) and the mountain station of the Main Astronomical Observatory (near Kislovodsk, USSR).

Since 1958 this method has been used at the physics Department of the Australian Solar Observatory, where Loughhead, Bray and others [3, 4] have used 35-mm cinecameras with the new photoheliograph to record fine details of the sun.

At the Solar Physics Department of the Main Astronomical Observatory at Pulkovo, cinematography for detailed studies of solar atmospheric disturbances were first used in 1957 under Professor Krat [5].

The observatories mentioned above mainly use either time-lapse photography (i.e. slow motion) or single frames at a rate of 24 per second on wide (35 mm) film. Cinematography at rates between 32 and 250 frames sec^{-1} is said to be "fast" and we use the term "high-speed cinematography" for frames at rates of more than 250 frames sec^{-1}. Ultra-high-speed photography in this case refers to ordinary photography with exposure times shorter than 0.001 sec and cinematograpy at rates higher than 250 frames sec^{-1} [6] is said to be "ultrahigh-speed". This terminology may be supplemented by the concept of astronomical cinematography or "astrocinematography". This is akin to the concept of microcinematography, in which the individual frames are taken through a microscope. In astrocinematography the camera objective is replaced by the optical system of the telescope. "Astrofilm" is the film obtained in astrocinematography.

In high-speed cinecameras highly sensitive 16-mm or 8-mm film is mostly used. The SKS-1 and SKS-1M are the most common in the USSR. As described in [7], the modified SKS-1M differs to some extent from the original SKS-1, having a guide roller and a filmguard for the takeup spool, as well as a redesigned time marker, viewfinder and focusing system. In addition, the design of the sprocket has been improved.

The SKS-1M is intended for the photographic study of fast moving objects. It has continuous film transport and does not have a mechanical shutter. The role of a shutter is performed (in the 16-mm camera) by a four-sided rotary prism.

The transport time of 30 m of film in the SKS-1M cinecamera versus motor voltage is shown in Figure 14-1. The frame rate for the same camera versus voltage is given in Figure 14-2. For the 30–127 V range (Figure 14-1 and 14-2), the filming by the SKS-1M is not done at a constant rate, since a considerable time is taken by the mechanism to reach constant speed.

The exposure time $t_{\exp}$ is inversely proportional to the frame rate n given in frames sec^{-1}, that is

$$t_{\exp} = k \frac{1}{n}, \qquad (14\text{-}1)$$

where k is a shutter constant which depends on the design of the camera. For the SKS-1M, $k = 0.2$, thus from Eq. (14-1) and for $n = 200$ frame sec^{-1} we have

$$t_{\exp} = 0.001 \text{ sec.}$$

For the time measurements on the film, the SKS-1M has a time marker in which a neon bulb MN-7 is used. This produces 100 flashes per second using a 50-Hz supply. The light of the neon bulb illuminates the photographic film along the edge, outside the frames. When the film is developed, the time marks can be seen in the form of a dashed line.

However, this time marker is suitable only for highly sensitive negative film, whereas in high-speed and ultrahigh-speed photography to study fine solar details we use the positive fine-grain film M 3 with low light sensitivity ($s = 0.7$ GOST units; $\gamma = 2.5$; $R = 95$ mm^{-1}) to obtain high-contrast photographs.

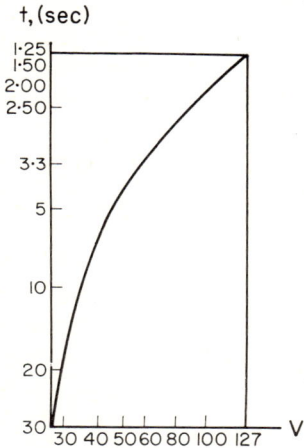

Figure 14-1 Transport time of 30 m of film in the 16-mm SKS-1M cinecamera versus ac motor voltage.

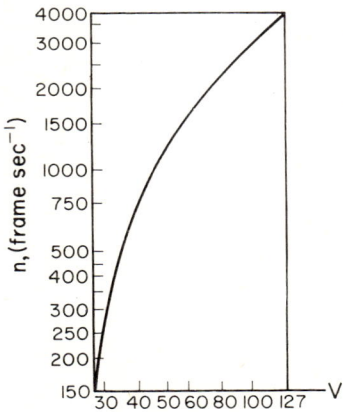

Figure 14-2 SKS-1M 16-mm cinecamera frame rate versus ac motor voltage.

The positive film M 3 is sensitive to the violet region of the spectrum, but not to the red. Since the neon bulb MN-7 gives a red glow, the factory-made time marker is unsuitable in this case.

We carried out experiments on time marking at the Solar Physics Department of the Main Astronomical Observatory in 1963. We use the SKS-1M camera with optical systems as described below, and with positive film.

The original design used an incandescent lamp and condensers.

2 TIME MARKER USING AN INCANDESCENT LAMP AND CONDENSERS

Let us first consider the optical layout for a time marker using an incandescent lamp and a single condensing lens (Figure 14-3). In this layout, the artificial light source is a 250-W incandescent reflecting-type lamp 7 which is on front of the lens 8 and is fed through a laboratory autotransformer (LATR). The light of the lamp 7 is modulated by the chopper 2 at a fixed frequency which depends on the number of slits in the chopper, and may vary between 50 Hz and a few thousand hertz. The mechanical chopper is made of Duraluminum mounted on the hysteresis motor (G-31) shaft 3. The G-31 drives the mechanisms at synchronous speed (3,000 rev min^{-1}) which is reached 10 sec after it is connected. The power at the shaft is 4 W.

The G-31 is connected to the ac line through a voltage stabilizer SN-300. The light beam from 7 falls on the single-lens condenser 8 with focal length 157 mm. After this, part of the beam passes through the 7-mm aperture 1 in the camera SKS-1M and is bent 90° by means of the rotating right angle prism 4 which has one reflecting surface. After passing through the aperture 5 in the cap of the prism case, it reaches the perforated area of the film 6.

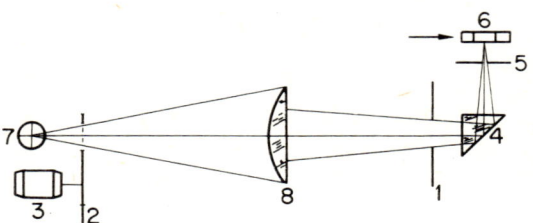

Figure 14-3 Schematic of a time marker system with incandescent lamp and single-lens condenser: 1, aperture in the SKS-1M camera (7-mm diameter); 2, disk shutter; 3, synchronous motor; 4, rotary prims; 5, aperture in the cap of the prism case; 6, perforated field of the film; 7, 250-W incandescent reflecting-type lamp; 8, single-lens condenser.

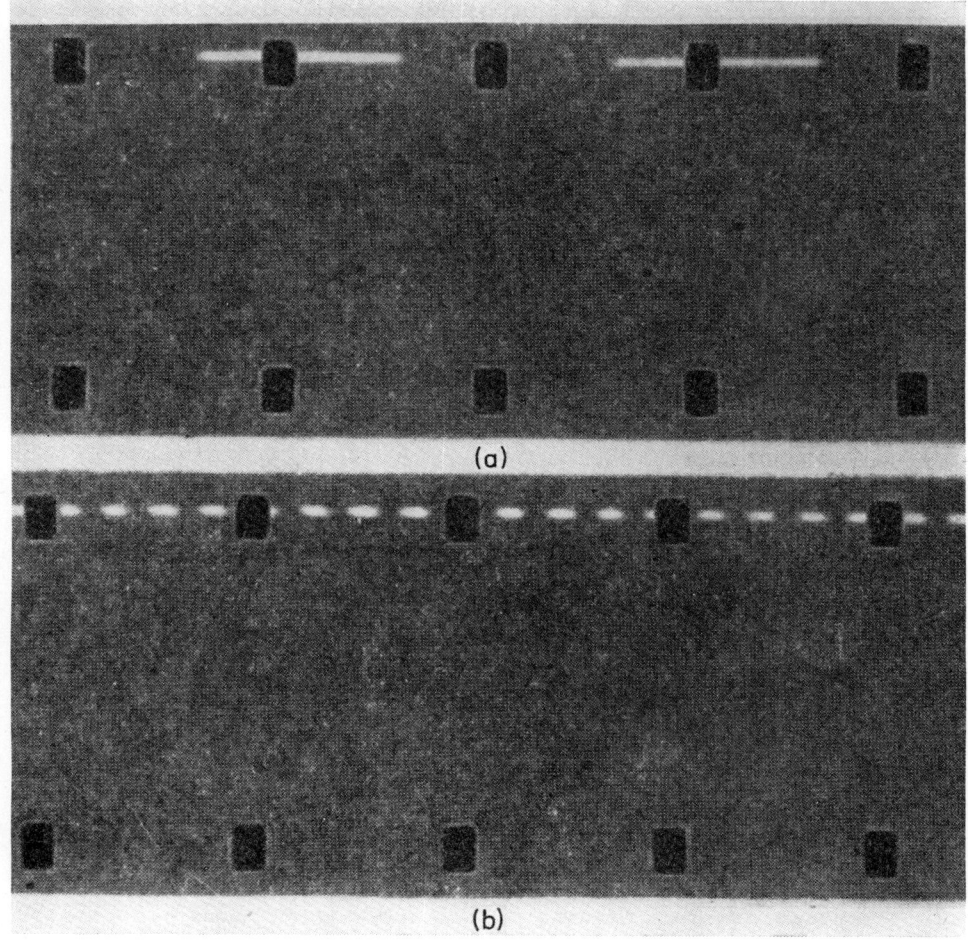

Figure 14-4 Time marks on the film edge along the perforation: (a) time marks 150 Hz and for 250 frames sec^{-1} (at the start of photographing); (b) time marks for the same frequency (in the middle of photographing).

The aperture 5 is circular in shape and it is 0.75 mm in diameter; the arrow in Figure 14-3 indicates the direction of the film motion.

Components 2, 3, 7 and 8 are mounted on the ways of the optical bench using special supports. Components 4 and 5 are inside the camera. After the film has been developed, the time marks are on the edge of the perforated side of the film having rectangular shape images of an average width of 0.5 mm. The length of the images varies from a few millimeters (at the start) up to 10 mm and more, increasing gradually from beginning to end.

Preliminary time marks were recorded on the 30-m of film (solar image was not recorded). Samples of time marks on two 16-mm films are shown in Figure 14-4.

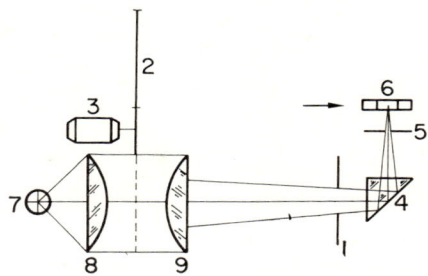

Figure 14-5 Schematic of the time marker system with incandescent lamp and two-lens condenser: 1, aperture in the SKS-1M camera, (7-mm diameter); 2, disk shutter; 3, synchronous motor; 4, rotary prism; 5, aperture in cap of prism case; 6, perforated film edge; 7, 250-W incandescent reflecting-type lamp; 8 and 9, two-lens condenser.

The single-lens condenser was next replaced by a two-lens condenser (Figure 14-5). In this system, the distance between the incandescent lamp and the plane of the film can be reduced, with the consequent increase in the aperture angle. The plano-convex lens *8* is chosen so that its diameter is approximately the same as that of the lens *9*, but with as small a focal length as possible. The focal length of lens *9* is chosen to be not less than the distance between *9* and the plane of the film *6*, and in our case is of the order of 27 cm. In addition, in a given case we may place the chopper between the condenser lenses to obtain much sharper time mark images. At high marking frequencies, the chopper may be placed closer to the camera between the condenser and the aperture *1*.

However, the best system to obtain the time marks on the film proved to be the one that uses sunlight.

3 TIME MARKER USING MIRRORS, AND SUNLIGHT

We first used a single-mirror system (Figure 14-6). The advantage of this system over the previous one is that it uses sunlight instead of an artificial light source. This system is used in cinematography of the solar disk edge.

But due to the fact that the distance between the mirror 8 and the tube (which replaces the objective of the SKS-1M) is comparable with the diameter of the solar image in the focal plane of the solar telescope, the scheme is not suitable for observations near the solar disk center.

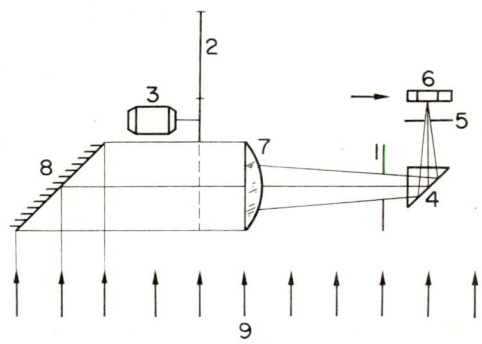

Figure 14-6 Schematic of the time marker which uses one mirror and sunlight: 1, aperture in the SKS-1M camera (7-mm diameter); 2, disk shutter; 3, synchronous motor; 4, rotary prisms; 5, aperture in cap of prism case; 6, perforated field of the film; 7, lens; 8, flat mirror; 9, solar rays.

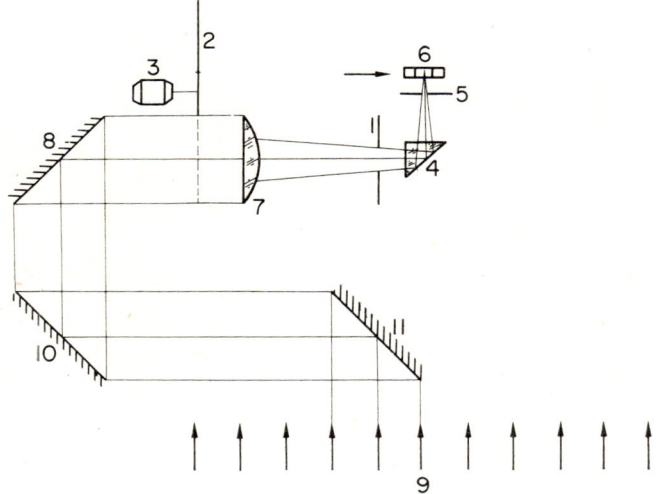

Figure 14-7 Schematic of the time marker using three mirrors, and sunlight: 1, aperture in the SKS-1M camera, (7-mm diameter); 2, disk shutter; 3, synchronous motor; 4, rotary prism; 5, aperture in cap of prism case; 6, perforated film edge; 7, lens; 8, flat mirror; 9, solar rays; 10 and 11, flat mirrors.

When pictures are to be taken any where on the solar disk, the best time marker uses three mirrors (Figure 14-7). Mirrors *8, 10* and *11* are aluminized and are in cells which can be rotated around two mutually perpendicular axes. The mirror cells are mounted on the optical bench ways. The adjustments in the horizontal and vertical directions makes it possible to use sunlight regardless of the distance between the first mirror *8* and the aperture *1* of the camera. Thus, this variant of the time marker does not have the drawbacks of the previous one.

To conclude, let us describe a time marker which uses a photomultiplier tube FEU, the viewfinder of the SKS-1M and the oscillograph model MPO 2.

In this case, the FEU is placed behind the eyepiece of the viewfinder and connected to the oscillograph. The number of frames is recorded by the oscillograph so that the time marking by the oscillograph is also the time marking for the SKS-1M camera. However, since this method is too awkward, it should mainly be used for testing the cinecamera to determine the transport time of the 30 m long film at different frame rates (function of the supply voltage).

The experiments performed at the Main Astronomical Observatory have confirmed that the SKS-1M camera can easily be adapted for high-speed cinematography to record the solar features. This is done as follows. The objective of the SKS-1M is removed and replaced by a Duraluminum light baffle (Figure 14-8). The tube has a slit to accomodate optical filters or a step attenuator. The configurations of the time marker could be arranged in accordance with the layouts indicated in Figures 14-3, 14-5, 14-6 and 14-7. Figure 14-8 shows the time marker used to generate the time marks shown in Figure 14-4. In this case, the distance between the aperture in the cap and the plane of the film is of the order of 1 mm.

The SKS-1M was mounted at the secondary focus of the horizontal solar telescope of the Main Astronomical Observatory. The average diameter of the solar image at this focus is 600 mm. In this case the plate scale is $3''.2 \text{ mm}^{-1}$. Using positive film with a sensitivity of the order of 0.7 GOST units high-speed solar cinematography (2,000 frames sec^{-1}), in integrated light, could be obtained. This observational technique has not been experimentally tested yet in monochromatic light, but it will be attempted in the near future. The time marks on the frames permit accurate timing of all short-term solar variations recorded on the film.

A fixed time scale and a good time marker is of great importance in high-speed solar cinematography.

Figure 14-8 SKS-1M camera with cover removed, and with new time marking system: 1, new time marking; 2, film; 3, tube; 4, slit for light filters; 5, screw for attaching time marking.

REFERENCES

1. V. A. Krat, *Zhurn. nauchn. i prikl. fotografii i kinematografii*, **1**, No. 4, 302 (1956).
2. G. P. Kuiper (Ed.), *The Sun*, University of Chicago Press (1953).
3. R. E. Loughhead and V. R. Burgess, "High resolution cinematography of the solar photosphere", *Austr. J. Phys.*, **11**, No. 1, 35 (1958).
4. J. R. Bray, R. E. Loughhead and D. G. Norton, "A 'seeing monitor' to aid solar observation", *The Observatory*, **79**, No. 908, 63, Corresp. Sect. (1959).

5. V. A. Krat, *Proceedings of the Conference on Stellar Scintillation* (Trudy Soveshcheniya po issledovaniya mertsaniya zvesd), 216 (1959).
6. *High-Speed Cinematography in Science and Technology* (Vysokoskorostnaya kinofotosemka v nauke i teknike), IL, Moscow (1955).
7. V. I. Lavrentev and V. G. Pell, *High-Speed Cinematography with the SKS-1 Camera* (Skorostnaya kinosemka kameroi SKS-1) "Iskusstvo", Moscow (1963).

CHAPTER **II-15**

The use of a horizontal long focal length telescope with coelostat for positional observations of the Moon using photography

N. G. RIZVANOV

Engelhardt Astronomical Observatory
Kazan

The horizontal telescope of the Engelhardt Astronomical Observatory was constructed in 1949 [1] to study the motions of the moon. The telescope has a fixed objective ($D = 200$ mm, $F = 8,000$ mm), a coelostat [2] with a gear drive mechanism and an auxiliary mirror. These are placed in a southern building, while the plateholder section of the telescope is in a nothern building; the light path area between the buildings is enclosed by a plywood corridor with corrugated black material on the inside wall.

At that time, the research program only required that the moon be photographed without the star background. The photographic method used for the study of the rotation of the moon and its surface is described in detail in the monograph by Khabibullin [1]; observations of this type can be performed fairly simply.

The problem becomes much more difficult when the telescope is used to obtain photographs of the moon and the stellar background simultaneously; this is required for the solution of a number of problems in astronomy and geodesy, such as the determination of the exact coordinates of the lunar features.

Photographs of the moon obtained at the Engelhardt Observatory using the Zeiss refractor ($D = 160$ mm, $F = 2,600$ mm) in conjunction with the Markowitz camera are used primarily for determining ephemeris time from measurements of points at the edge of the lunar disk; these photographs are not suitable for lunar cartography, due to the small image size in the photographs. When positional observations are made on a horizontal telescope, it is difficult to obtain sharp images of 9th and 10th magnitude background stars. With an $f/40$ objective, diffraction and atmospheric seeing cause the photographic image of the stars to reach from 0.5 to 1.0 mm in diameter. There are additional light losses in the two mirrors. Thus, in order to obtain high-contrast stellar images suitable for measurement, the exposure time to photograph the stellar field must be increased from three to ten minutes (when a refractor is used, the corresponding exposure time is of the order of 20 to 30 sec). At the same time, light scattering on the two mirrors increases the background on the photographic plate.

Under these conditions, the drive mechanism must operate faultlessly. Since guiding is not used, the rate of the drive mechanism is set for a star which is sufficiently close to the moon; remote control is used to adjust the drive mechanism governor from the northern building. A shutter is located on the front of the plateholder so that the moon may be photographed with short exposure times (from 0.1 to 0.3 sec). The shutter consists of two blades mounted on a small diameter metal rod. When the exposure begins, the rod under the action of a spring, is displaced perpendicularly to the optical axis of the telescope. Before the shutter is actuated one blade covers the film; after the shutter is triggered and the exposure completed the second blade takes the place of the first one.

This simple device enables us to obtain identical exposure time across the lunar image. The length of the exposure is determined by the distance between the blades. The time of the photograph and the length of the exposure are recorded by means of stop switches mounted at the ends of the rod and connected to a printing chronograph.

Since atmospheric turbulence and periodic errors in the drive mechanism may cause large random errors in the results obtained by this method, a Markowitz camera will be used in the future. In addition, it is proposed to replace the gear drive mechanism by an electric drive fed by a crystal oscillator. Measurement and processing of several photographs of the moon and stars obtained by the horizontal telescope in 1963 have shown that the accuracy of the plate reductions with good stellar images is perfectly satisfactory. In order to check the accuracy to which the coordinates of the lunar craters were determined from these photographs, we compared the positions of the moon found from edge point measurements by N. F. Bystrov using his photoelectric instrument [3]. The difference, which did not exceed 0."6 may to some extent be explained by the measurement and processing methods.

All the work in setting up the telescope, as well as the observations, were carried out in conjunction with K. S. Shakirov, the senior scientist at the Engelhardt Astronomical Observatory. I would like to take the opportunity to express my gratitude also to the engineer and mechanic A. I. Urmatsky at the Engelhardt Astronomical Observatory, who took an active part in the redesign of the telescope.

REFERENCES

1. Sh. T. Khabibullin, *Izv. AOE AN SSSR*, No. 31 (1958).
2. V. V. Vyazanitsyn and O. A. Melnikov, *Usp. Astr. Nauk*, **3** (1947).
3. N. F. Bystrov, "Automatic performance of astrometric measurements of lunar plates with a photoelectric device", *Astron. Zh.*, **34**, 146 (1962).

CHAPTER **II-16**

The use of short exposure photography for meteors

YE. N. KRAMER

Odessa Astronomical Observatory
Odessa

Photography for the study of meteors is one of the most accurate methods available and possibly the one to provide the largest amount of information, allowing the measurement of the height, velocity, brightness and deceleration of meteors. Using these data we can calculate the atmospheric parameters (density, temperature and pressure). Photographs of meteor can be used to obtain information about its mass, density and structure. However, visual, photographic and radar observations have shown that the phenomena associated with the meteor consists of complex processes occurring on the boundary between the meteor body and the surrounding medium. For example, the opinion is held by some that at least certain meteors, which are loose in structure, split up during flight. Furthermore, fast meteors are accompanied by afterglow.

Meteor photography, even with the use of modern methods [1, 2], has as a main drawback the fact that the meteor trail (or "photometeor", as Astapovich [3] has called it) is photographed, instead of the meteor itself. In this case, the image of the meteor is completely blurred. To obtain these photographs a shutter is placed in front of the objective (or elsewhere in the light path) which interrupts the image of the meteor trail on the film into separate portions or lines. If the angular velocity of the shutter is known, the velocity and deceleration of the meteor can be measured. Often, blurring may be detected between the broken lines of the photographed meteor trail; this is an evidence of a large size meteor.

In the meteor photograph given in Figure 16-1, the trail can be seen. It is easy to estimate the length of this trail, since the velocity of the meteor

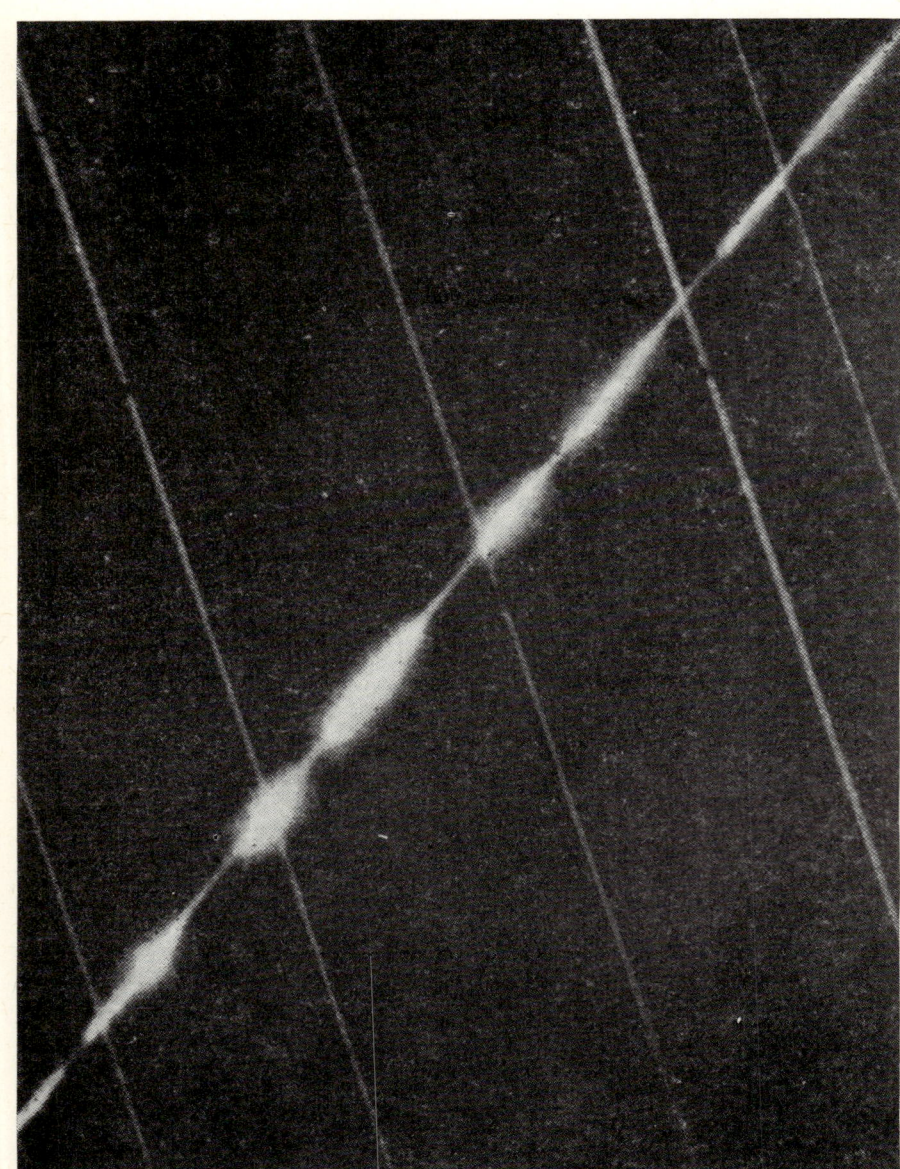

Figure 16-1 Short exposure time photography (rotating shutter) of a meteor.

is known (60 km sec^{-1} on the average). One trail image element corresponds to a complete revolution of the shutter (about 0.04 sec), and since the shutter duty cycle is 0.4, the length of the actual trail is about 1 km.

Note that the transversal dimensions of the meteor are usually not large, of the order of two to three meters. Indirect evidence indicates the existence of a meteor trail even when it cannot be seen on the photograph and the gaps between the image elements are quite sharp.

In Figure 16-2 we show two light curves of a single meteor [3]. The upper curve was obtained without the use of a shutter, and the lower with a rotating shutter in front of the objective. The first curve, is based on a photograph of the meteor and the trail thus, making it appear brighter. The second curve is based on a photograph of the meteor without the trail, thus, the meteor appears fainter. To obtain this photograph after the appearance of the meteor, the camera shutter cuts off the trail image so that it is not recorded photographically. Similar curves have been obtained at Kazan using photoelectric brightness measurements.

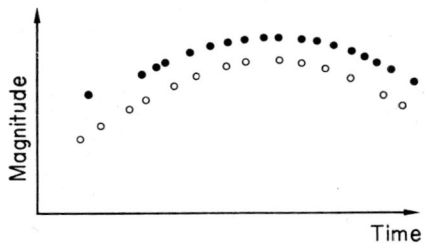

Figure 16-2 Light curves obtained for a single meteor. The black dots are the data from a photograph obtained with a camera without a shutter and the circles from a photograph with a camera having a shutter.

These facts indicate that the light of the meteor is not concentrated in a moving point, and that, while the cross section is small, the meteor trail measures hundreds of meters.

The optics that have been used in the photography of meteors enables to resolve objects at a distance of 100–150 km whose linear dimensions exceed 20–50 m. For an exposure of 1 msec the displacement of the meteor even at a speed of 60–70 km sec^{-1} will be an order less than its linear dimensions. It will lie within the resolution of the optics. Hence, short exposures of the order of 10^{-3} to 10^{-4} seconds will give a sharp meteor image. In the photographs obtained by this method the meteor looks brighter than the trail.

It is interesting to note that the division of a meteor into comparatively large fragments is almost impossible to record photographically by simply increasing the focal length of the camera, since an increase in focal length will reduce the field of view and therefore the chances of photographing the meteor. Moreover, even for focal lengths of one, or even two meters, the transversal distances between fragments will be less than the resolution of the camera. While the longitudinal distances between fragments will increase with time, due to the differences in air drag and will become larger than the resolution, the image of the separate fragments are superimposed on one another and cannot be distinguished by ordinary photographic methods. The use of short exposure makes it possible to separate the images of the individual fragments and thus to study the motion of each separately and measure its mass.

The use of short exposure meteor photography has not encountered any basic difficulties, as the method is widely used in other areas of research. However, at least two factors must be taken into account in this case.

Firstly, it is never known beforehand when a meteor will appear and so the system must be designed for long periods of operation (meteor patrols usually operate on every clear night). Therefore the mechanical design of the shutter must be highly reliable.

Secondly, the instantaneous image of a meteor lies along the path of its flight. Hence, the individual short exposure photographs must be separated on the film so that they do not overlap.

Several methods which consider these requirements may be proposed. They include the following.

(1) A sufficiently large rotating shutter in front of the objectives of the camera. The shutter consists of a disk from which a sector has been cut out; the sector has the same size as the cross-section of the cone of light gathered by the objective (in practice, no tmuch larger than the diameter of the objective). If R is the distance from the center of the objective to the center of rotation of the shutter and r the radius of the objective, then the effective exposure time is equal to $\eta = r/\pi R\mu$, where μ is the number of revolutions per second. If the velocity of the meteor is 50 km sec^{-1} and the linear dimensions of the trail is 2 km we must have $1/\mu \geqq 0.04$ sec in order that separate images are not superimposed. When $2r = 50$ mm and $\eta = 10^{-3}$ sec, then $R = 320$ mm. Since the radius of the shutter increases proportionally

to the aperture of the camera the assembly could become awkward, but this photographic method has the advantage of being simple; moreover, it uses the full clear aperture of the objective.

(2) Two movable gratings are placed in front of the camera objective. In each grating clear strips alternate with opaque strips (1 mm wide clear strip alternating with a 1 mm opaque strip). The gratings can be driven to obtain exposures of less than a millisecond, so that the time intervals between the images are of the order of 0.05 to 0.02 sec. A shutter such as this will give short exposures simultaneously over the whole photographic film, an aspect which is advantageous in meteor spectrography. The disadvantage are obvious: only 50% of the clear aperture of the objective is used, and the large accelerations which occur in the operation of the shutter components reduce its life and make it inadvisable to use it over long periods of time.

(3) A special focal plane shutter has been developed at the Odessa Astronomical Observatory and is now in operation. A shutter with one or two narrow clear sectors 5° to 10° in aperture rotates close to the focal plane of the objective, the focal length of which is 250 mm for some cameras and 750 mm for others. The shutter in one type of camera is driven by an asynchronous motor through a gear train unit; this is mounted outside the camera frame. In the other type of camera the shutter is driven directly by a synchronous motor. A hole is bored in the mounting flange of the aerial camera and through it the sleeve of the shutter is inserted. An opening of 20 mm diameter is cut out from the center of the 30 cm × 30 cm frame. The shutter rotates at 1500 rev min^{-1}. The systems are designed for exposures of 1/900 and 1/1,800 sec.

Under these observing conditions, star and sky background are very low, which allows a single frame to be exposed for about 10 to 15 hours over the course of several nights.

A similar system operates at the Tajik Institute of Astrophysics. A system using image converters EOP is under development at the Odessa Astronomical Observatory. On the one hand, this will allow fainter meteors to be photographed; and, on the other, short exposure photography can be somewhat simplified, since the signals may be modulated by electric or magnetic fields. In addition, the use of EOPs will make it possible to study meteors in the nonvisible as well as in the visible region of the spectrum.

REFERENCES

1. L. A. Katasev, *Photographic Methods in Meteor Astronomy* (Fotograficheskiye metody meteornoi astronomii), Gostekhizdat, Moscow (1957).
2. F. H. Whipple, *Harvard Meteor Program*, Harvard Reprint Series, Vol. II, No. 19 (1947).
3. I. S. Astapovich, *Meteor Phenomena in the Atmosphere of the Earth* (Meteornye yavleniya v atmosfere Zemli), Fizmatgiz, Moscow (1958).
4. N. I. Izrayetskaya, "An Effect of the Photometry of Meteors", *Byull. Kom. po kometam i meteoram* (to be published).

CHAPTER **II-17**

An astronomical dome made of plastic

T. E. KYUBAR

Tartu Observatory
Tartu

The basic design requirements for astronomical domes are: (1) reliable protection for the instrument; (2) suitability to the nature of the observations, and the absence of factors which might adversely affect the results; (3) low cost and technically simple manufacture; (4) durability and minimal maintenance costs.

The majority of existing dome designs fail to meet these requirements due to their massiveness and high construction costs. Probably the most difficult problems are the choice of a light, stable, easily assembled material for the dome skin and the design of the rotating support structure.

The present level of production of large components made of plastics (for boats, automobile bodies, etc.) and the weather resistance of these materials have suggested the possibility of using plastics in the construction of astronomical domes. The consequent reduction in the total weight of the dome will simplify the design of the drive mechanisms.

It was suggested by Yu. S. Streletskii and A. I. Kopylov (of the Main Astronomical Observatory) that an experimental dome made of plastic should be constructed at the Tartu Observatory (Estonia). Such a dome was built (see Figure 17-1) and completion took place during 1962. The general shape of the dome is spherical, with 5.25 m diameter and 1 m slit width. It is made of "glass-plastic" (resin PN-1 and fiberglass). The sections of the dome, shaped from plywood templates, were cemented using a resin; the thickness of the dome skin is 3 mm. The experimental dome has no steel frame. The sections could be made by any factory with a glass-plastic shop; the erection of the dome does not require any special skill. The dome rests on a ball bearing of 2 m diameter and is rotated by a low-power electric

Figure 17-1 Plastic astronomical dome at the Tartu Observatory.

drive; the shutter of the slit is manually operated. The total weight of the dome is about 400 kg.

On the basis of our experience accumulated during the construction of this experimental plastic dome, we would draw the following conclusions:

(1) The properties of the glass-plastic satisfies the main material requirements for the skin of the dome.

(2) A light weight dome on a bearing rotates smoothly and noiselessly.

(3) If the dome has a large diameter, it should have a steel or Duraluminum frame.

(4) At points of concentration of stresses, the sections should be bolted together, in addition to being cemented.

Subject Index

Achromate Class E. Fraunhofer 324
Air
  index of refraction 279
  turbulence 9, 208, 411
Algorithms 125, 126, 129,
Altazimuth telescope 73
  automatic azimuth guiding 73, 74
  automatic azimuth setting 73
  blind zone 74, 103
  coordinates converter 76, 77, 80, 81, 123, 124
Altazimuth telescope, design principles of digital control for 119
  digital automatic regulation system (SAR) 134, 135, 138
  guiding accuracy 138
  guiding-by-position 136, 137, 139, 140
  guiding-by-velocity 136, 137, 138, 139, 140
  tracking system (TsSS) and information theory 124, 128, 130, 137, 141, 142
Altazimuth telescope, combined control system for 107, 108, 109
  computer for dA/dt 110, 111, 114
  computer for dp/dt 112, 115
  computer for dz/dt 110, 111, 115
  positional drive 109, 110
Amplidyne 17, 21, 38, 62, 79, 82, 96, 97, 100, 385
Annealing 24
Apochromate 324, 328
  Zeiss A 324
Astrograph 185, 188, 190, 208
  double 406-mm 100
  Hamburg AG 340
  Pulkovo 230-mm 192
Astrocinematography 412

Astrofilm 412
Astrolabe 302
  Danjon 197, 207, 302
  prism 197, 207
  reflecting 207, 208, 210, 211
  absolute $Z_\odot$ measurements 208, 209
Astrometric
  determination 217
  instruments 199
  standards 337
Astrometry
  fundamental 207, 209, 211
  photographic 185, 331
Atmospheric
  dispersion 340
  extinction 170
  extinction service 172, 176
  mean refraction 170, 171
  mean refraction component correction 55
  mean refraction computer 20, 21, 172
  mean refraction constant 33, 143, 171
  mean refraction errors 172
  seeing 255
  transmittance 245, 246
Aurorae 312
Automatic observations 169

Binary
  code 36, 37, 55, 57, 58, 59, 60, 113
  numbers 48
Bits 34
Bolometer 303
Brightness equation 335, 336, 340

Camera for observation of artificial satellites (AES) 185
  description 186, 187

Camera (cont.)
  limiting magnitude 193
  optical compensator 185, 188, 189, 190, 192, 193
Camera, night aerial
  NAFA type 193
Cassegrain system 11, 12, 320, 321
Catadioptric telescope 317, 319, 328, 329
  astigmatism 327
  chromatic aberration 322, 323, 324
  coma 327
  confusion disk 320, 325, 326
  curvature of field 327
  description 319
  maximum useful magnification 325
  spherical aberration 325, 326
Catalog
  epochs of the 334
  observations 197
Chronograph 208, 218
  printing 189, 302
Cinecameras, solar high-speed 411, 412
  SKS-1 412
  SKS-1M 412, 413, 414, 416, 417
Cinematography, solar 416
  high-speed 411, 412
  ultrahigh-speed 412
Cluster
  M 13 183
  NGC 1664 408
  Pleiades 183, 184
Coelostat 280, 399, 400
Color equation 335, 336, 340, 342
Common method, testing mirrors by the 280, 281
Computer
  BESM 125
  Strela 125
Concentric optical system 311, 312, 313, 314, 315
  a two mirror with aplanatic lens 315
  double pass through the meniscus 311
  solid 312
  with a meniscus 313, 314
  with negative lens 314
Conoid 21, 22

Coordinates converter 85, 86, 87, 88, 89
  PK-I 91
  PK-II 91
  PK-III 91
  PK-IV 91
  PK-V 91, 92, 93, 94, 97, 98, 100, 103, 105, 108, 116
  PK-VI 98, 99, 100
  PK-VII-A 101, 102, 104, 105
  PK-VII-B 101, 102, 103, 104, 105
Coordinates, relationship between equatorial and horizontal 86, 87, 89, 90, 108
Coudé system 11, 12, 321
Counter, Afanaseyev-Platonov 302

Danjon astrolabe (see Astrolabe, Danjon)
Detectivity (D^*) 244
Dewar 244
Diffraction gratings 399, 400, 401, 402
Digitial Differential Analyzers 125, 134, 300
Digital integrators 125, 126, 127, 128, 133
Digital voltmeter 389
  Solartron LM-902 390
  ETsPV-1 390
  V2-8 390
Dihydrophosphate of ammonia 288
Diode, Zener 236
Dome 7, 9, 87, 88, 90
Dome, automatic synchronization of 85, 97, 300
  with asymmetrically mounted telescope 87, 88, 89
  with symmetrically mounted telescope 87, 88, 90, 91, 94, 99
Dome, phantom 23
Dome, plastic 431
  description 431
Doolittle method 331
Doppler shift 289

Earth
  atmosphere 9
  gas corona of 256
Electro-optics 273

Subject Index 437

Electrometer
  preamplifier   226
  tubes   227, 236, 390
Electronic camera   394
Electron telescope   269, 273
Epoxy resin ED-6   360
Equinoxial corrections   207

Fabry lens   225, 248
Fabry–Pérot etalon   256
Fabry–Pérot etalon, low order   395
  adjacent passbands   396
  description   395
  resolution   395, 397
  separators   395
Feedback
  flexible negative   80, 82, 96
  rigid negative   80, 96
Ferrites   63, 134
Filters, interference   273
Filters, interference polarization   287
  H_α solar observations   287
  IT-53, description   287
  IT-58, description   288
  K-line solar observations   287, 397
Filter, optical
  IKS-3 type   248
Flint
  heavy (see Glass TF 3 type)
  short (see Glass OF 3 type)
  light (see Glass OF 1 type)
Fluorite   322, 323, 328
Foucault knife edge tester   281, 285

Geodetic instruments   199
Germanium detector   244, 249, 250
  spectral response   245, 246
Ghosts, Lyman   399
Ghosts, Rowland   399, 401
  intensity   401, 402
  compensation in double pass monochromators   402
Glass
  BK 2   322, 323
  K-7   322
  K-8   297
  LK-5 (see Pyrex type)

Glass (cont.)
  OF 1   323
  OF 3   322, 323
  Pyrex type   16, 24
    internal strain   16
  TF 3   322, 323, 325
  TF 5   322
  TK 4   323, 325
Glass-plastic   431, 433
Gray code   37, 56, 57, 58, 59, 60

Hale telescope (see Telescope, Hale)
Hartmann
  constant   350
  criterion   342
  test   30, 328, 348
  test description   348
  test screen   349

Igdantin   368
  preparation   369
  properties   370, 373
  molding   369
Image converter   269, 272, 273, 275, 315, 393
  applications to direct photography   272, 273
  applications to spectroscopy   273, 274, 275, 276
Image converter, photocontact   255, 269, 273
  gain   255
  magnetic field effects, shielding   270, 271
  resolution   255
Image converter, cascaded   393, 394
  background   393, 394
Image intensifier   255, 257, 258, 273
  applications to spectroscopy   255
Infrared astronomy   243
Interferometer   200
Interferometer, stellar   73, 74, 80, 81
  coarse guiding   74, 76
  coarse guiding, circuitry   80, 81
  fine guiding   10
  photoelectric automatic guidance   74, 82

Interferometer (*cont.*)
  setting system   81, 82
Interferometer, Uversky   203
Interstellar
  absorption   249
  dust   243
Iris monochromator (*see* Microphotometer, iris)
Iris photometer (*see* Photometer, iris)
Islandic spar   288
Isophotometer   261, 262, 267

Jupiter
  infrared spectra   244, 251
  atmosphere   252

Lead sulfide detector   244, 249, 250
  NEP   244, 245
  spectral response   245, 246
Lens, aplanatic   315
Level, liquid   199
Level tester, screw   204, 205
Level tester, standard wedge   199, 201, 205
  description   199
  performance   202
  procedures   200
Light source, β emission excited   226
Linnik stellar interferometer (*see* Interferometer, stellar)
Lithium fluoride   12
Luminescence, upper atmosphere   258
Lyman ghosts (*see* Ghosts, Lyman)
Lyot system   287

Machine, grinding-polishing   24, 25, 27, 30
Magnetic amplifiers   22, 300
Maksutov system   311, 318, 328, 331
Maksutov, compensation shadow method by   31
Markowitz camera   185
Measuring engine, coordinate   370
  KIM3-type   348
Meniscus system   12, 208, 296, 311
Meridian circle   143
  observations   207

Meteor   315, 425
  light curve   427
  photography   425, 427, 428
  trail   425, 427
  spectrography   429
Meteor camera   428, 429
  focal plane shutter   429
  gratings   429
  image converter   429
  rotating shutter   428
Microphotometer   304
Microphotometer, iris   405, 406, 407
Microphotometer MF-2   265, 304, 405, 406, 407
  description   262, 263
  performance   265, 266, 267
Microphotometer, recording intensity   259, 260, 261
  requirements   262
  description   262, 263
  performance   265, 266, 267
Microphotometer, Rosenberg   260
Microphotometer, Zeiss   260, 261
Microscope
  KIM-3-type   406
  MMI-type   370
  UMI-21-type   361
Milliameter
  H-370 type   247
Mirror
  chromium   296
  flat   367, 371, 373, 374
  flat-concave   367, 371, 372, 373, 374, 375
  flat-convex   367, 371, 372, 373, 374, 375
  hyperbolic   367
  lightweight   370
  metallic   296
  modeling, frozen-in method   360
  ribbed   359, 360
  testing precision (*see* Testing precision mirrors) with central hole   362
Monochromator   303, 395
  double pass   399, 400
  interference   257
  single pass   399

Nasmyth system 11, 12, 321, 238
Nebula 315
  Crab
    infrared radiation 244, 248, 249
    magnetic field 249
    relativistic electrons 249
  emission 256
  diffuse, infrared spectra 274
  NGC 6618 396, 397
  NGC 7000 (North America) 256, 396, 397
  observation 274
  Orion, spectrum 274, 276
Night sky
  emission from 395
  H_α emission 256
  infrared spectrum 244, 246, 250, 276
  photograph 6

Optical systems, fast 311
Optimeter 201, 204
Orthoscopic 327
Oscilloscope ENO-I type 406

Palomar Photographic Star Atlas (see Sky Survey, National Geographic Society-Palomar Observatory)
Parallax 207
Permalloy, magnetic shielding made of 274
Photocathode
  multi-alkalai 269, 274
  cesium-oxide 269, 274
Photocell STsV-3 247
Photographic emulsion
  A-650 272
  Agfa-Astro 193
  Agfa-Rapid 202
  Ilford HP3 193
  Kodak OaO 193
  M-3 413, 414
  NIFKI-1 194
  NIFKI-2 193
  RF-3 287
Photographic Vertical Circle (FVK) 207, 208, 210

Photographic Zenith Tube (FZT) 197, 207, 213, 217, 222
  automatic observation cycle 213, 214, 217, 218
  control unit, description 214, 215, 217, 218, 219, 220, 221, 222
  electromechanical counter 215, 216, 219
Photography
  high-speed 412
  ultrahigh-speed 412
Photoheliograph 411
Photometeor (see Meteor trail)
Photometer, automatic stellar photoelectric 223, 224, 237
  automatic cycle and program unit 227
  electrometer preamplifier and integrator amplifier 226
  operating procedure 228, 229
  optical system, photometer head 224, 225
Photometer, differential 248
Photometer, iris 181, 183, 184, 300
  description 182
  performance 181, 183
Photometer MF-6 type 184
Photometer, photoelectric with digital printout 389
  description 389, 390, 391
Photometer, stellar 1—2.5 μ 243, 244
  description, circuitry 246, 247
Photometric system, UBV 237, 295
Photometry 8
  multicolor photoelectric 223, 224, 227
  stellar 248, 300
Photomultiplier tubes 304
  head-on 225
  EMI 6094 B 389
  EMI 9502 B 389
  FEU-16 234
  FEU-17 82, 156
  FEU-19 226
  FEU-25 263
  FEU-29 226
Piazzi–Smyth correcting plate 297, 335
Plateholder 9, 103, 295
  Ritchey 16, 224

Pleiades, constellation 345
Polarimeter, automatic, photoelectric 231, 237, 304
  description, circuitry 234, 235
  optical train 233
  specifications 231, 233
Polarization, amount of 231, 232, 233, 236, 238, 239
Polarization, angle of 231, 232, 233, 236, 238, 239
Polarization plane 238
Polarization, analog computer to obtain directly the amount of 237, 238
Printer, digital 390, 391

Radial velocity, stellar 291
Radio Telescope 100, 107, 300
Rayleigh criteria 12
Recorder, EPP-90 type 164, 227, 228, 263, 264
Ritchey plateholder (see Plateholder, Ritchey type)
Rock salt 328
Ross corrector 319
Rowland ghosts (see Ghosts, Rowland)

Saturn, infrared spectra 244, 251, 252
Schmidt system 311, 318, 328
Scintillation 161, 162, 164, 165, 166
  amplitude 162
  brightness 161
  color 161
  frequency spectrum 162, 165
Seeing 162, 163, 164, 165, 166, 167, 279
  frequency spectrum 161
  local 9
  stellar images 161
  photoelectric method to measure 161, 163
Semiapochromate, Sonnefeld AS 324
Sensitogram 267
Sensitometer 267
Sensitometry 262
Shaft angle encoder 36, 37, 55, 56, 57, 67
Silica, clear fused 12, 328

Sky Survey, National Geographic Society-Palomar Observatory 6
Slevogt–Richter system 311, 315
Slyusarev system 319
Solar chromosphere, H_α-line profile in the spectrum of 265, 266
Solar L_β emission 256
Solar spectra, high resolution 253
Solar Spectrum, Utrecht Photometric Atlas of the 261
Solar tower 253
Spectrogram 291
Spectrograph 395
  nebular 312
  prism 273
Spectrograph, slitless 289
  description 289, 290, 291, 292
  guiding 293
  reference fringes 289
Spectrograph, vacuum 296
Spectrometer 395
Spectrometer, stellar 1—2.5 μ 243, 244, 250, 251
  characteristisc 250, 251
  description 250
Spectrophotometer, photoelectric solar 399, 400, 402
Spectroscopy, interferometric 253
Speed stabilizer, electric motor 239
  description, circuitry 239, 240
Stars
  absolute coordinates 333
  binary 207
  brightness measurements 405
  position of 331
  proper motions of 331, 332, 337
  reference 333
  spectral class of 303
  variable 227
  α Lyrae 166
  XY Lyrae 250
  NCyg. 1948 183
  α Perseus 337
Stereocomparator 291
Sun, infrared spectra 251, 252
Support system, mirror 377
  back 378, 379, 380, 381, 382, 383, 384

Subject Index 441

Support (*cont.*)
  Lasell type 377, 378, 379
  lateral 378, 379, 380, 381, 382, 384

Talbot's bands 289, 290, 292
Telescope, amateur 296
Telescope, 200-mm (APR-31) 224
Telescope, 250-mm (AP-250) 151
Telescope, 320-mm meniscus 100
Telescope, 480-mm (AZT-14) 179, 180, 389
Telescope, 50-cm Maksutov 183, 184
Telescope, 70-cm meniscus; astrometric study 183, 237, 331
  brightness and color equation 335
  distortion 340
  Hartmann constant 342
  plate constants 334, 336
  plate inclination 341
  position accuracy 332
  plate scale 342
Telescope, 700-mm (PM-700) 32, 297
  building 71
  cross section 356
  Foucault test 346, 347
  front wave aberrations 350, 351
  Hartmann test 348, 349, 350
  optical systems 345, 350
  telescope mount deformation 353
Telescope, 1-m 100
Telescope, 2-m Universal (Tautenberg, DDR) 301
Telescope, 2.6-m Shain (ZTsh) 4, 19, 297
  axis
    polar 12, 13
    declination 12, 13
  control system 16, 20
  cross section 13
  dome 11, 18, 23
  dome windscreen 17, 18, 20, 22, 23
  drive specifications 17
  focusing mechanism 17
  grinding tool main mirror 28
  guiding 16
  lunar-planetary drive 21
  main mirror image quality 31

Telescope (*cont.*)
  main mirror manufacturing 11, 24, 26, 29, 30
  main mirror support system 14
  optical systems 11, 12
  plateholder 16, 17
  temperature considerations 18, 22
  sidereal motion 21
  slew motion 22
  tube 13, 14, 18
    flexure compensation 158
  worm-gear 13
Telescope, 6-m (BTA) 359
Telescope, 13-inch 396
Telescope, 50-inch 89, 90, 97, 98, 250, 251, 253, 312
Telescope, 82-inch McDonald 253
Telescope, 120-inch Lick Observatory 14
Telescope, 200-inch Hale 9, 14, 16
Telescope, Electronic Digital Control System (ETsUM) of the PM-700 32, 33, 34, 35, 36, 38, 39, 41, 45, 48, 52, 54, 55, 57, 62, 63, 64
  adders (Σ) 38, 49, 50, 51, 52, 53, 56
  basic characteristics and block diagram 33, 35
  Coarse setting control Units (UUGN-t and UUGN-) 36, 37, 62
  Code converter (PK) 37, 38, 57, 58, 59
  Coding unit 37
  Control console (UP) 36, 37, 48, 49, 65, 70
  Cycle of operation 35
  delai circuit (SZ) 54
  fine correction control units (UUTK-t and UUTK-δ) 36, 38, 60, 63
  frequency dividers (DCh$_1$ and DCh$_2$) 34, 39, 40, 44, 50, 63
  mean refraction correction analog computer (SRM) 37, 69, 70
  signal delayed pulse generator (GOZI) 34, 42, 44, 45, 46, 48, 51, 55, 56
  substractor ($\varDelta_1$ and $\varDelta_2$) 51, 54, 55, 56
  switching pulse transmitter (DPI) 36, 39, 40, 49

Telescope (cont).
  synchronization pulse generator (GSI)   34, 41, 42, 44, 45, 49
  universal arithmetic unit (UAU)   37, 55
  velocity control mechanism (MUS)   34, 63, 66, 68
Telescope, catadioptric (see Catadioptric telescope)
Telescope, chromospheric   287
Telescope for limited observing program   169
  programmed control   169
Telescope mount APSh-6   385
  photoelectric guiding   387
Telescope, phantom   23
Telescope tube flexure   33, 152, 295
  automatic compensator   146, 147, 151, 153, 154, 155, 156, 157
  compensation   143, 144, 159
  correction   145, 148, 149, 150, 152
  differential   158
Telescope, vertical, solar   399, 400
Testing precision mirrors, vacuum chamber for   279
  description   281, 285
  limitation testing in air   279, 280

Textolite   360
Time
  ephemeris   191, 192
  marker   414, 416, 417
  marks on film   415
  universal   192
Time counter, Neuchâtel   219, 222
Time service, standard   192
Transit instrument   143, 195, 197, 207
  observational results   195, 196
  photoelectric   195, 207
Turner, linear equations   334
Twilight flash   257, 258

Uran, objective   202

Vacuum chamber   279, 282, 283, 285
Valtzmassa   360, 364, 365, 368
  composition   360
Verticle Circle   207

Wind screen   85, 90, 100

Yale catalog   333

Zodiacal light   312, 315

Name Index

Abele, M. K.   xi
Abrazhevsky, B. P.   224
Achermann, V.   56, 72
Afanasyeva, P. M.   xi, 197, 205, 302
Aleksandrov, V. V.   xi, 296
Andreyenko, N. R.   204, 205
Argunov, P. P.   317
Aronson, M.   142
Astapovich, I. S.   425, 430
Avedisova, V. S.   181, 299

Babcock, H. W.   268
Baker, J. G.   318
Barabashov, N. P.   241
Barney, I.   343
Batranina, G. A.   365
Belonovsky, A. S.   72
Belyaev, N. M.   32, 375
Belyaev, Yu. A.   63, 72
Bigay, M. J.   254
Billings, D. E.   268
Birthley, W. B.   261, 267
Bolshakova, G. I.   345, 353
Borodina, G. V.   305
Bray, J. R.   411, 419
Breido, I. I.   193, 194
Brown, W. N.   261, 267
Bruckner, G.   268
Bruns, A. V.   xi, 224
Bruk, G. L.   20, 298
Bryushkova, Ye. I.   202
Bugoslavskaya, Ye. Ya.   343
Burgess, V. R.   419
Bystrov, N. F.   422, 423

Cashman, R. J.   254
Cavedor, M.   197
Chasmar, R. P.   253

Chretien, M. H.   318
Chuvayev, K. K.   x
Connes, P.   254
Cooper, R. H.   268
Couder, A.   18

Deutsch, A. N.   343
Dieckvoss, W.   343
Dimov, N. A.   xi, 303, 304
Ditchburn, R. W.   289, 294
Dixon, M. N.   343
Dobrogursky, S. O.   83, 106, 117
Dobroravin, P. P.   296, 299, 300, 301
Dobychin, P. V.   x, 296, 297
Dolgov, P. N.   197
Dolidze, M. V.   268
Dravskikh, A. V.   63
Drichko, N. M.   287

Elsässer, H.   167
Evans, J. W.   268

Fatchikhin, N. V.   193, 194
Feldbaum, A. A.   142
Felgett, P. B.   250, 254
Fishkova, L. M.   268

Galilei, Galileo   6
Gardiner, A. J.   226, 229
Gerasimow, F. M.   xi, 305
Gerasimova, T. S.   xi
Gershberg, R. Ye.   268
Gershtein, G. M.   180
Goreva, G. I.   143
Gourzadyan, G. A.   268
Grossvald, Ye. G.   359, 365, 367, 375
Gush, H. P.   254

Haffner, H. 409
Hall, J. S. 167
Hawkins 318
Heckmann, O. 343
Hilbert, D. 56, 72
Hiltner, W. A. 268
Hoag, A. A. 409
Hoffleit, D. 343
Houtgast, J. 261, 268

Idelson, N. I. 343
Ilin, V. G. xi
Ilyasov, U. I. 411
Ioannisiani, B. K. x, 6, 11, 143, 295, 296, 297, 298, 299, 357
Ioffe, S. B. 287, 305
Ivanov-Kholodny, G. S. 259, 265, 267
Izrayetskaya, N. I. 430

Johnson, H. L. 226, 229, 409
Jones, F. E. 253
Jones, R. B. 343

Kalinenkov, N. D. xi, 299, 300, 302, 304
Kalinin, V. I. 180
Kalinyak, A. A. 303
Kalpnyak, A. A. 290
Karkalev, V. I. 281
Karpinskii, V. N. xi, 399, 403
Kasanov, V. A. 83
Katasev, L. A. 430
Kazakov, A. V. 106, 117
Kernahan, J. J. J. 72
Khabibullin, Sh. T. 310, 421, 423
Kharadze, Ye. K. 304
Khatisov, A. Sh. 331
Khodorov, T. Ya. 117
Kholodny, G. S. I. (see Ivanov-Kholodny)
Kholopov, P. N. 300
Kies, C. C. 403
Kiladze, R. I. 332, 341, 343
Kirillov, G. V. 247, 253
Kiselev, A. A. 343
Klein, M. 142
Kohlschutter, A. 343
Konev, P. S. xi, 310
Kononovich, E. V. 259

Konshin, V. M. 20, 72
Koppel, R. 387
Kopylov, A. I. 431
Kopylov, I. M. x
Kopystyanskii, A. A. 403
Korbut, I. F. 204
Korkin, S. I. 268
Koroler, A. V. 63
Korotkov, S. V. 119
Kotlyar, L. M. 261, 268, 304
Kox, H. 343
Kozak, P. P. 399, 403
Kramer, Ye. N. 425
Kranjc, A. 261, 268
Krat, V. A. 411, 419, 420
Ksanfomaliti, L. V. 231, 241, 298, 299, 300, 303, 304
Kubeva, Z. H. 169
Kuiper, G. P. 251, 253, 254, 419
Kulikovsky, P. G. 302
Kundzin, A. P. 223, 224, 303
Kuprevich, N. F. 303
Kyubar, T. E. 385, 431

Laffineur, M. 268
Lapkin, Ye. M. 213
Lavrentev, V. I. 420
Lebedev, A. N. 106
Lee, R. H. 268
Lekhnitsky, S. G. 365, 376
Linfoot, E. H. 318
Linnik, V. P. 73, 281, 289
Lobanov, A. I. 224
Longhurst, R. S. 289, 294
Loughhead, R. E. 411, 419
Lovell, D. J. 254
Luncl, M. 250, 254
Lyot, B. 287

McMath, F. C. 49
Maiorov, F. V. 72
Maksutov, D. D. xi, 31, 267, 268, 296, 297, 302, 318, 365
Maleyev, P. N. 253
Mamedova, Z. N. 177, 299, 301
Markowitz, W. 194, 197
Martynov, D. Ya. 297, 303, 304, 409

Name Index

Mayer, U. 162, 167
Meggers, W. F. 403
Melnikov, O. A. 305, 343, 423
Mensky, B. M. 72
Miczaika, G. R. 254
Mikesell, A. H. 167
Mikhailov, A. A. 343
Mikhelson, N. N. x, 63, 72, 147, 177, 268, 295, 296, 299, 301, 304, 305, 355, 365, 377
Mikhkyura, V. N. 241
Minnaert, M. G. J. 261, 268
Morgan, G. 142
Moroz, V. I. 243, 253, 264, 303
Myao-Yun-Zhui 202
Myasnikov, V. A. 119, 298

Nabokov, M. Ye. 267, 268
Naumov, V. A. 217
Neant, M. 254
Nefedev, A. A. 310
Neplokhov, Ye. M. 298, 303
Nikolayev, P. V. 73, 83, 143, 158, 297, 298, 299, 301
Nikonov, V. B. x, 7, 224, 295, 297, 299, 301, 304, 305
Norton, D. G. 419
Novoseltsev, Ya. V. 106

Obrashevsky, B. P. 298
Oke, J. B. 261, 268
Omarov, S. Z. 331, 343
Oshurko, V. V. 23, 295, 296
Osokina, D. N. 375

Pannekoek, A. 267
Panov, D. Yu. 72
Pavlov, N. N. 195, 197, 297, 301, 302
Pell, V. G. 420
Penning 318
Phillips, J. G. 261, 268
Pilnik, G. P. 301, 302
Platonov, Yu. P. xi
Podobed, V. V. 301, 302
Polozhentsev, D. D. 207, 208, 302
Ponomarev, D. N. 213
Popov, G. M. 311, 315

Potter, Kh. I. 185, 194, 207, 301, 302, 303
Pronik, V. I. 268, 274

Richards, R. K. 72
Richter 318
Rizvanov, N. G. 421
Ross, F. E. 319
Rozanov, N. S. 365
Rozhnova, I. P. 161, 171, 299
Ryvkin, S. M. 253

Sabinin, Yu. A. 73, 83, 85, 106, 107, 117, 119, 143, 158, 161, 169
Saksina, I. F. xi
Savin, V. A. 279, 295, 301, 305
Schorr, R. 343
Schwarzchild, K. 318
Semenov, L. I. 204, 205
Serrurier, M. 14
Severny, A. B. 6
Shain, G. A. 7
Shakhovskoi, N. M. xi, 304
Shakirov, K. S. 423
Shcheglov, P. V. 255, 277, 295, 297, 303, 393, 395
Shefov, N. N. 258
Shelovitelev, P. I. 6
Shileiko, A. V. 142
Shkutov 370
Shreiber, V. G. 328
Shumakher, A. V. xi, 143, 147, 344
Siedentopf, H. 181
Sitnik, V. F. 296
Slavenas, I. V. 305
Slavenas, P. V. 304
Slevogt, H. 318
Slyusarev 319
Smith, R. A. 253
Solomonovich, A. Ye. 298, 300
Sonnefeld 319
Stanislavsky, B. I. 83
Stepanov, V. E. 403
Streletskii, Yu. S. xi, 185, 194, 207, 297, 357, 431
Sukharev, L. A. 199, 302, 303
Sukhov, V. B. xi

Sumin, V. S.   63
Suvorov, N. A.   375

Talbot, F.   289
Tammemyagi, O.   387
Tavastsherna, K. S.   359, 367, 374
Tikhomirova, L. N.   167
Timoshkova, G. M.   202
Titov, V. K.   83, 106, 117
Trumbachev, V. F.   375
Tsesevitch, V. P.   330
Tsimmerman, G. K.   205

Urasin, L. A.   405, 409
Urmatsky, A. I.   409, 423
Usanov, D. S.   296, 297

Vakkur, E.   387
Vanderkerkhov, E.   260, 261, 267
Vasilev, A. S.   205
Vasilev, O. B.   299

Veisman, V. K.   385, 389
Vinogradova, R. G.   167
Vitovsky, N. A.   253
Volkov, I. V.   277
Vyazanitsyn, V. V.   423

Whipple, F. H.   430
Williams, R. C.   268
Wilson, W.   254

Yasevichus, V. A.   179, 300
Yazev, A. I.   205
Yegorov, V. P.   85, 106, 107, 117, 298, 300, 301
Yesipov, V. F.   269, 277

Zavoisky, Ye. K.   394
Zhukova, L. N.   162, 167
Zhurkin, N. S.   20, 297, 298
Zverev, M. S.   207, 210, 211, 296